Springer Series in Optical Sciences   Volume 43

Edited by Theodor Tamir

Springer Series in Optical Sciences

X-Ray Microscopy

Proceedings of the International Symposium,
Göttingen, Fed. Rep. of Germany,
September, 14 – 16, 1983

Editors

G. Schmahl and D. Rudolph

With 262 Figures

Springer-Verlag
Berlin Heidelberg GmbH 1984

Professor Dr. GÜNTER SCHMAHL
Dr. DIETBERT RUDOLPH

Forschungsgruppe Röntgenmikroskopie, Universität Göttingen,
Universitätssternwarte, Geismarlandstraße 11
D-3400 Göttingen, Fed. Rep. of Germany

ISBN 978-3-662-13547-1   ISBN 978-3-540-38833-3 (eBook)
DOI 10.1007/978-3-540-38833-3

Library of Congress Cataloging in Publication Data. Main entry under title: X-ray microscopy.
(Springer series in optical sciences ; v. 43) 1. X-ray microscope – Congresses. I. Schmahl, G. (Günter),
1936- . II. Rudolph, D. (Dietbert), 1935- . III. Series. QH212.X2X2 1984 502'.8'2 84-1367

Preface

X-ray microscopy fills a gap between optical and electron microscopy. Using soft x-rays, a resolution higher than with visible light can be obtained. In comparison to electron microscopy, thick, wet, unstained specimens can be examined. This is especially advantageous for biological applications.

The intense synchrotron radiation of electron storage rings and the development of optical elements for soft x-rays render x-ray microscopy feasible for basic research. Wider applications will be possible in the future with the development of laboratory x-ray sources and microscopes.

In 1979 a conference on x-ray microscopy was organized by the New York Academy of Sciences and in 1981 a symposium on high resolution soft x-ray optics was held at Brookhaven. The present volume contains the contributions to the symposium "X-Ray Microscopy", organized by the Akademie der Wissenschaften in Göttingen in September 1983.

In their capacity as conference chairmen, the editors would like to thank the Akademie der Wissenschaften, especially Prof. H.G. Wagner, Secretary of the Academy, and Mr. J. Pfahlert for organizing the symposium. We are indebted to the Stiftung Volkswagenwerk for financial support. The symposium was held at the Max-Planck-Institut für Strömungsforschung. We are grateful for their hospitality and assistance during the symposium. Thanks are due to all authors and to the Springer Verlag for their combined efforts. We thank Dipl.-Phys. P. Guttmann, Dr. B. Niemann and Mrs. A. Marienhagen for their assistance during the final preparation of the manuscripts.

Göttingen, January 1984   *G. Schmahl   D. Rudolph*

Contents

Part III **X-Ray Detectors**

Part IV **X-Ray Microscopes**

Introduction

G. Schmahl and D. Rudolph

Forschungsgruppe Röntgenmikroskopie, Universität Göttingen
Geismarlandstraße 11, D-3400 Göttingen, Fed. Rep. of Germany

Low energy x-ray research has experienced a renaissance in the last decade due to new developments in physics. Cosmic x-ray sources are being explored through x-ray telescopes orbiting in satellites. Hot plasmas for fusion energy are studied with x-ray optical techniques. Microstructure fabrication with small dimensions is performed through x-ray lithography. Atomic and molecular as well as solid-state and surface physics phenomena are investigated by use of x-ray spectroscopy with tunable x-ray beams from synchrotron radiation sources. X-ray microscopy is performed using recently developed x-ray optics and electron storage rings as brilliant soft x-ray sources.

X-ray microscopy was started with considerable enthusiasm in the 1950's. In 1952, WOLTER [1] proposed that a natural contrast mechanism for biological specimens could be exploited because there is an order of magnitude difference between the absorption coefficients for water and protein in the soft x-ray region between 2.3 and 4.4 nm. Because for x-radiation the index of refraction does not change much in going from air or vacuum to solid materials, it is impossible to make refractive lenses. One can, however, build diffraction and reflection optics. Pioneering work in grazing incidence optics was done by KIRKPATRICK and BAEZ [2] using a system with two cylindrical concave mirrors and by WOLTER [1] using two confocal conicoidal surfaces. The application of zone plates as diffraction lenses for x-ray microscopy has been discussed by BAEZ [3]. MÖLLENSTEDT and coworkers [4] have performed first x-ray imaging experiments with zone plates. At the same time the development of shadow projection microscopes by COSSLETT and NIXON [5] and microradiography (contact microscopy) by ENGSTRÖM [6] were initiated. This initial enthusiasm, however, slowed down after it became evident that the achievement of high resolution was beset with considerable experimental difficulties.

The renewal of interest in x-ray microscopy is caused by the development of improved x-ray optics as zone plates, multilayered reflection optics and grazing incidence optics, the use of high resolution x-ray resists for contact microscopy and the development of intense soft x-ray sources.

In Part I of this volume modern developments of intense soft x-ray sources are described. Storage rings have, up to now, by far the highest spectral brilliance which can, in addition, be improved further by undulators. A high spectral brilliance is especially important for high resolution scanning x-ray microscopes and for x-ray holography. This is discussed in more detail in several papers. The development of pulsed laboratory sources seems promising for special applications where it is advantageous to use very short exposure times and where it is desirable to be independent of large synchrotron sources.

In Part II the developments of different kinds of x-ray optics are described. With these developments the whole range of the above-mentioned fields of low energy x-ray research is covered as far as imaging systems are necessary. For high resolution x-ray microscopy good condensers and especially high resolution optical elements are necessary. High resolution systems must be fabricated with very small tolerances. This can be obtained the easier,the smaller the systems are, which is well known from optical systems in the visible. Micro zone plates for x-ray microscopy have diameters of less than one hundred micrometers and hence have, up to now, the highest resolution of all x-ray imaging systems.

In Parts III, IV and V, x-ray detectors, x-ray microscopes, contact microscopy and applications are discussed. In Part VI, holography with soft x-radiation is described.

X-ray imaging microscopy is, up to now,developed to a resolution of 50 nm and exposure times of a few seconds. It gives the possibility to examine relatively thick speciemes, e.g. biological specimens in an aqueous state. X-ray microscopes with improved resolution of about 10 nm are under development. Scanning x-ray microscopes, which are under development in several laboratories, cannot make pictures with exposure times as short as with imaging microscopes. They have, however, the advantage that the radiation dose applied to the specimen is considerably lower. Contact microscopy has proved that biological specimens can be imaged with high resolution of about 10 nm and good contrast without staining. The high resolution with this method, however, can be obtained only for very thin layers which are in direct contact to the recording medium. Therefore people who are applying contact microscopy to solve special problems will switch over to imaging systems as soon as they are available with about 10 nm resolution and short exposure time.

The described applications and proposals of x-ray microscopy comprise biological and medical problems. Only one proposal deals with quantitative microanalysis with high resolution as a surface-sensitive technique for materials research. In future developments it can be expected that x-ray microscopy will be applied to many other fields, too.

References

1 H. Wolter: Annalen der Physik 10, 94-114 (1952)
2 P. Kirkpatrick and A.V. Baez: J.Opt.Soc.Am. 38, 766 (1948)
3 A. Baez: J.Opt.Soc.Am. 42, 756 (1952)
4 G. Möllenstedt, K.H. v. Grote, and C. Jönsson: "Production of Fresnel Zone Plates for Extreme Ultraviolet and Soft X-Radiation" in: X-Ray Optics and X-Ray Microanalysis, H.H. Pattee, V.E. Cosslett, and A. Engström Eds., Academic Press, New York (1963)
5 V.E. Cosslett and W.C. Nixon: Nature 168, 24 (1951)
6 A. Engström: "Quantitative Micro- and Histochemical Elementary Analysis by Roentgen Absorption Spectrography", Acta Radiologica Suppl. 63 (1946)

Part I

X-Ray Sources

1. The BESSY Soft X-Ray Source and Future Developments

G. Mülhaupt

Berliner Elektronenspeicherring-Gesellschaft für Synchrotronstrahlung mbh
(BESSY), Lentzeallee 100, D-1000 Berlin 33

1.1 Introduction

BESSY is a national synchrotron light source, covering the spectrum range down to $\lambda_c = 2$ nm [1.1,2]. Since July 82 it is used in regularly scheduled user shifts for three classes of experiments:

1) Basic and applied research in physics, chemistry, biology and medicine performed by universities and research institutions

2) Applied research and industrial developments in the semiconductor and microelectronic field performed by the Fraunhofer Gesellschaft and four German semiconductor companies

3) Radiometry performed by the PTB.

The user demands range from maximum photoflux (x-ray lithography), over maximal brightness, respectively brilliance (basic research), to very small emittances (metrology). The storage ring should also be prepared to have appropiate magnetopical conditions for installation of an undulator and a free electron laser experiment. In addition to that an operation mode is required for time-resolved spectroscopy to decrease the length of the individual light pulses down to $\tau = 10$ psec.

In order to fulfill the different experimental requirements the facility includes an 800 MeV electron storage ring with an extremely flexible magnet lattice, a 800 MeV separated function injector and a 20 MeV preinjector microtron.

1.2 BESSY Set Up

The central part of the BESSY building is the experimental hall (1600 m^2) housing the storage ring and the experimental areas for the three mentioned types of experiments (Fig.1.1). The x-ray lithography laboratory is kept dustfree (class 10000) and is air conditioned ($\Delta t = \pm 1^oC$). The radiometry laboratory is also air conditioned ($\Delta t = \pm 1^oC$) while the largest part of the hall, which is dedicated to basic research, has no air conditioning.

The base plate of the experimental hall is a 30 cm steel-enforced concrete layer on a 2 m highly densified sand bed. The base plate is completely decoupled from the remainder of the building.

There are some ten laboratories for the preparation of experiments available for users as well as office space and an elementary equipped mechanical self-service workshop.

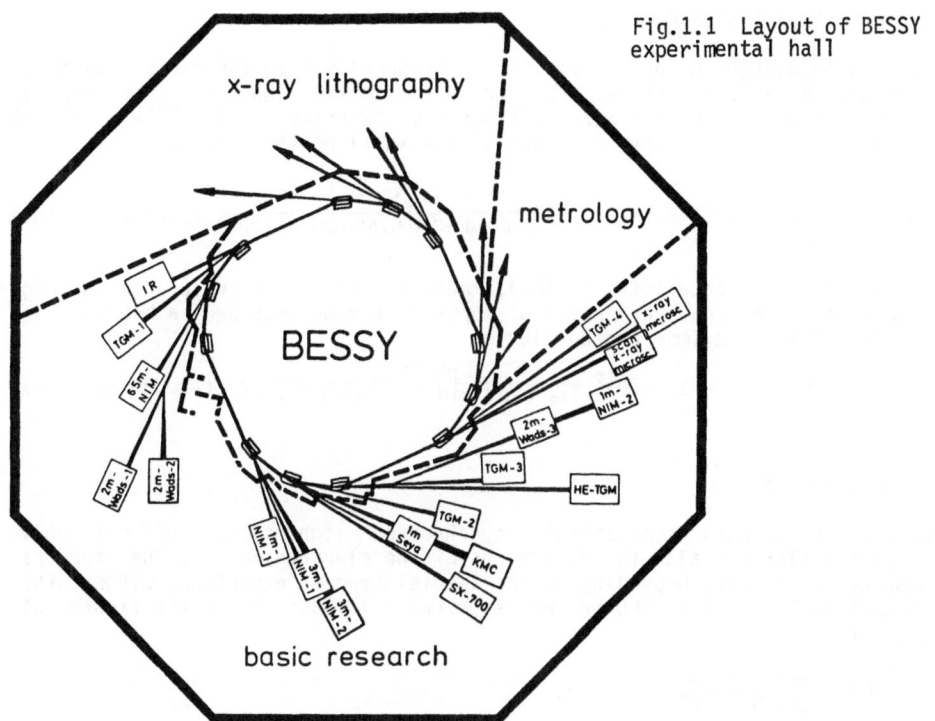

Fig.1.1 Layout of BESSY
experimental hall

The storage ring is filled from below via a transfer line, connecting the
storage ring with the separately placed injector. So all 2π of circumference
are available for experiments.

Today 18 beam lines are in operation, three in the x-ray lithography laboratory, two in the metrology laboratory and 13 in the basic research area,
most of them being equipped with monochromators covering the range from
several eV to 1.4 keV.

The facility operates normally in a one shift mode, delivering eight
hours net beam time per day to the users (5 day week). From the middle of
1984 on, a two shift operation is planned delivering twelve hours net beam
time per day.

1.3 Spectrum

BESSY is normally operated close to its maximum energy of 0.8 GeV. The photon spectrum of the synchroton radiation emitted out of the dipole magnets
has a critical wavelength $\lambda_c = 2$ nm. This has been chosen as an optimum to
fulfill two x-ray lithography requirements: to have as short a wavelength as
possible to minimize the exposure time and on the other hand to have sufficiently large wavelengths as not to deteriorate the contrast from the mask.

It is possible to increase the wavelength up to $\lambda_c = 16$ nm by decreasing
the energy to 0.4 GeV. A decrease of λ_c by increasing the energy above 0.8
GeV is not possible, since the dipole magnets run at 0.8 GeV already at a
magnetic field level of 1.5 T. Further increase in the magnetic field increases the field perturbations due to saturation effects to an unacceptable
level.

1.4 Spectral Photon Flux

At a given energy the photon flux n_γ is proportional to the stored electron current. The largest current achieved so far in the high flux optic (XRAY), which is mainly used for x-ray lithography experiments, is I_{XRAY}^{max} = 350 mA (design goal: 500 mA) corresponding to a maximal photon flux of

$$\text{Optic "X-RAY": } n_\gamma^{max} = 2.5 \cdot 10^{12} \frac{\text{photons}}{1\% \text{ band width,sec,mrad (hor)}} \quad (\text{around } \lambda \simeq 1\text{nm}).$$

In the low emittance optic (METRO), mainly used for metrological and high resolution experiments, a current of I_{METRO}^{max} = 150 mA has been achieved corresponding to a maximum photon flux of

$$\text{Optic METRO: } n_\gamma^{max} = 1.0 \cdot 10^{12} \frac{\text{photons}}{1\% \text{ band width,sec,mrad(hor)}} \quad (\text{around } \lambda \simeq 1\text{nm}).$$

1.5 Brilliance

For most of the more sophisticated synchrotron light experiments not only the photon flux but also the dimensions of the electron beam at the tangent point are important. Depending on the special type of experiment either the spectral light density SLD or the spectral brilliance SB is the figure of interest:

$$\text{SLD} = \frac{n_\gamma}{(2.35)^2 \sigma_x \sigma_z}$$

$$\text{SB} = \frac{n_\gamma}{(2.35)^2 \sigma_x \sigma_z \ 2.35 \sqrt{\sigma_{z/nat}'^2 + \sigma_{z/electr.}'^2}}$$

σ_x, σ_z: standard deviation of the hor. resp. vert. electron beam distribution

$\sigma_{z/nat}'$: standard deviation of the angular distribution of the photons emitted from a single electron

$\sigma_{z/electr.}'$: standard deviation of the angular distribution of the electrons in the stored beam.

Both quantities are given together with several other characteristic figures in Table 1.1. Comparing these figures with other existing or proposed synchrotron light sources one has to keep in mind that low energy machines always have a larger $\sigma_{z/nat}'$ than high energy machines, but usually due to the much shorter beam lines the acceptable horizontal radiation angle in low energy machines is more than an order of magnitude larger than in high energy machines.

The geometrical beam size depends on the stored current (Fig.1.2). In the case of the low emittance beam it seems to be clear that the beam is unstable above a current level of ~15 mA if no additional Landau damping is provided. In our case the additional Landau damping is provided by the stored electron current. If one excites a stored beam of more than 15 mA nonadiabatically to amplitudes larger than the beam size (which will destroy the

6

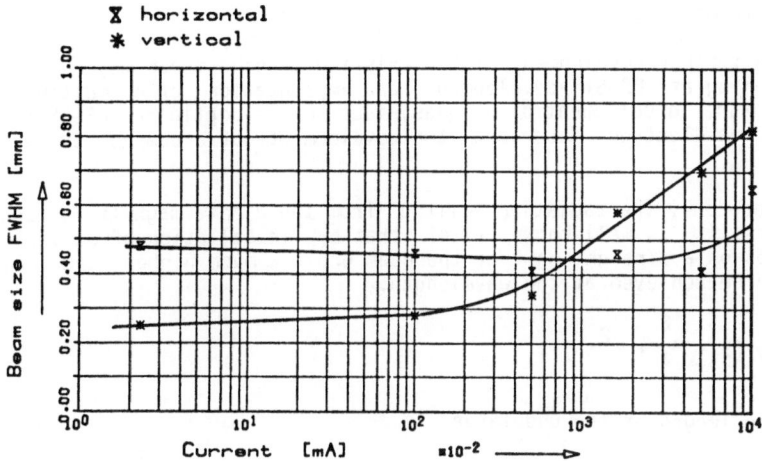

Fig.1.2 Beam size in optic METRO vs stored electron current

stabilizing Landau damping due to the trapped ions), the current will immediately shrink down to the level of ~15 mA.

    To supply the same amount of Landau damping to the beam by external octo-
pole fields as the trapped ions do, the dynamic aperture of the machine would
shrink down to unacceptable low values. It is therefore doubtful whether
stored currents of several hundred milliamps with emittances of even an order of magnitude smaller than in BESSY´s METRO optic (what has been proposed
[1.3]) could be kept stable.

Table 1.1 Parameter list of available BESSY optics

	BESSY(METRO) theor.	BESSY(METRO) achieved	BESSY(XRAY) theor.	BESSY(XRAY) achieved
Electron energy [GeV]	0.8	0.8	0.8	0.8
Radius of bending magnets [m]	1.78	1.78	1.78	1.78
λ_c [nm]	2.0	2.0	2.0	2.0
Electron current [mA]	200	150	500	350
Number of 1 keV photons per sec, mrad(hor),1‰ bandwidth	$1.3 \cdot 10^{12}$	$1. \cdot 10^{12}$	$3.3 \cdot 10^{12}$	$2.3 \cdot 10^{12}$
σ_x [mm]	0.18	0.2	0.9	0.9...1.2
σ_z [mm]	0.08	0.15...0.35	0.6	0.4...1.4
$(2.35)^2 \, \sigma_x \sigma_z$ [mm^2]	0.08	0.16...0.38	3.0	2.0...9.3
Natural σ_z' [mrad]	0.32	0.32	0.32	0.32
Electron σ_z' [mrad]	0.03	0.05...0.11	0.3	1.4
Total σ_z' [mrad] x 2.35	0.75	0.76...0.79	0.45	3.4
<Spectral light density>	$1.6 \cdot 10^{13}$	$2.6 \cdot 10^{12}$	$1.1 \cdot 10^{12}$	$2.5 \cdot 10^{11}$
<Spectral brilliance>	$2.1 \cdot 10^{13}$	$3.3 \cdot 10^{12}$	$2.4 \cdot 10^{12}$	$7.3 \cdot 10^{10}$

1.6 Future Developments

Apart from the further development of the existing machine three major projects are running at BESSY: development and implantation of a multipole wiggler/undulator, development of a quasi-isochronous optic in BESSY for an infrared free electron laser and the development of a compact storage ring.

a) Up to now only the radiation emitted from the dipole magnets is used for experiments. It is well known [1.4] that by installation of periodic magnetic fields on a straight part of the beam orbit a much higher spectral brilliance can be achieved at the wavelength:

$$\lambda_n = \frac{\lambda_o}{n \cdot 2\gamma^2} (1 + \frac{k^2}{2} + \Theta^2 \gamma^2)$$

λ_o = period length of the undulator

n = harmonic number

k = $0.0934 \cdot B_o \cdot \lambda_o = \gamma \cdot \delta$, B_o [kГ], λ_o [cm]

δ = max. deflection angle of the electron in the undulator structure

B_o = max. magn. field in the beam path.

If K $\leq$ 1 interference effects will strongly enhance the radiation emitted at λ_n by a factor of $(2\,N)^2$ (undulator regime), while with K>>1 the radiation of the different periods of the structure will just incoherently add (wiggler regime), so that the radiator power will be N times the power radiated from a single period of magnetic field B_o.

It has been decided to build at BESSY a hybrid structure out of permanent magnets enhanced by soft iron yokes with λ_o= 7 cm and N = 36 [1.5].

The K can be varied from K = 0.5 to K = 7.3. The corresponding data are given in Table 1.2.

Table 1.2 Data of the proposed multipole wiggler at different values of K:
800 MeV, 200 mA, L = 2.52 m, λ_o = 7 cm, N = 36

	K = 0.5	K = 1	K = 2	K = 7.3
B_o [kG]	0.76	1.53	3.06	11.2
g [cm]	7.4	5.2	3.7	1.5
E_1 [eV]	77.1	57.8	28.9	3.1
λ_c [A]	383.2	190.4	95.2	26.0
ε_c [eV]	32.4	65.2	130.4	477.
δ [mrad]	0.32	0.64	1.3	4.7
$\alpha_x$tot [mrad]	1.4	1.6	2.5	6.8
$\alpha_y$tot [mrad]	1.3	1.3	1.3	1.3
P_t [watt]	1.2	4.8	19.2	257
$P(E_1)$ [watt]	0.028	0.064	0.064	at
$N(E_1)$ [$\frac{phot.}{sec}$]	2.3×10^{15}	6.9×10^{15}	1.4×10^{16}	$\Delta\lambda/\lambda = 0.01$

b) The free electron laser experiment is a project of the Free University of Berlin (FUB) with a completely different approach to all other free electron laser projects.

The basic idea is as follows: as long as the electron beam current is completely uniform with respect to the optical potential formed by the radiation field inside the optical cavity of the laser, as many electrons will give energy to the radiation fields as will take up energy out of the radiation field. A net transfer of energy from the electrons to the radiation field requires a density modulation of the electron beam compared to the optical field. This is normally achieved during one single passage through a long multipole structure by the action of the radiation field on the beam by energy modulation.

But after one turn in the storage ring, the density modulation is completely smeared out due the momentum compaction factor of the storage ring:

$$\frac{\Delta L}{L} = \alpha \cdot \frac{\Delta p}{p} \text{ with } 0.02 < \alpha < 3 \text{ in low energy machines.}$$

In the FUB project the idea is to decrease the momentum compaction factor down close to zero to conserve the density modulation for several turns. It is an interesting side effect that by shrinking the momentum compaction factor also the electron bunch length decreases.

Up to now the BESSY storage ring has been tuned to $\alpha = 0.004$ corresponding to a bunch length of $\tau_\ell \sim 25$ psec. But a further factor of 2 in decreasing of α is required at least.

Table 1.3 Parameter compact source (COSY)

λ_c	=	12 A ($\hat{=}$ $\varepsilon_c = 1$ keV)
B_{max} (on e^- orbit)	=	5 T
Stored energy in the magnetic field	=	12 MJ
ρ	=	0.382 m
n	=	0.5 - 0.7
Circumference	=	2.4 m
Circumferential time	=	8.00 nsec
E_{max}	=	560 MeV
HF frequency (h=1)	=	125 MHz
U_0	=	23.3 keV
Injection energy E_i	=	100 keV
Injection field	=	0.003 T
Transitron (betatron/synchrotron)	=	5 - 10 MeV
Transitron field	=	0.05 - 0.09 T
$\sigma_x \cong \sigma_z$	=	0.8 mm
Stored current	=	0.3 A
Number of beam lines	=	10

c) The largest project at BESSY is the development of a compact storage ring designed to deliver soft x-ray radiation to x-ray lithography users. The compact storage ring should have a price below 5 - 8 Mill. DM and should have sufficiently small geometrical dimensions as to fit into an existing infrastructure of microelectronic fabrication.

The main idea to minimize the geometrical dimensions and the electron energy needed is to use a weak focussing magnet excited by superconducting coils [1.6]. Further ideas like full current injection by a linac and the use of an NBS-type cavity came up in a feasibility study by a Munich group [1.7], which was stimulated by the Fraunhofer Gesellschaft.

BESSY is now building up a prototype compact storage ring under contract with the Fraunhofer Gesellschaft. The main parameters of this machine are given in Table 1.3. A still open question is whether to use a 10 MeV linear accelerator for injection or to use the guide field itself to form a betatron preaccelerator. In the case of the betatron preacceleration one could save about one quarter of investment costs, could decrease drastically the space needed to build up the machine and would decrease significantly the problems of radiation protection.

To clear this question BESSY is building up a prototype betatron and it is planned to be ready for tests at the end of January 1984. The hardware of the superconducting magnet should be ready for tests in Berlin by the middle of 1985 so that the first injection tests could be made in the last quarter of 1985.

References

1.1 D. Einfeld et al: IEEE Trans. Nucl. Sc. NS-26, No. 3, 3801 (1979)
1.2 D. Einfeld, G. Mülhaupt: Nucl. Instr. Meth., 172, Nos. 1,2, 55 (1980)
1.3 R. Sah: Proposal ALS, Lawrence Berkeley Lab. 1982
1.4 e.g. A. Winick, S. DONIACH: Synchrotron Radiation Research, Plenum Press, N. Y. 1980
1.5 W. Gudat, E. Umbach, BESSY TB 39/1983
1.6 BMFT-Bericht, Anwendung der Synchrotronstrahlung in der Halbleiter-Lithographie 1977
1.7 M. Jahnke et al.: "Klein Erna", Int. Bericht Techn. Univ. München 1981

2. Recent and Future Developments at L.U.R.E. (Orsay)

Y. Petroff

L.U.R.E., Bat. 209 C, Université Paris-Sud
F-91405 Orsay, L.P. CNRS 008, France

2.1 Introduction

The French synchrotron radiation center (L.U.R.E.) uses at the moment two storage rings: A.C.O. (0.546 GeV), which is fully dedicated to synchrotron radiation, and D.C.I. (1,8 GeV), used only at 25% (the other 75% are for high energy physics).

A third machine, Super A.C.O., (0,8 GeV) is under construction.

In this talk I will not describe the experiments concerning the rings but discuss the future developments on the machines (wigglers, undulators and free electron lasers).

2.2 A.C.O.

A.C.O. (Anneau de Collisions d'Orsay) is one of the oldest storage rings. It was built in 1965 and was used extensively until 1977 for e^+-e^- collisions. Since then it has been completely dedicated to synchrotron radiation. Thirteen experiments (on three beam lines) are installed around the ring. They cover different fields: atomic and molecular physics, solid state and chemistry, applied physics and microscopy.

Typical parameters are given below:
Energy: 150 to 540 MeV. For most of the experiments we work at 540 MeV. The low energies (150 to 240 MeV) are used only for the free electron laser
Injection Energy: 240 MeV by a Linac
Particles: e^- at the moment
Intensity: 150 mA at 540 MeV
500 mA at 240 MeV
Circumference: 22m
Radio frequency: 27.2361 MHz
Number of bunches: 1 or 2
R.M.S. bunch length: 1 ns
Beam life time: 8 hours at 100 mA
14 hours at 60 mA
Transverse dimensions: $1,2 \times 1.5$ mm^2 for large current at 540 MeV. They strongly depend on the intensity in the electron beam.

The developments that we are planning are the following:

1) Free electron laser

A.C.O. has only one straight section available for undulators. It has been used to install a free electron laser experiment based on an undulator having the following characteristics:

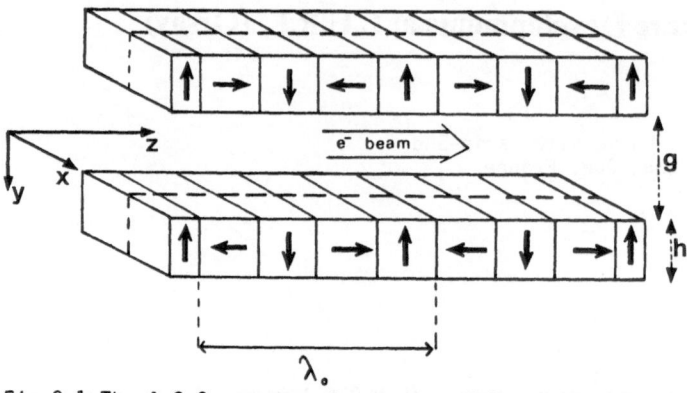

Fig.2.1 The A.C.O. permanent magnet undulator uses a configuration of magnets rotated by $\pi/2$ as proposed by K. HALBACH [2.1]. In this sketch of a 2 - period undulator the first and last magnets are adjustable half-periods

Full length: 1.33 m
Number of periods: 17
Period: 7.8 cm
Maximum field: 0.31 Tesla.

The magnet structure is the arrangement defined by HALBACH [2.1] (Fig.2.1). In this configuration the maximum field available, B_0, is given by

$$B_0 \text{ (Tesla)} = 1.426 \cdot B_r \cdot e^{-\pi g/\lambda_0} \tag{2.1}$$

with B_r the remanent field (0.85 T in our case), g the undulator gap and λ_0 the period.

The wavelength λ of the spontaneous emission is given along the axis of the undulator by (in S.I. Units)

$$\lambda = \frac{\lambda_0}{2\gamma^2} \cdot (1 + \frac{K^2}{2}) \tag{2.2}$$

with $K = \frac{e \cdot B_0 \cdot \lambda_0}{2 \cdot \pi \cdot m \cdot c}$,

e and m are the electron charge and mass, c is the speed of light in vacuum and γ is the total electron energy divided by mc^2.

Due to the small gain measured with this undulator (2.10^{-4}) it was transformed in an optical klystron, which is a device consisting of two undulators separated by a dispersive section as shown in Fig.2.2.

The maximum magnetic field in this section is 0.53 Tesla [2.2].

Laser oscillation was first demonstrated in June 1983 around λ= 6500 A. However, the limited straight section length (1.3m) available on A.C.O. will limit the gain to a few parts in 10^{-3} per pass. This will allow oscillation

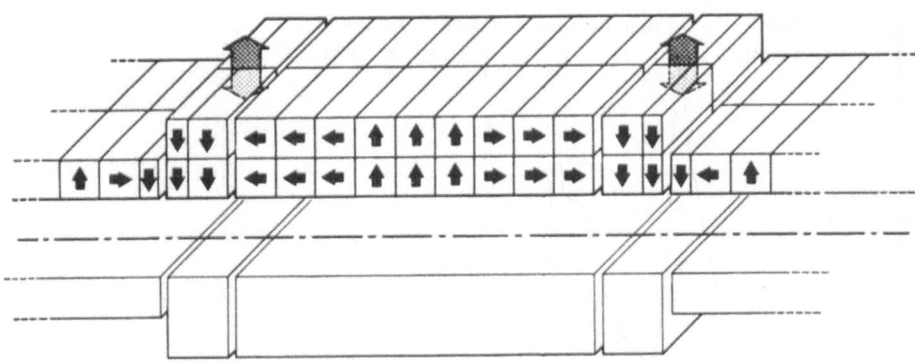

Fig.2.2 Magnetic structure of the Orsay optical klystron

only in the range 3000-8000 A. This is the reason why we have decided to study only the physics of the laser on A.C.O. and to build a 5m undulator for D.C.I. to try to obtain a laser between 1200 A and 5000 A.

2) Production of intense pulsed coherent radiation between 2000 and 200 A.

It is possible to bunch a relativistic electron beam on an optical wavelength scale using a powerful Nd : YAG Laser in conjunction with an optical klystron. This bunched beam will produce coherent synchrotron radiation not only at the laser frequency but also at the harmonics [2.3]. The light will be produced in short pulses with a low repetition rate given by the external laser. Conversion efficiencies varying from 10^{-3} for the 3rd harmonic to 10^{-7} for the higher harmonics are expected.

We hope to be able to perform these experiments in the next few months.

3) Use of the undulator in the V.U.V.

The emission of the undulator has been measured in the V.U.V. spectral range with an electron energy of 536 MeV and a small toroidal grating monochromator [2.4], without any entrance optics. One can see in Fig.2.3 a typical spectrum showing the harmonics (3 to 10). It can be seen that the spectral widths are broadenend by the angular dispersion of the electron beam. It appears also that the harmonics of even rank are like the odd harmonics, although they should vanish in the forward direction in the case of a linear undulator.

This is probably due not only to the increased emittance of the beam at 540 MeV but also to some misalignment. The harmonics of rank higher than 7 are very small in comparison with the theory. This is due to the spectral behavior of the monochromator which falls off for wavelengths smaller than 200 A (grazing angle on the grating 19⁰). At 240 MeV harmonics up to the 23rd have been observed.

The gain in brightness compared to a bending magnet of A.C.O is of the order of 230.

A high resolution toroidal monochromator (<15 meV between 10 and 250 eV) is under construction and will be installed before the end of 1983. This experiment (undulator and monochromator) will be used for:

13

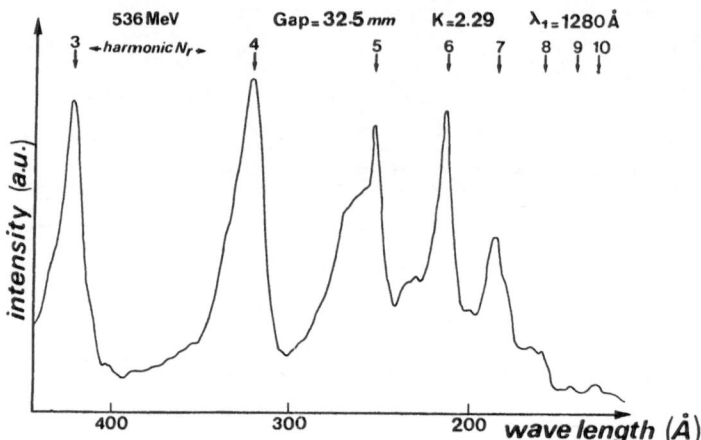

Fig.2.3 Spectral distribution of the light emitted in the V.U.V. region by electrons of 536 MeV for an undulator gap of 325 mm (K ∿2.2) at θ = 0. The spectrum has not been corrected for the monochromator spectral response. The decrease in intensity below 200 A is due to the large grazing incidence angle of the grating (19°)

    a) spectroscopy in atomic, molecular and solid state physics,

    b) obtaining coherent radiation with a coherence length

$$l_{coh} = \lambda^2/\Delta\lambda \sim 1mm.$$
This will be used for holography and holographic microscopy.

4) Use of e^+ instead of e^-

The use of e^+ instead on D.C.I of e^- has improved both the lifetime and the stability of the beam. Previous experiments of A.C.O. by PETIT and LEVEL [2.5] have shown that σ_x and σ_z are almost independent of current intensity. (Fig.2.4).

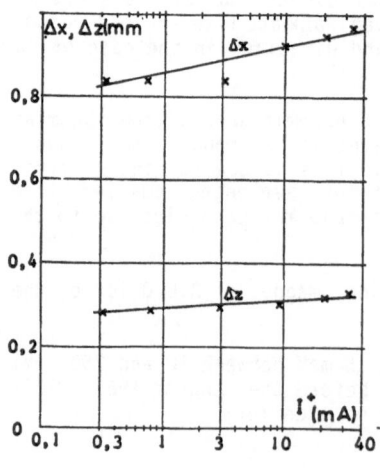

Fig.2.4 Variations of the transverse dimensions of the positron beam as function of the current for A.C.O. at 540 MeV [2.5].

2.3 D.C.I.

D.C.I. is used at 75% for high energy physics and 25% dedicated to synchrotron radiation (80 shifts of 12 hours per year). Ten experiments (on two beam lines) are installed around the ring. The fields covered are: topography, protein crystallography, Compton and Raman scattering, fluorescence, x-ray imaging, E.X.A.F.S., small angle scattering, diffuse and anomalous scattering. The useful range of photons is 3.5 (due to a beryllium window) - 25 KeV.

Typical parameters are given below:

Energy: 600 MeV to 1.8 GeV (for synchrotron radiation it is used mostly at 1.72 GeV).
Injection energy: 1 GeV by a Linac.
Particles: e^+. We use positrons instead of electrons because the lifetime is better (by a factor of 2) and the beam more stable.
Intensity: 300 mA.
We show in Fig.2.5 the typical currents injected during synchrotron radiation shifts for the period Oct. 81 - Oct. 82.
Number of bunches: 1
Beam Lifetime: 30 hours at 300 mA.
50 - 60 hours at 190 mA.
Transverse dimensions: 2 x 7 mm^2 for large current at 1.72 GeV (this is an average value, the size depends on the position in the ring).

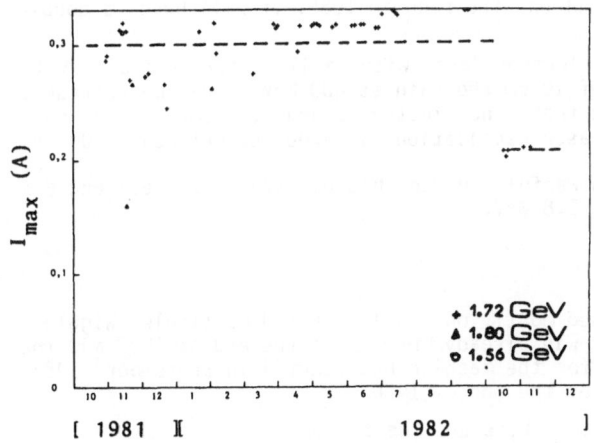

Fig.2.5 Intensity of the stored beam in D.C.I. between October 1981 and October 1982 for the synchrotron radiation shifts

The developments underway are the following:

1) Construction of a superconductive wiggler

A design study for a 5 poles - 5 Tesla wiggler has been carried out [2.6] and the device is now under construction at the C.E.A. (Saclay). It will be inserted in 1985 in one of the straight sections. Six experiments will share

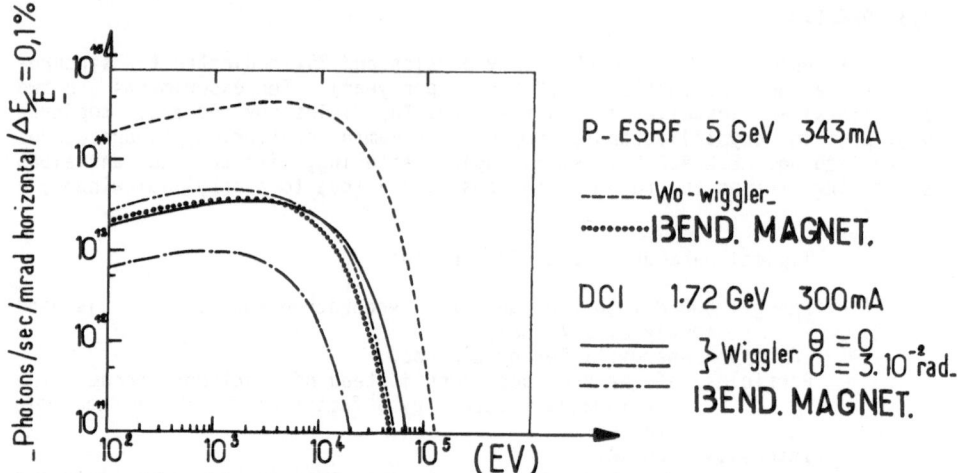

Fig.2.6 Calculated spectra for a 5.T wiggler (0.3 A) on D.C.I. and E.S.R.F.

the 60 mrad. The calculated spectra for a bending magnet and the wiggler are shown in Fig.2.6.

2) Free electron laser and undulator

We are starting the construction of a 5 meter permanent magnet undulator that we are planning to install on the ring in 1985. It will have a double purpose:

- to obtain a free electron laser with a 17 meter optical cavity. With 50 periods of 10 cm the gain at 400 MeV should be between 8 and 20%, assuming that the electronic density remains the same as at 600 MeV. Laser oscillation is expected between 5000 and 1200 A.
- to give a very powerful source between 120 A(100 eV) and 6 A ($\sim$ 2 KeV) for E = 1.8 GeV.

2.4 Super A.C.O.

This is a 800 MeV machine based mostly on undulators and multipole wigglers [2.6] and using positrons. It was already financed at the end of 1982 and the construction of the building for the machine has started in September 1983. The first beams are expected by the end of 1986.

The expected characteristics are the following:

E = 300 - 800 MeV.
I_{max} $\sim$500mA (20 mA per bunch)
Emittance: ε_x = 3.8 10^{-8}m x rad
λ_c = 19 A
Circumference: 72.041 m. Eight straight sections will allow to install six undulators or multipole wigglers (see Fig.2.7)
RMS bunch transverse dimension:
- bending magnets σ_x = 0.144 mm
σ_z = 0.165 mm
- undulators σ_x = 0.575 mm
σ_z = 0.2 mm

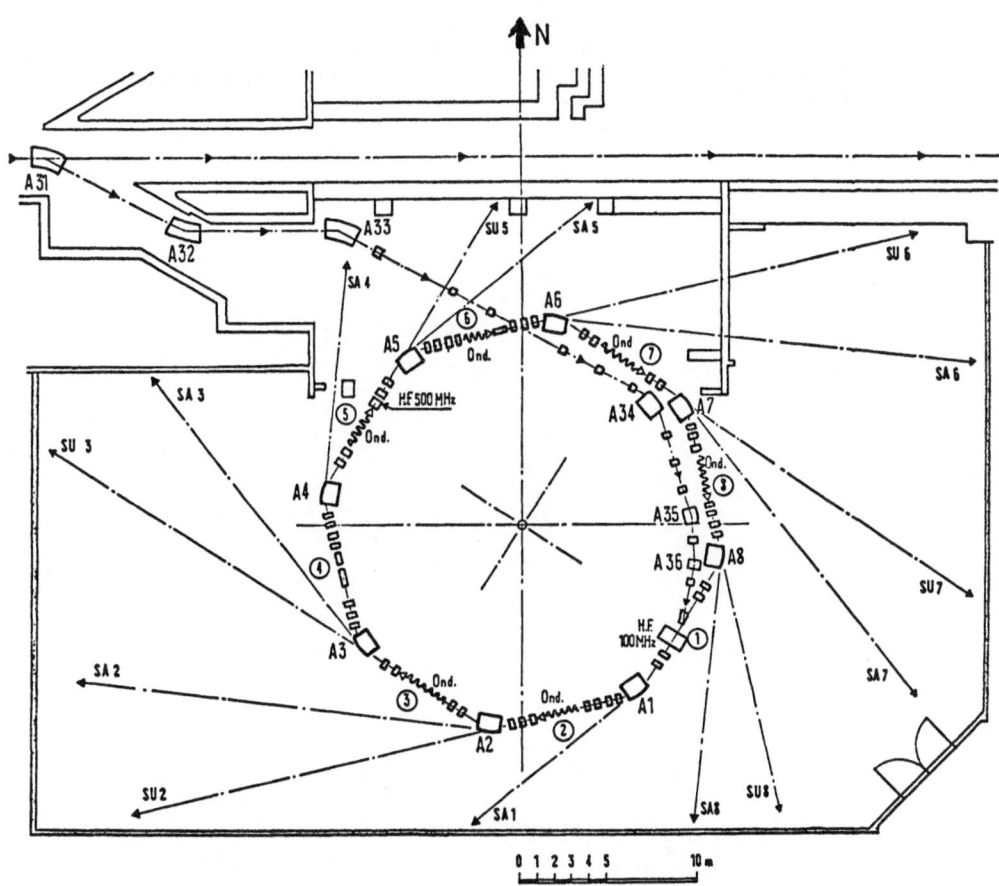

Fig.2.7 Overview of Super A.C.O. indicating the position of the six undulators (or hybrid wigglers) and the 8 beam lines on the magnets (24 experiments)

Expected lifetime: 10 hrs
Time structure: There will be two cavities with 100 and 500 MHz. For the 100 MHz cavity the number of bunches will go up to 24.
Fundamental frequency: 4.161 MHz
Delay between bunches: $\dfrac{240}{n}$ (ns) $n = 1 - 24$

Bunch length σ_1 (mm) = 32 to 50
For the 500 MHz cavity n_{max} = 125
Bunch length σ_1 (mm) = 9 to 12 .

References

2.1 K. Halbach: Nucl. Inst and Methods, 169,1 (1980)
2.2 M. Billardon, D.A.G. Deacon, P. Elleaume, J.M. Ortega, K.E. Robinson, C. Bazin, M. Bergher, J.M.J. Madey, Y. Petroff and M. Velghe: Proceedings of the Bendor Free Electron Laser Conference, Sept. 1982. Journal de Physique CI - 29 (1983), P. Elleaume p. 333, same volume

2.3 R. Coisson and F. De Martini: ibid., p. 163

2.4 L.H.T. 30 produced by Jobin - Yvon (France)

2.5 Petit - Level: 1976 L.A.L. internal report (unpublished)

2.6 D. Raoux: (unpublished)

3. X-Ray Emission from a 1 kJ Plasma Focus

G. Herziger

Institut für Angewandte Physik, Technische Hochschule Darmstadt
Schloßgartenstraße 7, D-6100 Darmstadt, Fed. Rep. of Germany

3.1 Introduction

Several devices capable of producing dense, high temperature plasma are presently investigated as possible sources in the soft x-ray range [3.1-6].
Among these are the laser-induced plasma, the gas puff systems, the sliding
spark, and the plasma focus. A plasma focus device produces a small volume
of high-temperature dense plasma, that is formed by pinch compression on the
focus axis in front of the center electrode (Fig.3.1). The plasma focus is
of roughly cylindrical shape frequently fluctuating in its local position
and geometrical shape. With respect to the generation of soft x-rays these
fluctuations are disadvantageous as the radiation source suffers from corresponding fluctuations in position, size and brightness. Improvement of the
reproducibility of plasma position and geometry has been achieved in a modified version of the plasma focus which differs from standard devices by an
external control of the ignition phase [3.8]. The modifications resulted in
an enhanced brightness and improved reproducibility of the x radiation.

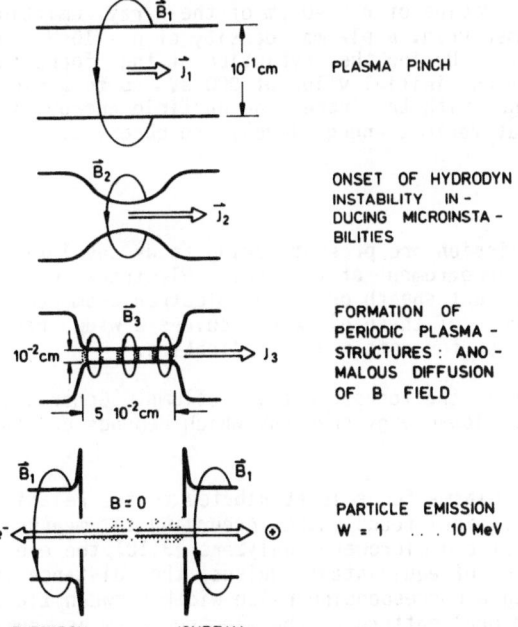

$\vec{B}_1$, $\vec{j}_1$, 10^{-1}cm	PLASMA PINCH
$\vec{B}_2$, $\vec{j}_2$	ONSET OF HYDRODYN INSTABILITY IN - DUCING MICROINSTA - BILITIES
$\vec{B}_3$, J_3, 10^{-2}cm, $5 \cdot 10^{-2}$cm	FORMATION OF PERIODIC PLASMA - STRUCTURES : ANO - MALOUS DIFFUSION OF B FIELD
$\vec{B}_1$, $B \approx 0$, $\vec{B}_1$, e^-, ⊙	PARTICLE EMISSION $W = 1 \ldots 10$ MeV
ELECTRON BEAM, IONBEAM	

Fig.3.1 Schematic evolution of
the plasma focus

3.2 Description of Device

Our experiments have been performed with a 1 kJ Mather type [3.7] plasma focus. Most of the experience has been with hydrogen gas, although x-ray emission has been investigated with several other gases, including argon and neon. The temporal evolution of the plasma focus during the period of most powerful x-ray emission is shown in Fig.3.1. Starting with the final pinch the drift energy of the particles is about 800 eV, the plasma is of nearly cylindrical shape and its contour is moving perpendicular to the focus axis. As the plasma temperature is about 20 eV, the plasma is far beyond thermal equilibrium. By onset of a hydrodynamic instability of m = 0 type the plasma gets increasingly compressed down to a final radius $r \simeq 50$ µm. During compression a constant current of $I = 2 \cdot 10^5$ A is enforced through the converging plasma by the external circuit. The further evolution is distinguished by onset of microinstabilities which modulate the plasma density periodically.

Due to the periodic plasma structure the diffusion velocity of the magnetic field is enhanced. The particles travelling in a cyclotron drift motion are accelerated by the rapidly penetrating magnetic field from an initial energy of 800 eV up to a final energy of several MeV. The magnetic energy stored in the surroundings of the compressed plasma is converted to kinetic energy of the particles within a few nanoseconds. When the magnetic energy is consumed by the acceleration process the force balance does not hold any longer and electrons and ions are simultaneously emitted into opposite directions. According to the measured number of emitted particles there is experimental evidence that almost every particle in the collapsing volume is accelerated and emitted, so leaving an evacuated volume that interrupts the current. The instant of particle emission indicates the final focus phase at which the soft x-ray emission is finished.

By x-ray pinhole diagnostics a radius of $r \simeq 40$ µm of the x-ray emitting part of the plasma volume was measured. A plasma density of $n \simeq 10^{20}$ was determined by standard diagnostics. During the evolution of the focus the particle energy increases from an initial value of 800 eV up to a final value of several MeV. In accordance with the increasing particle energy different x-ray emission processes at various energy levels are observed.

3.3 X-Ray Emission

Two different sources of x-ray emission are present during focus evolution. The first results from electron bombardment of the center electrode either by the electrons moving in the current sheath or by the electron beam emitted from focus. The energy spectrum of these electrons covers a wide range (keV - MeV) and is of minor interest for soft x-ray application.

The second x-ray source is from the focused plasma column. Generally, this source is characterized by a low-energy spectrum which depends on the filling gas.

The hard x-ray emission of the plasma focus is attributed to the relativistic electron beam emitted in the final focus phase. According to measurements with Čerenkov detectors [3.9] and microwave analyzers [3.10], the electron beam consists of a sequence of equidistant pulses, the distance of which is between 10 and 30 ps with a corresponding pulse width between 1 and 3 ps. In Fig.3.2 the measured temporal pattern of the electron beam is compared with model calculations [3.11].

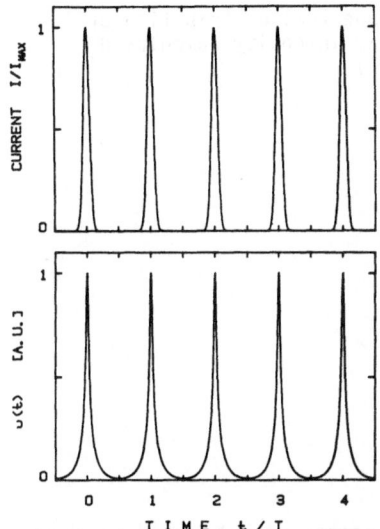

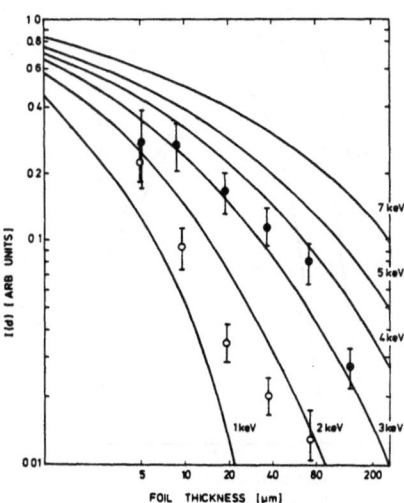

Fig.3.2 Temporal structure of the electron beam emitted from the plasma focus
Below: Model calculations of the temporal structure of the electron density during the final focus phase. Pulse distance is in the range of 10-30 ps, pulse width in the range of 1-3 ps

Fig.3.3 Soft x-ray intensity versus thickness of Al foils. Black dots indicate end-on measurements (axis of diagnostics parallel to focus axis) white dots indicate side-on measurements. Full lines are theoretical curves calculated from (3.1). Parameter is the electron temperature

A second type of quasithermal x radiation is observed at the beginning of the focus phase. From absorption filter measurements this radiation appears to be caused by electron bremsstrahlung in the ion field. Best fit of the experimental data is obtained if the electron temperature T is chosen in the range (Fig.3.3):

$$kT = 3-4 \text{ keV (end on)}$$
$$kT = 1-2 \text{ keV (side on)}.$$

According to Fig.3.3 the x-ray intensity depends on the angle of emission, which is a strong indication that the plasma is not in thermal equilibrium. Despite that, we have assumed thermal equilibrium in order to obtain a rough estimation on the spectral distribution of the x-ray emission (Fig.3.4). With this assumption the power radiated per unit volume in a wavelength interval is [3.12]

$$dP = 6.01 \cdot 10^{-30} \lambda n_e \Sigma(n_i Z^2) T^{-1/2} \lambda^{-2} \exp(-12.4/\lambda T) d\lambda \tag{3.1}$$

where T is in units of keV, g is the Gaunt factor and λ the wavelength in Angstroms. It should be noted that the spectral distribution shown in Fig. 3.4 is a crude approximation, as there are further experimental indications that the x-ray emission is dominated by processes other than those of thermal equilibrium and the mechanisms producing the x rays are not fully understood. According to our experiments the intensity of x-ray emission increa-

21

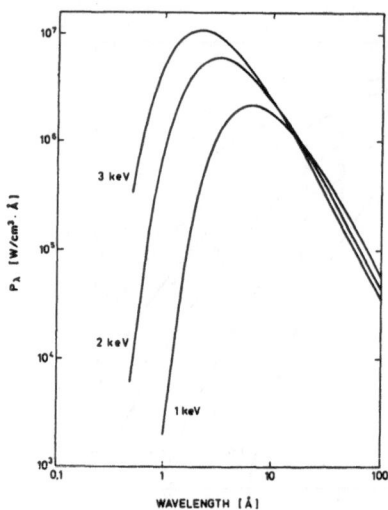

Fig.3.4 Spectral distribution of soft x-ray intensity calculated from (3.1)

ses considerably when adding gases of higher atomic weight to the hydrogen plasma, in which case line radiation becomes dominant. The quantity of additives is up to the present limited to values below 10%. Larger amounts of additives or even experiments in pure gases of high atomic weight require a new adjustment of the focus parameters according to the scaling parameter:

$$a = (\mu_0 I^2/8\pi^2 r_0^2 \rho_0 u^2)^{1/2} = \text{const} \tag{3.2}$$

where r_0 is the radius of the center electrode, ρ_0 the gas density, and u the steady-state velocity of the plasma sheath during the run down phase [3.13].

The angular distribution of the quasithermal x radiation has been investigated with a multipinhole x-ray camera which supplies three-dimensional images. End on and side on power measurements have been performed for additional proof. According to our preliminary results the quasithermal x radiation is nearly isotropic in space. Repetitive three-dimensional imaging in turn offers a possibility to determine the geometry, position, and reproducibility of the x-ray source.

A third type of x rays, which is anisotropic and monochromatic with a quantum energy of about 1 keV, is additionally emitted during the focus phase into a cone of angle ϕ around the focus axis (Fig.3.5). Wavelength and angular distribution have been measured by a method of multiple filter absorption [3.14]. The linewidth $d\lambda$ has been determined by a von Hamos spectrometer. The measured value $d\lambda/\lambda = 0.003$ corresponds to the resolution limits of the apparatus.

We ascribe the generation of this anisotropic and monochromatic x radiation to the interaction of a relativistic electron beam with the periodically varying potential of the modulated electron density which appears in the focus plasma at this time. Periodic structures of the electron density with a period length of 40 - 100 μm (Fig.3.6) have been observed during the compression phase by optical Schlieren diagnostics [3.15]. However, with proceeding focus evolution the period length is decreasing to values below the resolution power of optical diagnostics, so that in the final focus phase the structurized electron density has to be deduced from a measurement of the anisotropic x radiation itself.

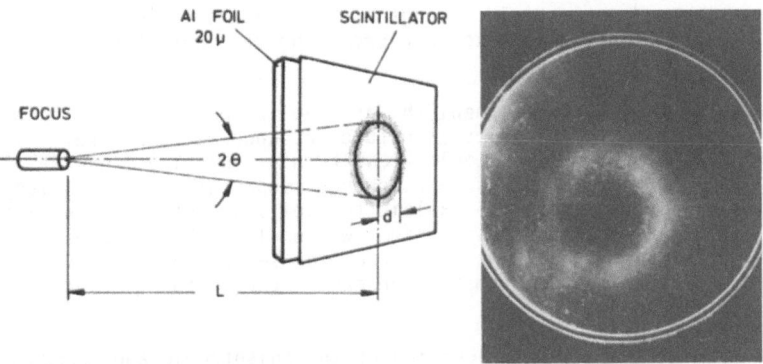

Fig.3.5 Geometry of the emission of the anisotropic soft x rays
Right: Scintillator image

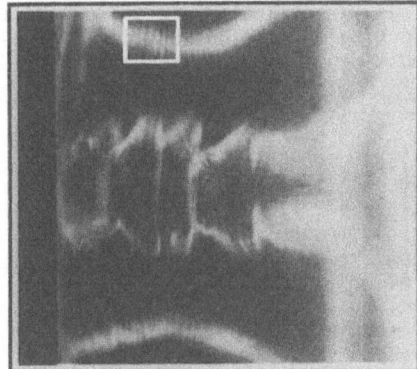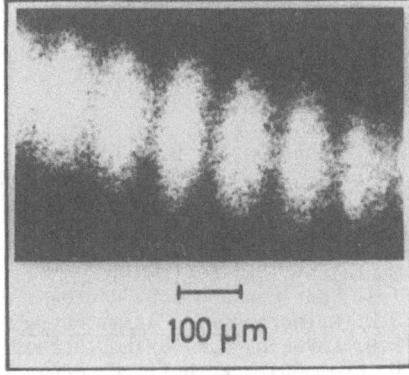

100 µm

Fig.3.6 Periodic structures in the electron density during focus evolution
observed by Schlieren diagnostics

A relativistic electron beam of monochromatic particle energy $W = \gamma mc^2$, moving through a periodic potential, of period length Λ emits electromagnetic radiation at a wavelength $\lambda = \Lambda/2\gamma^2$. According to the relativistic motion of the electrons the radiation is emitted into a cone of angle $\Theta = 1/2\gamma$. By measuring wavelength λ and emission angle Θ of the x radiation, the period length Λ and the relativistic factor γ were determined to be

$$\Lambda = 17 \text{ nm} ; 1 < \gamma < 4. \tag{3.3}$$

The measured value of Λ is close to the Debye length of the plasma. Assuming the occurrence of periodic density structures as a manifestation of strongly oscillating plasma instabilities the type of instability can be determined by a simultaneously performed measurement of period length and frequency. From correlated measurements, combinations of period length and frequency were found, which can be attributed either to a lower hybrid drift instability during the collapse phase, or to a cyclotron drift instability during the final focus phase. The latter instability constitutes a powerful means to accelerate particles in a plasma as its phase velocity is increasing with increasing magnetic field strength. For these obvious reasons we conclude the cyclotron instability to be a promising candidate for particle acceleration.

The low value of the relative linewidth of the x-ray emission points to a well-ordered and uniform density structure. As uncorrelated x-ray emission of individual beam particles would result in a relative linewidth in the order of $d\lambda/\lambda \simeq 1$ we have to assume a coherent superposition of a large number of emission acts in order to explain the low value of the bandwidth of the x radiation. From this consideration we conclude that the electron density has a well-ordered periodic structure over several hundreds of period lengths and that this order remains constant during the time of x-ray emission.

The assumed mechanism for the generation of the anisotropic and monochromatic x radiation is very similar to that of a free electron laser, so that the question arises whether laser action might be achievable in a plasma focus. The measured total power density of 10^{16} W/cm^2 would exceed the necessary threshold pump power for a laser of that kind but at present it is unknown which part of the total power density contributes to the coherent pumping process. A coherent coupling of the single emission acts is imaginable by distributed feedback from the periodic density modulations.

References

3.1 N.P. Economou and D.C. Flanders: Vac. Sci. Technol. 19, 868 (1981)
3.2 I. Bailey et al.: Appl. Phys. Lett. 40, 33 (1982)
3.3 D.J. Nagel et al.: Jp. J. Appl. Phys. 17, 472 (1978)
3.4 R.A. McCorkle and H.J. Vollmer: Rev. Sci. Instrum. 48, 1055 (1977)
3.5 G. Dahlbacka et al.: Proc. Conf. on Low Energy X-Ray Diagnostics, Montherey, Calif. (1981)
3.6 I.S. Pearlman and I.C. Riordan: J. Vac. Sci. Technol. 19, 1190 (1981)
3.7 J.W. Mather: Dense Plasma Focus, in Plasma Physics B, edit. R.H. Lovberg, H.R. Griem, Academic Press, New York, London (1971)
3.8 H. Krompholz, W. Neff, F. Rühl, K. Schönbach, and G. Herziger: Phys. Lett. 77 A, 246 (1980)
3.9 W. Neff, H. Krompholz, F. Rühl, K. Schönbach, G. Herziger: Phys. Lett. 79 A, 165 (1980)
3.10 K. Schmitt, H. Krompholz, F. Rühl, G. Herziger: Phys. Lett. 95 A, 239 (1983)
3.11 H. Krompholz and G. Herziger, in: Chaos and order in nature, ed. H. Haken, Springer Series in Synergetics, Vol. 11 (Springer, Berlin, 1981)
3.12 S. Glasstone and R.H. Lovberg: Controlled Thermonuclear Reactions Van Nostrand, Princeton, New Yersey (1960)
3.13 G. Herziger et al.: Phys. Lett. 71 A, 54 (1979)
3.14 G. Herziger et al.: Phys. Lett. 64 A, 390 (1978)
3.15 R. Haas et al.: Phys. Lett. 88 A 03 (1982)
3.16 F. Asbeck, diploma thesis, Inst. für Angewandte Physik, TH Darmstadt (1976)

4. Laser Produced Plasma VUV and Soft X-Ray Light Sources

M. L. Ginter

Institute for Physical Science and Technology, University of Maryland
College Park, MD 20742, USA

4.1 Introduction

Vacuum ultraviolet (VUV) and soft x-ray (SXR) radiation is available from well-established laboratory (rotating anode, sliding sparks, etc.) and facility-based (synchrotron, etc.) light sources, as well as from more developmental laser-driven (plasma, direct and indirect generation, etc.) light sources. Hence, the experimenter often is confronted with the problem of choosing the best light source to use for a particular application. This process often begins and ends with a rapid survey of tabular information such as the sample comparision of properties for light sources from each of the three previously mentioned groupings in Table 4.1, which contains data abstracted from a 1980 survey report [4.1] on VUV sources and applications. While the tabulated interrelationships in Table 4.1 remain roughly the same today (the "laser, indirect generation" and "laser generated plasmas" values now appear to have been over and under estimates, respectively), it should be emphasized that any such comparative listing of source properties can be extremely misleading. For example, if one wishes to perform experiments using VUV radiation that requires the use of source images on small apertures or irradiation with large solid angles, then the high intensities quoted

Table 4.1 Properties [4.1] of Representative VUV Sources

Demonstrated, 1980 (Projected, 1985) Units	Photon Energy keV	Rep. Rate Hz	Peak Flux Phot. sec. MHz	Peak Intensity Phot.(1%BW) sec. mr^2	Average Intensity Phot.(1%BW) sec. mr^2
Synchrotron, Bending Magnets	<100 (200)	10^6	10^8 (10^{10})	10^{17} (10^{19})	10^{13} (10^{15})
Synchrotron, Undulators	< 10 (100)	10^6	10^{10} 10^{13}	10^{19} 10^{22}	10^{16} 10^{19}
Laser, Indirect Generation	< 10^{-1}	10^2	10^{19}	10^{21}	10^{15}
BRV	< 1	10^1	10^{11}	10^{13}	10^7
Laser-Generated Plasmas	< 1	10^2	10^{11}	10^{14}	10^8
Condensed Gas Spark	< 10^{-1}	10^2	10^{10}	10^{15}	10^{11}

for wigglers and undulators in actuality are not available to the experiment because these sources are extremely extended (several meters) and have very low divergences. In addition, there is the natural tendency for persons developing a particular type of light source to choose systems of units which present their source most advantageously in tabular comparisons. The use of "Peak Flux" (photons x second^{-1} x megahertz^{-1}) is especially advantageous to sources generating high VUV outputs in very narrow wavelength intervals (i. e., "Lasers, Indirect Generation" in Table 4.1) while "Average Intensity" (photons in a 1% wavelength bandwidth x seconds^{-1} x milliradians^{-2}) would be a much more advantageous choice for sources with broad range wavelength outputs (i. e., "Synchrotron,..." in Table 4.1). Hence, the choice of an optimum light source will not be the same for all types of experiments, and such an optimized choice can be made only by a detailed matching of a particular experimental apparatus to the available sources' specific rather than general output parameters. Unfortunately, there are few direct comparisons of the VUV outputs from different classes of sources so that it is difficult to make really optimized comparisons. Hence, one tends to be satisfied with a best guess based on the limited data available.

Having made the obligatory disclaimers implicit in the preceding discussion, it is practical to describe the principal features of a repetitively pulsed, laser-produced plasma source of VUV and SXR radiation [4.2-5] in use at the University of Maryland and to indicate a few applications of the source to lithography and spectroscopy. The basic properties of laser plasma sources have been reviewed recently [4.6], with the conclusion that laser-produced plasmas have been shown to be light sources of exceptional intensity in the VUV (∿6 eV-100 eV) and SXR (∿100 eV-5 KeV) spectral regions. The experiments emphasized in subsequent sections were conducted using plasmas produced by focusing lasers onto metal targets with irradiances in the ∿10^{12}-10^{14} W/cm^2 range to optimize [4.7,8] radiative outputs with energies less than ∿1 KeV.

4.2 Light Source

The principal light source utilized in our recent experiments [4.2,5] is compact and highly mobile. It consists of a cubic vacuum chamber 8.25 cm on a side which contains a cylindrical target attached to a stepping-motor-driven screw [4.2,5]. This source cube can be attached conveniently to almost any experimental chamber or spectrometer and the entire source (including stepping motor and logic circuitry, power supplies, and vacuum attachments but excluding a driver laser) weighs less than 15 kg and packs into a box 0.4 m x 0.6 m x 0.3 m.

VUV and SXR emitting plasmas are produced by focusing the output from a laser onto a point on the target's surface. The target can be rotated to provide a fresh area of surface to the laser-driven pulse after any specified number of laser pulses. Target ablation (see below) and subsequent surface contamination sometimes makes it advantageous to view the plasma plume at nearly right angles to the incident laser beam. While we have used this light source chamber with low pulse repetition rate (less than 10^{-1} Hz) lasers such as ruby and CO$_2$, emphasis here will be on summarizing results [4.4] obtained using a 10 Hz driver. This represents a first step toward laser-produced plasma SXR light sources with pulse repetition rates in the 10^2 - 10^3 Hz range and with very high average intensities.

4.3 Experimental Results

Light from an International Laser Systems (ILS) Nd:YAG laser producing 25 nsec pulses of 1.06 μm radiation with energies and repetition rates variable up to 0.8 J and 10 Hz, respectively, was focused by a glass lens (f= ~10 cm) through a pyrex window onto the source's target cylinder [4.4]. Radiation emitted by the plasma passed from the target chamber [4.2] into an experimental chamber with the plasma viewed nearly perpendicular to the laser beam. The laser–produced single-mode, TEM_{00}, pulses which could be tightly focused to provide target irradiances in the 10^{11}-10^{13} W/cm^2 range, and targets of Yb, Hf, Pb, Al , Sn, U, C (graphite), steel (Fe), teflon, and brass were utilized [4.4].

Diagnostic observations of the light emitted from the source were made [4.4] in the ~30-1.4 nm (~40-800 eV) range using grazing-incidence (1-m, 1200 l/mm grating) and rudidium acid phthalate crystal spectrographs with Kodak 101 film to record spectra. Short wavelength (~2.5-80 nm band pass) pinhole camera images of the plasma were recorded on 101 film, and time histories of the SXR emissions were recorded using an x-ray diode (XRD) filtered so that its response was predominately in the 14 to 4.3 nm (90 to 284 eV) region [4.4].

It was found [4.4] that SXR intensities measured using the filtered XRD were not strongly dependent on focal conditions, which is similar to previous observations [4.2,3,9] at longer VUV wavelengths. There was a very slow drop in XRD signals, marginally detectable after several thousand plasma pulses, which very likely is due to a slight coating of the detector window by evaporated materials (see below). Also, the XRD signals for multiple plasma pulses produced from the same target position indicated that between 10 and 100 pulses of ~0.5 J each can be used at each target location if per pulse variations in the SXR output of ~25% are acceptable to the experimenter.

Figure 4.1 illustrates typical VUV emissions observed from the source using low, intermediate, and high atomic number (Z) target materials. Relatively low Z materials such as aluminum exhibit well-defined line spectra, which increase in complexity through intermediate Z elements such as Fe. However, on reaching high Z elements such as Yb and Pb the observed emissions consist of continua overlaid by weak and sometimes sparse line spectra. The Yb spectrum in Fig.4.1 is characteristic of the now well-known rare earth continua [4.10-13] which apparently result [4.10,11] from various combinations of bremsstrahlung, bound-free transitions, etc., and from open $d^n f^m$ configurations dominating the electronic structures of the ionization stages

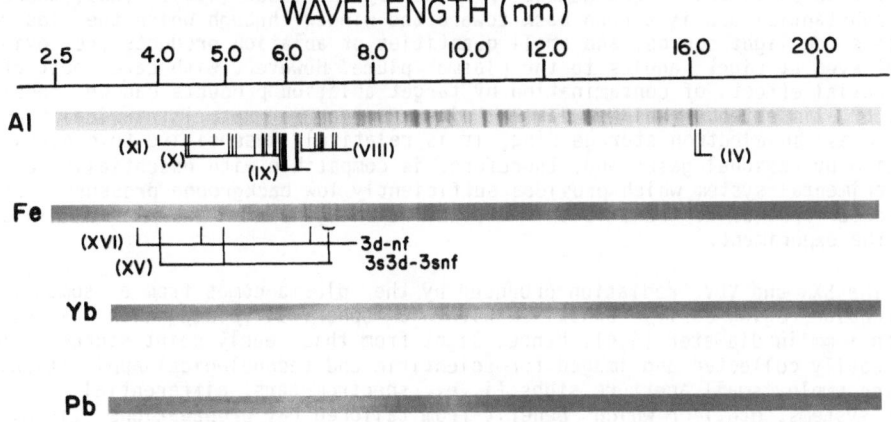

Fig.4.1 Typical spectra from the plasma source at 10 Hz pulse rates

produced in these plasmas. The spectra in Fig.4.1 are very similar to spectra generated using the same light source and ruby lasers with pulse energies up to ∿4 J/pulse.

Figure 4.2 illustrates typical variations of the source's SXR output with laser pulse energy for several elements determined using the filtered XRD [4.4]. The per pulse laser energy threshold for production of measurable XRD signals was in the 25-50 mJ range. Actual data points are included for Yb and Fe only. The illustrated points represent averages of 5 or more pulses, while the error bars represent one standard deviation from the average.

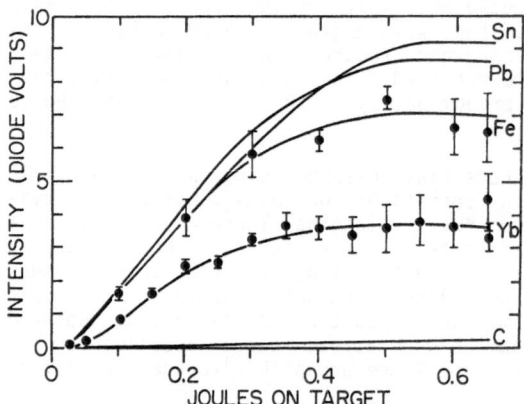

Fig.4.2   Dependence of SXR output on laser pulse energy and target
          materials for 10 Hz pulse rates

Damage craters produced by the laser are largest in materials which vaporize readily and crater depths increase with increasing per pulse energy. Target debris and vaporization products generated by the laser pulse are a significant consideration in experimental applications which employ a laser plasma light source. These target ablation products tend to be concentrated in a cone approximately concentric with the principal plasma cone and are proportional to the target damage produced by the laser [4.4]. Thus, there is substantial debris thrown back toward the window through which the laser enters the light source, and small quantities of ablation products are evident even at right angles to the plasma plume. However, with care most of potential effects of contamination by target ablation products can be reduced or eliminated. While a laser plasma source is not intrinsically as clean as an electron storage ring, it is relatively insensitive to contamination by residual gases and, therefore, is compatible with essentially any experimental system which provides sufficiently low background pressures of absorber species to transmit the desired wavelengths from the plasma plume to the experiment.

The SXR and VUV radiation produced by the plasma comes from a somewhat irregular region easily contained within a spherical volume which is less than 1 mm in diameter [4.4]. Hence, light from this nearly point source can be easily collected and imaged for scientific and technological applications which employ small aperture stops (i. e., spectrometers, differential pumping systems, etc.) or which benefit from tailored ray propagations (i. e., parallel light for lithography, condensed light for microscopy, etc.) as has been demonstrated [4.4].

Acknowledgments

This work was partially supported by the National Science Foundation.

References

4.1 M. Ginter, T. Namioka: "Report of U.S.-Japan Seminar (S-MPC-0104) on Production and Applications of High Power Levels in the Extreme Ultraviolet", Sendai, Japan 1980 (National Science Foundation, Washington, 1981)

4.2 G. O'Sullivan, P. Carroll, T. McIlrath, M. Ginter: Appl. Opt. 20, 3043 (1981)

4.3 G. O'Sullivan, J. Roberts, W. Ott, J. Bridges, T. Pittman, M. Ginter: Opt. Lett. 7, 31 (1982)

4.4 D. Nagel, C. Brown, M. Peckerar, M. Ginter, J. Robinson, T. McIlrath, P. Carroll: "Repetitively-Pulsed Soft X-Ray Plasma Source", Appl. Opt. (in press)

4.5 P. Carroll, E. Kennedy, G. O'Sullivan: "Table-Top EUV Continuum Light Source", IEEE - J. Quant. Elect. (in press)

4.6 T.J. McIlrath: "Laser Produced Plasmas as VUV and X-Ray Light Sources", Laser Focus (in press)

4.7 G.L. Stradling: "Time Resolved Soft X-Ray Studies of Energy Transport in Layered and Planar Laser-Driven Targets", Lawrence Livermore Lab. report UCRL-53271, 1982, available from Nat. Tech. Infor. Sev., U. S. Dept. of Comm., Springfield, VA 22161

4.8 D. Babonneau, D. Billon, J. Bocher, G. Di Bona, X. Fortin, G. Thiell: "X-Ray Emission Experiments at 1.06 μm and 0.35 μm", Proc. 6th Int. Workshop on Laser Interactions and Related Phenomena, Monterey, CA, Oct. 1982

4.9 M. Kuhne: Appl. Optics 21, 2124 (1982)

4.10 G. O'Sullivan: J. Phys. B 16, 000 (1983)

4.11 P.K. Carroll, G. O'Sullivan: Phys. Rev. A25, 275 (1982)

4.12 G. O'Sullivan, P.K. Carroll: J. Opt. Soc. Am. 71, 227 (1981)

4.13 P.K. Carroll, E.T. Kennedy, G. O'Sullivan: Appl. Opt. 19, 1454 (1980)

5. Spectral Radiant Power Measurements of VUV and Soft X-Ray Sources

M. Kühne and B. Wende

Physikalisch-Technische Bundesanstalt, Institut Berlin, Abbestraße 2-12
D-1000 Berlin 10

5.1 Introduction

In the last decade the use of VUV and soft x-ray radiation has steadily been growing, not only in fields of basic research like atomic , molecular , solid state , surface and astrophysics, but also in applied physics and technology like controlled nuclear fusion research, x-ray microscopy and x-ray lithography. In principle, an electron storage ring dedicated to the generation of VUV and soft x-ray radiation meets nearly all requirements in the research areas mentioned above but disadvantages exist especially in the fields of applied physics and technology: (i) high capital and operating costs are disadvantageous, (ii) all instrumentation has to be transported to and operated at the site of the storage ring, (iii) the ultra – high vacuum requirements are incompatible with some types of applications, (iv) especially in the soft x-ray range the usable emission angle perpendicular to the plane of the electron orbit is rather limited. Therefore, different laboratories started with the development of alternative sources for VUV and soft x-ray radiation (particularly for x-ray lithography and microscopy) which are more compact, less expensive to construct and to operate and which are optimized for the specific application. The interest is concentrated mainly on pulsed plasma sources,e.g., [5.1-4] of small emitting areas having high spectral radiant power like vacuum sparks, laser-produced plasmas, plasma focus devices and plasma pinches. Presently the quoted radiant properties of the different sources can hardly be compared amd improvements can be judged only with certain reservations because reliable (absolute) spectral radiant power measurements are missing in the extreme VUV and soft x-ray range. Therefore, in the radiometric laboratory of PTB at the Berlin electron storage ring BESSY an instrumentation was developed particularly suited for spectral radiant power measurements under grazing incidence. In this paper we describe the method and the corresponding uncertainties. The quality of the method is studied by measuring the radiant properties of a laser-produced plasma.

5.2 Method for Spectral Radiant Power Determination under Grazing Incidence Utilizing Electron Storage Ring Radiation

The unknown spectral radiant power of a source is determined by comparison with the calculated spectral radiant power of an electron storage ring (Fig. 5.1). An ellipsoidal mirror images the tangent point of the storage ring or the source under investigation into a monochromator under conditions that the radiant flux is not limited by the entrance slit size. For the two sources the same surface element of the mirror and the same angle of incidence are used. A common aperture stop A determines the spectral radiant power of either the electron storage ring or the source under investigation.

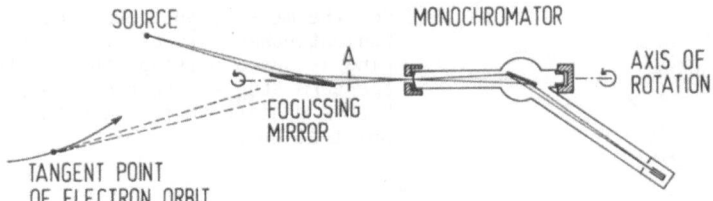

SOURCE   MONOCHROMATOR

A

AXIS OF
ROTATION

FOCUSSING
MIRROR

TANGENT POINT
OF ELECTRON ORBIT

Fig.5.1 Instrumentation for comparing under grazing incidence the unknown
spectral radiant power of a source with the calculated spectral
radiant power of an electron storage ring

To allow for corrections due to the different degree of polarization of both
sources (synchrotron radiation is highly polarized near the electron orbit
plane), the monochromator can be rotated around its optical axis defined by
the center of the entrance slit and the center of the grating. The photocurrent of the detector at the exit slit of the monochromator is i^{sr} for the
storage ring and i^{so} for the source under investigation. The quantity i is
labeled $i_\parallel$ when the plane of incidence of the monochromator (defined by the
direction of the incoming radiation and the grating normal) is in the plane
of the electron orbit, and is labeled $i_\perp$ when the plane of incidence is perpendicular to the orbit. Then (5.1) and (5.2) hold for the unknown spectral
radiant power ϕ_λ^{so} of the source under investigation.

$$\phi_\lambda^{so} = \frac{i_\perp^{so}}{i_\perp^{sr}} \cdot \phi_\lambda^{sr} \cdot F_\perp(\lambda) \tag{5.1}$$

$$\phi_\lambda^{so} = \frac{i_\parallel^{so}}{i_\parallel^{sr}} \cdot \phi_\lambda^{sr} \cdot F_\parallel(\lambda) . \tag{5.2}$$

By applying the SCHWINGER formula [5.5] the spectral radiant power ϕ_λ^{sr} of
a storage ring (for definition see Fig.5.2) is given by

$$\phi_\lambda^{sr}(\lambda) = \phi_\parallel^{sr} + \phi_\perp^{sr} = \frac{2}{3} \cdot \frac{e}{\varepsilon_0} \cdot R^2 \cdot I \cdot \frac{b}{d} \cdot \frac{1}{\lambda^4} \cdot \frac{1}{\gamma^4} [\ldots]$$

$$[\int_{\psi_0-\frac{a}{2d}}^{\psi_0+\frac{a}{2d}} [1+(\gamma\psi)^2]^2 \cdot K_{2/3}^2(\xi) \, d\psi + \int_{\psi_0-\frac{a}{2d}}^{\psi_0+\frac{a}{2d}} [1+(\gamma\psi)^2](\gamma\psi)^2 \cdot K_{1/3}^2(\xi) \, d\psi] \tag{5.3}$$

with $\gamma = \frac{E}{m_0 c^2}$ and $\xi = \frac{2\pi R}{3\gamma^3 \lambda} [1 + (\gamma\psi)^2]^{3/2} .$

E is the energy of electrons with rest mass m_0 and charge e moving in an
orbit with the radius of curvature R at the tangent point of observation
and representing an electron beam current I. $\phi_\parallel^{sr}$ is the spectral radiant
power with the electrical vector of the radiation field oscillating in the
electron orbit, and $\phi_\perp^{sr}$ that one with the electrical vector perpendicular
to the orbit. c is the speed of light in vacuum and ε_0 the permittivity of
vacuum. $K_{2/3}$ and $K_{1/3}$ are Bessel functions of the second kind.

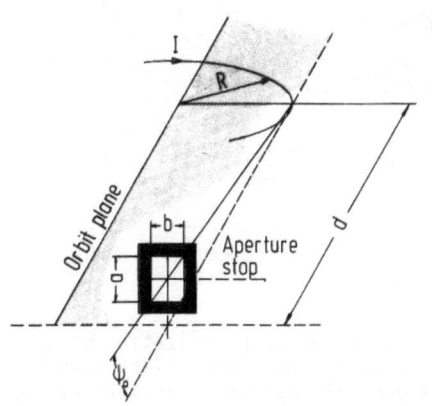

Fig.5.2 For the measurements the spectral radiant power ϕ_λ^{sr} of the storage ring is used passing through the aperture stop A with the size a x b at a distance d from the tangent point at an angle ψ_0

$F_\perp(\lambda)$ and $F_{\shortparallel}(\lambda)$ (in (5.1) and (5.2)) are wavelength–dependent polarization correction functions accounting for the polarization characteristics of the monochromator detector system P^{mo} and the focussing mirror P^{mi}, and accounting for the different degree of polarization P^{sr} and P^{so} of the two sources:

$$F_\perp(\lambda) = \frac{1 + P^{sr} \cdot P^{mo} + P^{mi}(P^{sr} + P^{mo})}{1 + P^{so} \cdot P^{mo} + P^{mi}(P^{so} + P^{mo})} \qquad (5.4)$$

$$F_{\shortparallel}(\lambda) = \frac{1 - P^{sr} \cdot P^{mo} + P^{mi}(P^{sr} - P^{mo})}{1 - P^{so} \cdot P^{mo} + P^{mi}(P^{so} - P^{mo})} . \qquad (5.5)$$

For unpolarized sources ($P^{so} = 0$ and $P^{sr} = 0$) or sources of identical degree of polarization ($P^{so} = P^{sr}$), $F(\lambda) = 1$ holds. The methods for determining the quantities P^{sr}, P^{mo}, P^{mi} and P^{so} entering in $F(\lambda)$ are shortly discussed.

The degree of polarization $P^{sr}(\lambda)$ of the storage ring radiation is calculated according to (5.3) using the usual definition

$$P^{sr}(\lambda) = \frac{\phi_{\shortparallel}^{sr} - \phi_\perp^{sr}}{\phi_\perp^{sr} + \phi_{\shortparallel}^{sr}} . \qquad (5.6)$$

The polarization characteristics of the monochromator detector system $P^{mo}(\lambda)$ is determined according to (5.7) by rotating the monochromator and measuring the corresponding photocurrents $i_{\shortparallel}^{sr}$ and $i_\perp^{sr}$ generated by the electron storage ring radiation:

$$P^{mo}(\lambda) = \frac{i_\perp^{sr} - i_{\shortparallel}^{sr}}{i_\perp^{sr} + i_{\shortparallel}^{sr}} \cdot \frac{1 + P^{sr} \cdot P^{mi}}{P^{sr} + P^{mi}} . \qquad (5.7)$$

If the measurement is performed with storage ring radiation of $P^{sr} = 1$ at $\psi_0 = 0$, according to (5.3) the following simple expression holds

$$P^{mo}(\lambda) = \frac{i_\perp^{sr} - i_{\shortparallel}^{sr}}{i_\perp^{sr} + i_{\shortparallel}^{sr}} , \qquad (5.8)$$

meaning that $P^{mo}(\lambda)$ can be determined without knowing $P^{mi}(\lambda)$.

Knowing P^{mo} the polarization characteristic of the focussing mirror $P^{mi}(\lambda)$ is determined by repeating the rotation of the monochromator using a source of unpolarized radiation (labeled "up" in (5.9)), e.g.,a plasma source under approximately thermal equilibrium:

$$P^{mi}(\lambda) = \frac{i_\perp^{up} - i_\parallel^{up}}{i_\perp^{up} + i_\parallel^{up}} \cdot \frac{1}{P^{mo}} . \tag{5.9}$$

In the case when the mirror is used under grazing incidence at short wavelength, P^{mi} is small compared to 1 and can be calculated with sufficient accuracy from n and k values of the reflecting surface material.

Finally it has to be considered that the source under investigation is emitting radiation with an unknown degree of polarization $P^{so}(\lambda)$. With $P^{mo}(\lambda)$ and $P^{mi}(\lambda)$ known $P^{so}(\lambda)$ is determined by rotating the monochromator and measuring the corresponding photocurrents of the detector $i_\perp{}^{so}$ and $i_\parallel{}^{so}$ for the source under investigation

$$P^{so}(\lambda) = (\frac{i_\perp^{so} - i_\parallel^{so}}{i_\perp^{so} + i_\parallel^{so}} - P^{mo} \cdot P^{mi}) / (P^{mi} - P^{mo} \frac{i_\perp^{so} - i_\parallel^{so}}{i_\perp^{so} + i_\parallel^{so}}) . \tag{5.10}$$

Thus far the measurements or the calculations of those quantities entering in the basic equations (5.1) and (5.2) have been described. This covers the general case of a radiometric comparison between a source standard and a source under investigation with unknown spectral radiant power and unknown degree of polarization. In the following sections some experimental details are reported and an estimation of uncertainties is given.

5.3 Instrumental Details

The instrumentation for the spectral radiant power measurements of VUV and soft x-ray sources (Fig. 5.3) is located in the radiometric laboratory of PTB at BESSY [5.6]. The ellipsoidal focussing mirrors made out of Al with an electroplated polished Ni surface is located 15 000 mm from the tangent point of the storage ring. Using an angle of incidence of 86°, the radiation is focussed into the entrance slit of the TGM in a distance of 930 mm from the mirror. The ellipsoidal mirror can be rotated inside the mirror chamber so that alternatively the source under investigation or the stored electron beam is imaged into the entrance slit of the TGM using the same surface area of the mirror and same angle of incidence. Due to the smaller distance of 5000 mm between the mirror and the source (compared to the storage ring) the image is formed for the source in a distance from the mirror 130 mm greater than for the storage ring. The aperture stop (A in Fig.5.1), typically between 6 x 6 mm^2 to 1 x 1 mm^2, is located directly behind the mirror in the common optical path of both sources.

The TGM is connected to the mirror chamber by means of bellows and a differential pumping system which allows the TGM to rotate around its optical axis maintaining a free optical path under UHV conditions. The bellows are needed to allow the TGM to slide forwards and backwards by 130 mm in order to position the entrance slit within focus of the ellipsoidal mirror for both the storage ring and the source under investigation. Consequently the storage ring irradiates a larger surface area on the grating than the source (see estimation of uncertainties in section 5.4). The TGM has an entrance arm length of 1000 mm, a fixed angle between the arms of 162° and

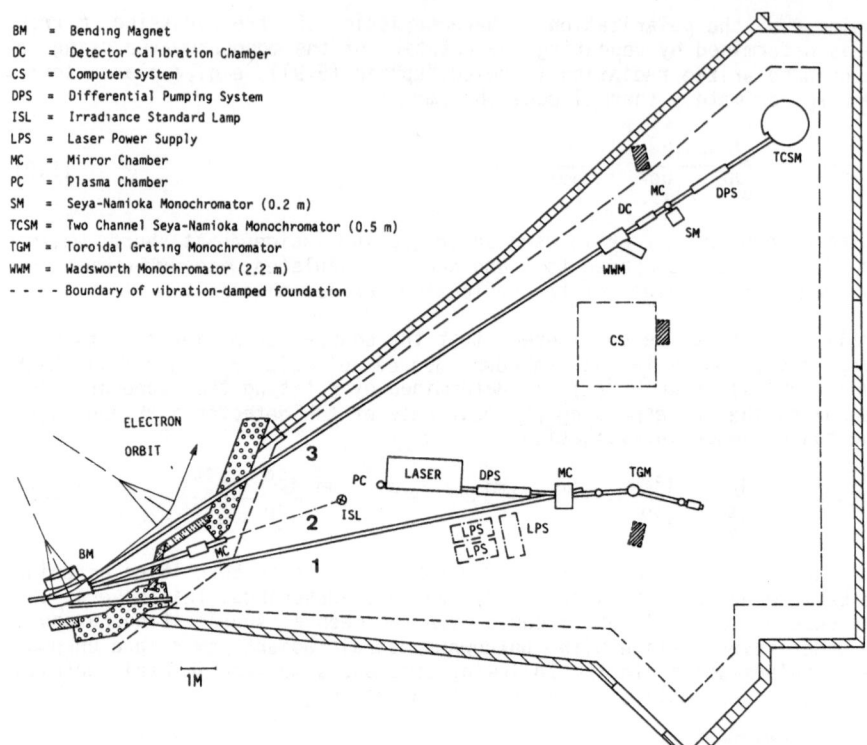

BM   = Bending Magnet
DC   = Detector Calibration Chamber
CS   = Computer System
DPS = Differential Pumping System
ISL = Irradiance Standard Lamp
LPS = Laser Power Supply
MC   = Mirror Chamber
PC   = Plasma Chamber
SM   = Seya-Namioka Monochromator (0.2 m)
TCSM = Two Channel Seya-Namioka Monochromator (0.5 m)
TGM = Toroidal Grating Monochromator
WWM = Wadsworth Monochromator (2.2 m)
- - - - Boundary of vibration-damped foundation

Fig.5.3 Radiometric Laboratory of PTB at BESSY. Beam line No. 1 is used
for spectral radiant power measurements of VUV and soft x-ray
sources. The main parts of the beam line are the bending magnet
chamber BM, the focussing mirror chamber MC, the toroidal grating
monochromator TGM, the differential pumping system DPS, and the
source under investigation, here a laser-produced plasma consisting of the plasma chamber PC and a laser. The differential pumping
system is used to connect the high-vacuum plasma chamber to the
ultra-high vacuum beam line

an exit arm length of 1415 mm. It is equipped with 3 ion etched gold coated
gratings with a diffracting area of 30 x 80 mm having 1800, 600, or 200
lines/mm to cover wavelength ranges from 4 to 12 nm, 12 to 36 nm or 36 to
100 nm respectively. The three gratings are interchangeable under vacuum
conditions. The resolution for all gratings at all wavelengths is better
than 400.

5.4 Estimation of the Characteristic Uncertainty of the Method

According to (5.1) the uncertainty of Φ_λ^{so} is determined by the uncertainty
of the ratio i^{so}/i^{sr} and by the uncertainty of the product $\Phi_\lambda^{sr} \cdot F(\lambda)$. The
uncertainty of $\Phi_\lambda^{sr} \cdot F(\lambda)$ is independent of the type of source under investigation and is considered therefore as a characteristic uncertainty of the
method of spectral radiant power measurement presented in this paper. In
Tab.5.1 those quantities are compiled which contribute to the uncertainty of
Φ_λ^{sr} and $F(\lambda)$. It was found $\Delta\Phi_\lambda^{sr}/\Phi_\lambda^{sr} = 3.7 \cdot 10^{-2}$ and $\Delta F(\lambda)/F(\lambda) = 7.3 \cdot 10^{-2}$.

Table 5.1 Uncertainties* of the spectral radiant power of the storage
ring ϕ_λ^{sr} and of the polarization correction function $F(\lambda)$

(a) contributions to the relative uncertainty of ϕ_λ^{sr}

distance from tangent point to aperture stop	d	$7 \cdot 10^{-4}$
width of aperture	b	$1.4 \cdot 10^{-2}$
height of aperture	a	$1.4 \cdot 10^{-2}$
angle between electron orbit and center of aperture	ψ_o	$2.2 \cdot 10^{-2}$
electron energy and radius of curvature of the orbit	E, R	$2 \cdot 10^{-2}$
electron current	I	10^{-3}

relative uncertainty of ϕ_λ^{sr} (addition in quadrature) $\dfrac{\Delta\phi_\lambda^{sr}}{\phi_\lambda^{sr}} = 3.7 \cdot 10^{-2}$

(b) contributions to the relative uncertainty of $F(\lambda)$

polarization of storage ring radiation	P^{sr}	$2 \cdot 10^{-2}$
polarization characteristic of the monochromator detector system	P^{mo}	$5 \cdot 10^{-2}$
polarization characteristic of the focussing mirror	P^{mi}	$5 \cdot 10^{-2}$

relative uncertainty of $F(\lambda)$ (addition in quadrature) $\dfrac{\Delta F(\lambda)}{F(\lambda)} = 7.3 \cdot 10^{-2}$

* The uncertainties listed in Table 5.1 represent the status as of summer
1983 (the first year of BESSY operation) where both the electron storage
ring and the radiometric instrumentation were still under test. The
achievable uncertainties under later routine operation conditions will be
significantly smaller.

In addition, the contribution to the uncertainty has to be considered
which is caused by the different sizes of the irradiated grating surface due
to the different distances of the electron storage ring and the source under
investigation from the focussing mirror. Using different sized apertures A
the uncertainty caused by possible local efficiency variation on the grating
surface (the irradiated areas differ by 29 %) was determined to less than
$5 \cdot 10^{-2}$. Combining the discussed three contributions in quadrature (ϕ_λ^{sr},
$F(\lambda)$, and the different irradiation of the grating surface) an overall un-
certainty of $\Delta\phi_\lambda^{so}/\phi_\lambda^{so} = 9.6 \cdot 10^{-2}$ follows, which is considered as the characteristic uncertainty of the method.

Additional uncertainties caused by a nonlinear response of the detector,
higher diffraction orders of the grating and stray light, which enter in the
uncertainty of the ratio i^{so}/i^{sr}, depend on the type of the source under
investigation and therefore have to be determined for each measured source
separately (see section 5.5).

5.5 Spectral Radiant Power Measurements of a Laser-Produced Plasma

Generation of the laser-produced plasma:
As a first source to be investigated a laser-produced plasma (LPP) was se-
lected. The Nd:YAG/glass laser system used for the experiment consists of a
Q-switched YAG oscillator, a first YAG amplifier, a spatial filter for TEM_{oo}

operation, a second and a third glass amplifier, a beam splitter for a power/energy meter and a Faraday isolator system. The laser operates at $\lambda = 1064$ nm with a maximum output energy of 6 Joules in TEM_{00} with pulse durations of typically 15-20 ns (FWHM) and a beam divergence of < 200 μrad. The laser pulses were directed through a glass window into a target chamber by a focussing optic with an effective focal length of 130 mm (Fig.5.4). The focus diameter on the target was less than 100 μm. A glass plate was used inside the target chamber to protect the inner side of the window from possible contamination by vaporized target material. This glass plate was routinely checked for change in transmission and replaced accordingly. As a target to the incoming laser beam, a sheet of tungsten was used under an angle of 45^0. The plasma radiation was observed under 90^0 to the laser beam direction. The target chamber was pumped by a turbo pump to a pressure of $5 \cdot 10^{-6}$ mbar.

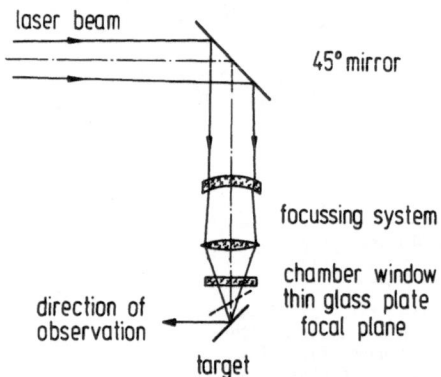

laser beam

45° mirror

focussing system

direction of observation

chamber window
thin glass plate
focal plane

target

Fig.5.4 Generation of a laser-produced plasma

Radiant properties of the LPP: The radiation of the LPP was measured at discrete wavelengths between 7 nm and 100 nm with a resolution of $\lambda/\Delta\lambda \simeq 150$ (Fig.5.5). The photocurrent of the electron multiplier EMI D233B at the exit slit of the TGM was recorded by a 500 MHz transient digitizer and the data were processed and stored into a HP 1000 computing system. Simultaneously the energy of the laser pulse was recorded. The quantities p^{sr}, p^{mo}, p^{so} were determined as given by (5.4) to (5.8) to $p^{sr}(\lambda) = 0.82...0.95 \pm 0.1$ (dependent on the wavelength), $p^{mo}(\lambda) = 0.00...0.15 \pm 0.05$, and $p^{so}(\lambda) = 0 \pm 0.05$. p^{mi} was calculated from n and k values of Ni [5.7] to $p^{mi}(\lambda) = -0.02...-0.11 \pm 0.05$. The negative sign notes that the plane of incidence coincides with the electron orbit plane.

For presenting the results of the spectral radiant power measurements of the plasma we chose the spectral radiant power time integrated over one plasma radiation pulse $Q_\lambda = \int \Phi^{so}_\lambda(\lambda,t)dt$, equivalent to spectral radiant energy or to number of photons $N_\lambda(\lambda)$ (Fig.5.5). Concerning the systematic uncertainty of the measurements we relate to the characteristic uncertainties of the method discussed in section 5.4. In addition those uncertainties (nonlinearity of the detector response, higher diffraction orders, stray light) have to be considered which are determined by the specific radiant properties of the source under investigation and which enter in the uncertainty of the ratio i^{so}/i^{sr} in (5.1) and (5.2). $\Delta(i^{so}/i^{sr})/(i^{so}/i^{sr})$ was estimated to be ≤ 0.1. Combining this uncertainty with the characteristic uncertainty of the method, an overall systematic uncertainty of 0.14 was found for the photon numbers $N_\lambda(\lambda)$ given in Fig.5.5.

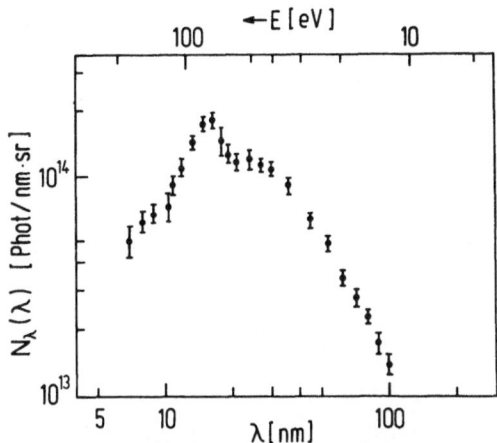

Fig.5.5 Example of radiant measurement of a laser-produced tungsten plasma
(laser pulse energy 800 mJ). $N_\lambda(\lambda)$ is the number of photons emitted during one plasma pulse within a bandwidth of 1 nm related to
a solid angle of 1 sr. The uncertainty bars represent the standard
deviation of 6 independent measurements using for the data evaluation (5.1) as well as (5.2). The systematic uncertainty is estimated at 14%

5.6 Conclusion and Future Development

It has been demonstrated that spectral radiant power measurement of VUV and
soft x-ray sources can be performed using stigmatic imaging optics under
grazing incidence and electron storage ring radiation. The instrumentation
described has a short wavelength limit around 5 nm. A further beam line and
the corresponding instrumentation is under construction to extend the wavelength range down to 0.5 nm (based on the principle shown in Fig.5.1). The main
part of the instrumentation will be a concave grating Rowland-type spectrometer under extreme grazing incidence.

References

5.1 M. Kühne and J.L. Kohl: Appl. Opt. 16, 1786-1788 (1977)
5.2 R.D. Bleach and D.J. Nagel: Appl. Phys. 49, 3832-3841 (1978)
5.3 M.L. Ginter, this volume, no.4
5.4 G. Herziger, this volume
5.5 J. Schwinger: Phys. Rev. 75, 1912-1925 (1949)
5.6 M. Kühne, F. Riehle, E. Tegeler, B. Wende: Nucl. Instr. and Meth. 208,
399-403 (1983)
5.7 H. Behren and G. Ebel, Ed.: Physic Data 18-1, Fachinformationszentrum
Karlsruhe (1981)

Part II

X-Ray Optics

6. Optimum Zone Plate Theory and Design

R. O. Tatchyn

Stanford Synchrotron Radiation Laboratory, Stanford University
Stanford, CA 94305, USA

Transmission zone plates composed of materials describable by the complex index of refraction $\hat{n}$ ($\hat{n} = (1 - \delta) + ik$) modulate both the amplitude and phase of the light passing through them. In this monograph, a general variational approach to the design of zone plate profiles which maximize system transfer functions like $I^{(m)}/I_{IN}$ or $I^{(m)}/I_{OUT}$ (where I_{IN} and I_{OUT} are the total input and output powers and $I^{(m)}$ is the diffracted power of the mth order) is presented in terms of the real and imaginary components of $\hat{n}$. The variational problem, as formulated here, is shown to generate some important classical zone plate configurations as special solutions and to lead naturally to the derivation of the most general zone plate configurations possible. Some specific optimum solutions are derived for gold in the soft x-ray range and tabulated for convenient reference.

6.1 Introduction

In recent years, various suggestions for zone plate configurations have been advanced for use in the VUV and soft x-ray ranges. At least two of these, the first by KIRZ (K) [6.1] and the second by CEGLIO and SMITH (C and S) [6.2], dealt explicitly with materials whose translucency to x rays depends on the complex index of refraction $\hat{n}$($\hat{n} = (1 - \delta) + ik$). Both of these articles examined the effects of specific thickness profiles of the zone plate material on the throughput efficiency into the diffracted orders, with K also investigating the effects of some nonuniform profiles on improving the diffracted output efficiency $I^{(1)}/I_{OUT}$, and C and S attempting to find, on the basis of intuitive arguments, the specific profile which would maximize the diffracted power into the first order. In this monograph, a more general approach will be employed to maximize the diffraction efficiencies of zone plates: the desired diffraction efficiency will be expressed as a functional of the zone plate thickness profile, and a variational analysis will be used to identify the optimum profile which maximizes it. As will be shown, some well-known zone plates will appear naturally as solutions to this variational analysis (for limiting values of k or δ), and the most general efficiency-maximizing profiles will be obtained as well. Once the solutions are derived, some specific numerical examples will be tabulated for gold in the 100-600 eV range.

6.2 Definitions

For the analysis included in this monograph, the following definitions will be useful (refer to Fig.6.1):

$\hat{n}(r) = (1 - \delta(r)) + ik(r)$ ≡ Complex index of refraction as a function of the zone plate radial demension r. δ is the phase-delay constant and k is the attenuation constant.

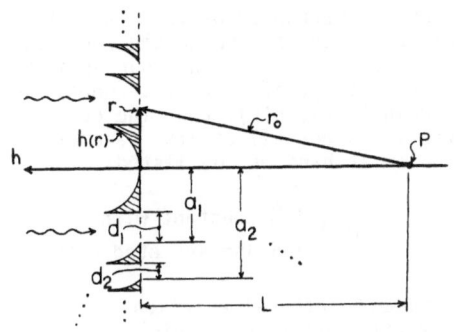

Fig.6.1 Side view of an optimum zone plate profile with focal length L. The incoming wave on the left is a monochromatic unit plane wave originating at infinity. The first order (m=1) focus is at P. Each zone i has radius a_i and an open-aperture area of decremental radius d_i. The zone plate geometry is defined by arbitrarily making its right face flat and by specifying its thickness as a function of r with the thickness profile function h(r). All dimensions are in units of "# of wavelengths" and $(L^2 + a_i^2)^{1/2} - L = 1$

r ≡ Zone plate radial dimension, measured in units of "# of wavelengths."

L ≡ Distance from the zone plate center to the point of observation. The focal length L is measured in units of "# of wavelengths."

a_i ≡ The radius of the ith zone, measured in units of "# of wavelengths."

d_i ≡ The open aperture dimension in the ith zone in units of "# of wavelengths."

$r_o = (L^2 + r^2)^{1/2}$.

m ≡ Number of the diffracted order. Positive m correspond to the real foci, and negative m correspond to the virtual foci.

h(r) ≡ The thickness profile of the zone plate material as a function of r, measured in units of "# of wavelengths."

$I^{(m)}/I_{IN}$ ≡ Throughput efficiency. Ratio of power diffracted into the mth order divided by the total input power.

$I^{(m)}/I_{OUT}$ ≡ Output efficiency. Ratio of power diffracted into the mth order divided by the total output power.

T_R ≡ Set of real numbers.

λ ≡ Wavelength of source light.

6.3 Formulation

In setting up the variational problem, we are free to define and maximize any suitable system transfer function which can be expressed as a functional of the zone plate geometry and ħ(r).[1] For the analysis in this monograph, the entire effect of the zone plate geometry will be represented by the parameter h(r), which is the thickness of the material presented to the light

---

1) Note that although we are restricting ourselves here to maximizing our transfer functions in terms of h, we could just as well maximize them with respect to any of the other parameters listed in the choices above.

at radius r. This reduction of the geometry is warranted by the fact that interface effects like refraction and reflection may be disregarded for most materials in the soft x-ray range [6.3], and it also means that our optimum solutions will be thickness profiles only and not specific shapes. Choosing the appropriate mathematical relation between $(h(r)$, $\hat{n}(r))$ and the desired system transfer function, we are also free to select all of its parameters as well. The principal choices for the problem at hand may be listed:

1. System transfer function $\begin{cases} I^{(m)}/I_{IN} \\ I^{(m)}/I_{OUT} \\ \infty \text{ others} \end{cases}$ 2. $\hat{n}(r) \begin{cases} \hat{n} \text{ constant with } r \\ \hat{n} \text{ unrestricted with } r \end{cases}$

3. Relation between zone plate geometry and system transfer function $\begin{cases} \text{Fresnel-Kirchhoff integral relation} \\ \text{Full-scale vector-integral relation} \\ \text{Others possible} \end{cases}$

4. Refraction $\begin{cases} \text{disregard} \\ \text{include} \end{cases}$ 5. $h(r) \begin{cases} h(r) \text{ predetermined} \\ h(r) \text{ unrestricted} \end{cases}$

6. Zone area geometry $\begin{cases} \text{standard zone plate geometry (see Fig.6.2)} \\ \infty \text{ others} \end{cases}$

7. $k \begin{cases} = 0 \\ \epsilon \text{ TR} \\ = \infty \end{cases}$ 8. $\delta \begin{cases} = 0 \\ \epsilon \text{ TR} \\ = \infty \end{cases}$ 9. Source $\begin{cases} \text{plane wave at } \infty \\ \infty \text{ others} \end{cases}$

There are clearly an infinite number of different cases that may be analyzed, and other choices for other parameters could have been listed as well. For this monograph, however, we will be interested in only a limited number of cases selected from the above listings. These are given in Table 6.1 below. The table assumes (1) the scalar Fresnel-Kirchhoff relation [6.4], (2) the standard zone plate geometry (Fig.6.2), (3) $\hat{n}$ constant with r, (4) a plane wave source at infinity, and (5) no refraction in the zone plate material. The constants have been partitioned as shown to correspond to materials which are transparent, translucent, and opaque, respectively.

Table 6.1 Zone plate cases individually suitable for optimization (see text). Separate columns for cases like (k=0, δ=0), (k= ∞, δ=0), etc., have not been included (1) due to their obvious physical interpretations and (2) to reduce redundancy

	Maximize	$k = 0$ $\delta \epsilon$ TR	$k \epsilon$ TR $\delta = 0$	$k \epsilon$ TR $\delta \epsilon$ TR	$k = \infty$ $\delta \epsilon$ TR
$h(r)$ Constant	$I^{(m)}/I_{IN}$	Case 1	Case 2	Case 3	Case 4
	$I^{(m)}/I_{OUT}$	Case 5	Case 6	Case 7	Case 8
$h(r)$ Unrestricted	$I^{(m)}/I_{IN}$	Case 9	Case 10	Case 11	Case 12
	$I^{(m)}/I_{OUT}$	Case 13	Case 14	Case 15	Case 16

In setting up the appropriate form of the Fresnel-Kirchhoff integral [6.4], we will determine the phase and amplitude of the electric field at each point on the zone area by tracing each incoming ray through the zone material and thereby describing its phase and amplitude relative to all the other rays by the factor $\exp(-2\pi i h(r)[\delta - ik])$. This is a valid procedure whe-

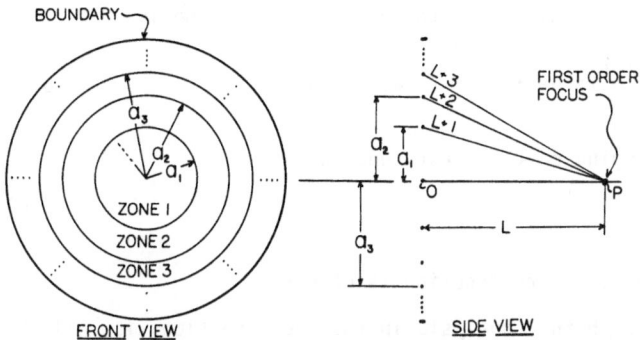

BOUNDARY

L·3
L·2
L·1

FIRST ORDER
FOCUS

a_3
a_2
a_1

a_2
a_1
r_0

P

L

a_3

ZONE I
ZONE 2
ZONE 3

FRONT VIEW

SIDE VIEW

Fig.6.2 The standard zone plate geometry, with all dimensions in units of "# of wavelengths." The successive radii are determined by: (1) $a_1^2 = 2L + 1$, (2) $a_2^2 = 4L + 4$, (3) $a_3^2 = 6L + 9$), etc. If $L \gg 1$, then each zone's area is approximately the same

never the zone radii are at least several wavelengths long and when the total zone plate radius is much greater than the zone plate thickness, conditions very well fulfilled by most zone plates in the soft x-ray range. In addition, we will restrict L to be sufficiently far away from the zone plate so that the cosine factor in the integrand of (6.1) may be neglected [6.4]. It is important to note that all of these restrictions in no way limit the validity and universality of the variational approach set forth in this monograph and are invoked primarily to obtain closed-form solutions in the present analysis.

The choice of zone geometry is depicted in Fig.6.2.

In the scheme shown, the successive zone radii are given by: (1) $a_1^2 = 2L + 1$, (2) $a_2^2 = 4L + 4$, (3) $a_3^2 = 6L + 9$, (4) $a_4^2 = 8L + 16$, etc., where L is the first-order focal length. For $L \gg 1$, all zone areas are approximately equal, and the higher real orders (m=2,3,4,...) are focused approximately at L/2, L/3, L/4, etc. This choice of zone dimensions is necessary for the decomposition of the integral in (6.1) below into a factor depending on the first zone geometry only and a factor depending only on the number of zones (6.2). In this choice of zone geometry, therefore, each zone contributes equally to the amplitude at P.

The desired relation, subject to all the previously stated choices and restrictions is, finally,

$$I^{(m)} \cong \left| \iint_\Sigma \frac{e^{2\pi i r_0}}{r_0} \, ds \right|^2 . \tag{6.1}$$

For m = 1, we can write (6.1) as

$$I^{(1)} \cong A \left| \int_0^{a_1-d_1} 2\pi \frac{e^{2\pi i (r_0 - h(r)[\delta - ik])}}{r_0} \, r \, dr + \int_{a_1-d_1}^{a_1} \frac{e^{2\pi i r_0}}{r_0} \, 2\pi r \, dr \right|^2 \tag{6.2}$$

where A is a function of the number of zones but not of h(r), and the constraint on L is $(L^2 + a^2)^{1/2} - L = 1$ (see Fig.6.1).

43

In all cases, the power emerging from the plate's first zone is

$$I_{OUT}^1 = \int_0^{a_1-d_1} e^{-4\pi h(r)k} 2\pi r dr + \pi(a_1^2 - (a_1 - d_1)^2) \tag{6.3}$$

and the total power impinging on the first zone is

$$I_{IN}^1 = \pi a_1^2 \quad . \tag{6.4}$$

Note that both (6.2) and (6.3) are functionals of $h(r)$.

In order to proceed with the analysis in the next section, it will be convenient to rewrite the quantity in absolute value signs in (6.2) as $(C^2 + S^2)$, where

$$C = \left[\int_0^{a_1-d_1} \frac{2\pi e^{-2\pi hk}}{r_0} \cos 2\pi ((L^2 + r^2)^{1/2} - h\delta + B)rdr \right.$$

$$\left. + 2 \sin \pi((L^2 + a_1^2)^{1/2} - (L^2 + (a_1 - d_1)^2)^{1/2}) \right] \tag{6.5}$$

$$S = \left[\int_{a_1-d_1}^{a_1} \frac{2\pi e^{-2\pi hk}}{r_0} \sin 2\pi((L^2 + r^2)^{1/2} - h\delta + B)rdr \right] \tag{6.6}$$

and

$$B = -1/2((L^2 + a_1^2)^{1/2} + (L^2 + (a_1 - d_1)^2)^{1/2}) \quad . \tag{6.7}$$

6.4 Analysis for Predetermined $h(r)$

If we predetermine $h(r)$ to be constant, the quantity $C^2 + S^2$ becomes

$$\frac{4\pi}{a_1^2} I_{(first\ zone)}^{(1)} = C^2 + S^2 = 4 \left[e^{-4\pi hk} \sin^2(\pi((L^2 + (a_1 - d_1)^2)^{1/2} - L)) \right.$$

$$+ 2e^{-2\pi hk} \sin(\pi((L^2 + (a_1 - d_1)^2)^{1/2} - L))\sin(\pi((L^2 + a^2)^{1/2}$$

$$- (L_1^2 + (a_1 - d_1)^2)^{1/2})) \times \left\{ \cos(2\pi(h\delta - \frac{1}{2}(L - (L^2 + a_1^2)^{1/2}))) \right\}$$

$$\left. + \sin^2(\pi((L^2 + a_1^2)^{1/2} - (L^2 + (a_1 - d_1)^2)^{1/2})) \right] \quad . \tag{6.8}$$

We can now maximize $I^{(1)}$ with respect to both d_1 and h by taking the ordinary derivatives of it with respect to these parameters and setting them to zero. This will yield

$$(L^2 + (a_1 - d_1)^2)^{1/2} - L = (L^2 + a_1^2)^{1/2} - (L^2 + (a_1 - d_1)^2)^{1/2} = \frac{1}{2} \tag{6.9}$$

and

$$e^{-2\pi hk} = \frac{\cos(2\pi h\delta - \gamma)}{\cos \gamma} \quad ; \quad k\delta \neq 0, \infty \tag{6.10}$$

where

$$\cos \gamma = \frac{k}{(k^2 + \delta^2)^{1/2}} \quad . \tag{6.11}$$

Once the optimum h, h_0, is found from (6.10), the maximum efficiency $(I^{(1)}/I_{IN})_{MAX}$ is given by

$$(I^{(1)}/I_{IN})_{MAX} = \frac{1}{\pi^2} (\sin^2 2\pi h_0\delta)(1 + (\frac{\delta}{k})^2) \quad ; \quad k, \delta \neq 0, \infty \quad . \tag{6.12}$$

Given (6.8) - (6.11), we can fill in the cases in the first row of Table 6.1. We first see that relation (6.9) holds for all values of k and δ. For $k = \infty$ (opaque material), we therefore clearly end up with the Fresnel zone plate (see Tables 6.1 and 6.2). For other limiting values of k and δ, (6.8) should be consulted directly.

Table 6.2 Optimum profiles corresponding to the cases given in Table 6.1 (for m=1). Some well-known classical zone plates are seen to appear as optimum solutions

	Maximize	$k = 0$ $\delta \in TR$	$k \in TR$ $\delta = 0$	$k \in TR$ $\delta \in TR$	$k = \infty$ $\delta \in TR$
h(r) Constant	$I^{(1)}/I_{IN}$	Rayleigh Phase Plate (Case 1)	Fresnel Zone Plate (Case 2)	Generalized Phase Plate (Case 3)	Fresnel Zone Plate (Case 4)
	$I^{(1)}/I_{OUT}$	Rayleigh Phase Plate (Case 5)	---- (Case 6)	---- (Case 7)	---- (Case 8)
h(r) Unrestricted	$I^{(1)}/I_{IN}$	Blazed Rayleigh Phase Plate (Case 9)	Fresnel Zone Plate (Case 10)	Generalized Blazed Zone Plate (Case 11)	Fresnel Zone Plate (Case 12)
	$I^{(1)}/I_{OUT}$	Blazed Rayleigh Phase Plate (Case 13)	---- (Case 14)	Generalized Output-Efficiency Zone Plate (Case 15)	---- (Case 16)

For $\delta = 0$, we see that h should be infinitely thick to provide complete attenuation in the material-filled areas, resulting once again in the Fresnel zone plate (Case 2). For $k = 0$, the equations enforce $h = 1/(2\delta)$ which yields the well-known Rayleigh phase plate. For k, $\delta \in TR$, we obtain the "generalized phase plate," whose thickness will be given by (6.10) for a given (k, δ) and which has also been previously analyzed by KIRZ [6.1].

To solve Cases 5-8 in Table 6.1, we need to take

$$\frac{d}{d(d_1, h)} \left\{ \frac{a_1^2(C^2 + S^2)}{4\pi^2(a_1^2 - (a_1 - d_1)^2(1 - e^{-4\pi hk}))} \right\} = 0 \quad . \tag{6.13}$$

For $k = 0$, the optimum solution is again the Rayleigh phase plate. The specific solutions for Cases 6-8 will not be presented in this monograph due to lack of space, but their determination, based on (6.13), is straightforward.

6.5 Analysis for Unrestricted h(r)

To find the optimum unrestricted functions $h_0(r)$ we will take the variation of the system transfer functions $I^{(1)}/I_{IN}$ and $I^{(1)}/I_{OUT}$ (which are functionals of $h(r)$) with respect to $h(r)$ and set them equal to zero [6.5]. This will yield two classes of functions which will make the system transfer functions stationary with respect to $h_0(r)$. Some two of these functions will maximize $I^{(1)}/I_{IN}$ and $I^{(1)}/I_{OUT}$.

The two stationary conditions may be expressed as

$$\frac{\delta(I^{(1)})}{\delta h(r)}\bigg|_{h(r) = h_0(r)} = 0 \qquad (6.14)$$

and

$$\frac{\delta(I^{(1)}/I_{OUT})}{\delta h(r)}\bigg|_{h(r) = h_0(r)} = 0$$

or

$$\left(\frac{\delta I^{(1)}}{\delta h(r)} - \Gamma \frac{\delta I_{OUT}}{\delta h(r)}\right)\bigg|_{h(r) = h_0(r)} = 0 \qquad (6.15)$$

where

$$\Gamma = \frac{I^{(1)}}{I_{OUT}}\bigg|_{h(r) = h_0(r)} . \qquad (6.16)$$

In solving the perturbations, we will use the identity

$$\left(\frac{\delta}{\delta h(r)}\left\{\left[\int R(h(r))dr\right]^2 + \left[\int I(h(r))dr\right]^2\right\} = 0\right) \iff$$

$$\left(\int \left[\left\{\int R(h(r))dr\right\}\frac{\partial R}{\partial h} + \left\{\int I(h(r))dr\right\}\frac{\partial I}{\partial h}\right]dr = 0\right) . \qquad (6.17)$$

Given this relation and the expressions (6.2-7), (6.11) and

$$\cos \nu = \frac{C}{(C^2 + S^2)^{1/2}} \qquad (6.18)$$

we straightforwardly derive the stationary condition for extremization of $I^{(1)}/I_{IN}$ to be

$$-4\pi^2(C^2 + S^2)^{1/2}(k^2 + \delta^2)^{1/2}(e^{-2\pi hk}/r_0)$$

$$\times \cos(2\pi((L^2 + r^2)^{1/2} - h\delta + B) + \gamma - \nu) = 0 \qquad (6.19)$$

and the stationary condition for extremization of $I^{(1)}/I_{OUT}$ to be

$$\cos(2\pi((L^2 + r^2)^{1/2} - h\delta + B) + \gamma - \nu) = \frac{4\pi \cos \gamma \, \Gamma r_0 e^{-2\pi hk}}{a_1^2(C^2 + S^2)^{1/2}} \quad . \tag{6.20}$$

For (6.19), the maximizing solutions will occur when the argument of the cosine is set equal to $\pi/2 \pm M\pi$, where M is any integer. If we specify the boundary condition

$$h_0(0) = 0 \tag{6.21}$$

the optimum profile in the first zone will be given by

$$\begin{cases} h_0(r) = 0; & (a_1 - d_1(opt)) < r < a_1 \\ h_0(r) = (1/\delta)((L^2 + r^2)^{1/2} - L); & 0 \le r < (a_1 - d_1(opt)) \end{cases} \tag{6.22}$$

where $d_1(opt)$ may be derived from the transcendental equation

$$e^{-2\pi \frac{k}{\delta} (F+1)} = \frac{-\sin(2\pi F - \gamma)}{\sin \gamma} \tag{6.23}$$

and the relation

$$F = (L^2 + (a_1 - d_1)^2)^{1/2} - (L^2 + a_1^2)^{1/2} \quad . \tag{6.24}$$

Once the optimum F, F_{opt}, is found from (6.23), the maximum possible efficiency for the optimum zone plate will be given by (cf. (6.12))

$$(I^{(1)}/I_{IN})_{MAX} = \frac{1}{\pi^2} (\sin^4 \pi F_{opt})(1 + (\tfrac{\delta}{k})^2) \; ; \; k, \delta \ne 0, \infty \quad . \tag{6.25}$$

At this point, we can examine the solutions for Cases 9-12 in light of (6.22-25). When $k = 0$, (6.23) and (6.24) state that $d_1(opt) = 0$, and we obtain what may be called the "blazed Rayleigh phase plate". For $k = \infty$ and $\delta = 0$, just as in the case of constant $h(r)$, the solutions generate the Fresnel zone plate. For the case most relevant to the soft x-ray range, where k, $\delta \varepsilon$ TR, we find that $d_1(opt)$ varies with both k and δ, and the profile and dimensions are specified by (6.22-24). The structure in this case (Case 11) may be called the "generalized blazed zone plate".

For (6.20), a closed-form solution is, in general, impossible. Solutions may be generated on the computer by setting the boundary condition

$$h_0(0) = 0 \tag{6.26}$$

and using the slope relation derived from (6.20)

$$\frac{dh_0}{dr} = \frac{\frac{r}{r_0}\left[\frac{1}{r_0} + 2\pi \tan(2\pi((L^2 + r^2)^{1/2} - h_0\delta + B) + \gamma - \nu)\right]}{2\pi\delta \tan(2\pi((L^2 + r^2)^{1/2} - h_0\delta + B) + \gamma - \nu) + 2\pi k} \tag{6.27}$$

to construct h_0 and to find the optimum d_1, $d_1(opt)$.

For k = 0, however, we can clearly see that the solution for Case 13 will also be the blazed Rayleigh phase plate and, for $k = \infty$, the solution for Case 16 will be identical to that for Case 8. For $\delta = 0$, we see from (20) that $h_o(r)$ will vary as $\ln ((L^2 + r^2)^{1/2})$. The structure corresponding to this case (Case 15) may be referred to as the "generalized output-efficiency zone plate".

It should be pointed out that the above results constitute a complete solution to the same problem addressed by KIRZ [6.1] in his comprehensive monograph. Numerical computations of h_o for this case indicate efficiencies $(I^{(1)}/I_{OUT})_{MAX}$ of as much as 82% for gold in the low end of the soft x-ray range ($\sim$150 eV).

6.6 Numerical Examples

In this section, we will tabulate optimum parameters and efficiencies for a gold zone plate ($\hat{n}$ constant with r) designed to maximize $I^{(1)}/I_{IN}$ (i.e., Cases 3 and 11). Some other useful numerical parameters and figures of merit for some of the constant-thickness zone plates (Cases 1-8) may also be found in KIRZ [6.1]. The numbers for Cases 3 and 11 are given below in Table 6.3 and have been derived from recent optical constant compilations [6.6]. Needless to say, the numbers are not to be taken as final, as the true values of the optical constants used in this table may yet prove to be significantly different in the future [6.3].

Table 6.3 Optimum parameters and figures of merit for gold zone plates with constant and unrestricted profiles over the range 100-600 eV. Text equations (6.10-12) and (6.23,24) have been used for the computations.

eV	$-F_{opt}$	$(I^{(1)}/I_{IN})_{MAX}$ (h(r) Unrestricted)	$(I^{(1)}/I_{IN})_{MAX}$ (h(r) Constant)	k	δ	k/δ
100	0.288	0.27	0.173	0.0454	0.111	0.409
120	0.2416	0.38	0.22	0.0178	0.071	0.25
140	0.238	0.39	0.22	0.0109	0.045	0.24
150	0.25	0.36	0.21	0.0101	0.037	0.273
160	0.209	0.31	0.19	0.0101	0.03	0.33
180	0.309	0.23	0.15	0.0107	0.021	0.51
200	0.337	0.19	0.13	0.0115	0.017	0.676
240	0.36	0.16	0.12	0.0113	0.013	0.87
280	0.3621	0.16	0.12	0.0097	0.011	0.88
300	0.372	0.15	0.116	0.009	0.0091	0.99
350	0.368	0.15	0.119	0.0073	0.008	0.91
500	0.368	0.15	0.118	0.0047	0.005	0.942
600	0.358	0.16	0.123	0.0034	0.004	0.85

(Note for "h(r) Constant" column: $F_{opt} = -0.5$)

It is of interest to compare these numbers to those derived by C and S [6.2]. In that paper, the authors attempted to specify intuitively a profile function which they felt would maximize $I^{(1)}$ and, on that basis, ended up with less-than-optimum throughput efficiencies. For example, one of their conclusions was that it would be of no value to blaze a zone plate when "$k/\delta \gtrsim 0.2$," whereas we see from Table 6.3 that doing so can actually yield efficiency gains of from 30 to 80% over the "generalized phase plate" (Case 3) - over the entire 100-600 eV range.

6.7 Higher Zones and Higher Orders

The reader will have noticed that for the optimum unrestricted profile our solutions specified only the first zone (6.22) and were presented for m = 1. This was done to conserve space and because the extension to higher zones and orders is easily made. Using (6.19), the general profile for the ith zone becomes (cf. (6.22))

$$
\begin{cases}
h_{0_i}(r) = 0; \ (a_i - d_i(opt)) < r < a_i \\
h_{0_i}(r) = \frac{1}{\delta} ((L^2 + r^2)^{1/2} - (L^2 + a_{i-1}^2)^{1/2}); \\
\quad a_{i-1} < r < (a_i - d_i(opt)); \ a_0 = 0 \ .
\end{cases}
\tag{6.28}
$$

For the constant-profile zone plates, Cases 1-4, the following relations hold:

$$
\begin{cases}
(L^2 + a_i^2)^{1/2} - (L^2 + (a_i - d_i)^2)^{1/2} = \frac{1}{2} \\
(L^2 + (a_i - d_i)^2)^{1/2} - (L^2 + a_{i-1}^2)^{1/2} = \frac{1}{2} ; \ a_0 = 0 \ .
\end{cases}
\tag{6.29}
$$

For the higher orders, the entire analysis is still valid, with a more general constraint imposed:

$$
(L^2 + a^2)^{1/2} - L = m; \ m = 1, 2, 3, \dots \ .
\tag{6.30}
$$

Similar extensions are obtained for all the other parameters and cases.

6.8 Summation

In our foregoing discussion, we have attempted to formulate the problem of maximizing system transfer functions which may be expressed as functionals of zone plate geometries in the most general way possible. As a result, none of the specific choices and restrictions made in the foregoing analyses limit or reduce the generality of our approach. For example, instead of a plane wave source at infinity, we might have included a point source a finite distance away in the Fresnel-Kirchhoff integral [6.4] (6.1). For that matter, the full-scale vector analog of the Fresnel-Kirchhoff integral could have been employed. With either choice, we could have selected a zone geometry that would have enabled us to separate our integral diffraction relation (as with (1) and (2) in the text), or we could have selected a more arbitrary geometry. Zone plate geometries on curved surfaces could have been considered, and more general optical effects like reflection and refraction or n̂ nonconstant [6.5] with r could have been included. With each of these refinements, the analysis might well become more complex, but it should always be possible to extract the solutions with a computer.

One interesting follow-up investigation that comes to mind would be to try to determine what restrictions on h(r) or what choice of transfer function in the presented formalism would lead to the Gabor zone plate [6.1] (which is clearly an extremum solution). Extending the present analysis to include refraction could also yield important results, as this becomes a strong effect in the VUV range for many metals [6.3]. Much work remains to

be done in finding zone area geometries (note that arbitrary transfer functions can also be maximized with respect to the zone area geometry Σ) and
appropriate transfer functions which would lead to optimal matching of sources like synchrotrons or undulators (wigglers) to desired image points P.
Finally, the maximization of other types of the system functions (such as,
e.g., the system resolution) is also an important area for future investigation.

In summary, it should be clear from these observations, as well as from
the multiplicity of choices listed in Section 6.3, that there is ample opportunity for the extension and further application of the analysis developed in this monograph.

Acknowledgments

The author is indebted to Paul Csonka for his interest and participation in
developing the analysis in this monograph. Portions of this work were performed at the Stanford Synchrotron Radiation Laboratory which is supported
by the National Science Foundation through the Division of Materials Research in cooperation with the Department of Energy.

References

6.1 J. Kirz: "Phase Zone Plates for X-Rays and the Extreme UV," J. Opt. Soc.
Am. 64, 301-309 (1974)
6.2 N.M. Ceglio and H.I. Smith: Proc. VIII Int. Conf. on X-Ray Optics and
Microanalysis (Boston, 1977)
6.3 R. Tachyn: Ph.D. Thesis, Stanford University, 1982, Chapter III
6.4 J.W. Goodman: Introduction to Fourier Optics, (McGraw-Hill, New York
1968, Chapter III)
6.5 R. Tatchyn, P.L. Csonka, and I. Lindau: "Optimization of Planar Metallic
Non-Refracting Transmission-Grating Profiles for m^{th} Order Intensity
Maximization in the Soft X-Ray Range", J. Opt. Soc. Am. 72, 1630-1638
(1982)
6.6 H.J. Hagemann, W. Gudat, and C. Kunz: "Optical Constants from the Far
Infrared to the X-Ray Region: Mg, Al, Cu, Ag, Au, Bi, C, and Al_2O_3,"
Rep. No. 74/7, Deutsches Elektronen-Synchrotron, Hamburg, West Germany
(1974)

7. Planar Techniques for Fabricating X-Ray Diffraction Gratings and Zone Plates

H. I. Smith, E. H. Anderson, A. M. Hawryluk, and M. L. Schattenburg

Department of Electrical Engineering and Computer Science
Massachusetts Institute of Technology, Cambridge, MA 02139, USA

Planar techniques employed in fabricating Fresnel zone plates and diffraction gratings are reviewed briefly, with emphasis on recent developments.

7.1 Introduction

In some respects, the fabrication of gratings and zone plates is simpler than the fabrication of the complex structures used in modern integrated electronics. Pattern description can be very concise and there are no requirements for multilevel alignment. On the other hand, the requirements of x-ray optics for structures with linewidths less than 100 nm and for spatial-phase fidelity (i.e. each element of a pattern must be located at its assigned position to within a fraction of the finest spatial period of the structure) present special challenges generally not encountered in other applications of submicrometer structures.

This paper presents a brief review of recent developments in planar fabrication techniques that have relevance to x-ray optics. We will not attempt to provide a thorough review of the field but, instead, will discuss selected topics that are new or we feel are especially important, and try to provide adequate references to current literature. Other reviews the reader may find useful are given in [7.1-6].

7.2 Multilayer-Resist Techniques

An important new development in planar fabrication is multilayer-resist techniques. A large variety of such techniques has been described in the literature [7.6-10]. Figure 7.1 summarizes their essential features. The basic idea is to separate the exposure and "resistance" functions of a resist. As depicted in Fig.7.1, exposure is carried out in a thin top layer. A high-aspect-ratio structure is then obtained in the lower layer, either by directional reactive ion etching or by exposure with x rays or deep UV. (The latter has been called the portable-conformable-mask technique [7.9].) Figure 7.1 is not all inclusive; many techniques follow a similar theme but a different procedure. In any case, multilayer techniques have greatly expanded lithographic flexibility and the range of certain lithography tools. For example, optical projection lithography at submicrometer linewidths has a small depth of focus, and thus demands a flat exposure surface. On surfaces with considerable topography, planarization can be achieved via multilayer resist techniques. The lower layer can also include a dye to suppress substrate back reflection [7.11]. In electron-beam lithography, multilayer-resist techniques lessen problems created by electron backscattering. Also, very thin resist layers become practical.

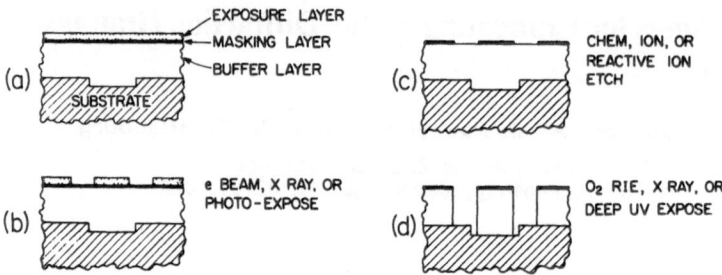

Fig.7.1 Schematic illustration of multilayer resist techniques - principles and variety of approaches

7.3 Scanning Electron-Beam Lithography

Scanning electron-beam lithography (SEBL) is an especially powerful tool for exposing Fresnel zone patterns [7.12,13]. A wide variety of zone plate spe- cifications can be achieved, changes are easily made in software, and the distortions of the electron-beam system, which would normally cause large phase errors, can be detected by moire techniques [7.14] and corrected [7.12]. In the work at M.I.T. Lincoln Laboratory on zone plate fabrication [7.12] (Fig.7.2), an SEBL system was used to generate an x-ray lithography mask. This mask was then used with C_K x-ray lithography to replicate the zone plate pattern in thick resist. (Such a two-stage process has signifi- cant advantages, as discussed below in Section 7.5). After x-ray exposure and development, the final zone plate structure was produced by gold electropla- ting.

Fig.7.2 Scanning electron micrograph of a free-standing gold zone plate. The gold thickness is 1.3 μm and the minimum zone width is 300 nm [7.12]

D. Kern and co-workers at IBM have developed further the methods of SEBL zone plate fabrication [7.13], using both single-layer and multilayer resist techniques. The latter, after reactive-ion etching, yielded 180 nm-wide zones in 1.4 μm-thick resist. They have also exposed patterns on x-ray masks and achieved good quality structures with minimum zone widths of 100 nm. The SEBL system used by KERN et al. was designed for research at 100-nm linewidths and below [7.15]. It includes proximity-effect correction [7.16] and a specially designed pattern generator which provides convenient Cartesian-to-polar-coordinate transformations. These features are important for the zone plate work.

In order to fabricate zone plates capable of approaching the resolution limits of x-ray microscopy it will be necessary to push toward the limits of electron-beam lithography. Electron scattering within the resist and back from the substrate has long been recognized as a serious problem in SEBL, although it can be overcome in most cases by proper control of exposure dosage and resist contrast. Recently, a rather simple method for modeling the effect on exposure of electron scattering was described [7.17]. The method applies to the exposure of periodic grating patterns with a ratio of linewidth W to period P of 1/2. A so-called resist-contrast function K is defined as

$$K = (E_A - E_B)/(E_A + E_B) \qquad (7.1)$$

where E is the normalized energy density absorbed in the resist; E_A is at the middle of a "line" and E_B is at the middle of a "space". The quantity E includes contributions from the input beam, from the forward scattering of the input beam, and from substrate backscattering [7.18]. Contributions from all lines of the grating are included. Electron scattering is calculated using a two-Gaussian model [7.19]. Figure 7.3 plots the resist contrast function K versus linewidth W assuming W/P = 1/2, for two values of η_E (the ratio of the volume integrated exposure by the backscattered electrons to that by the forward-scattered electrons) [7.18].

For a high-contrast resist, such as PMMA, a K value of 0.4 would probably be adequate if precise control of exposure and development parameters is

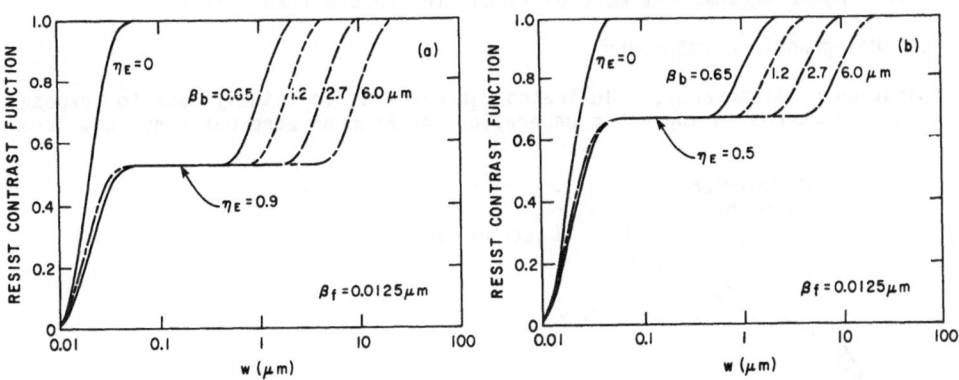

Fig.7.3 Resist contrast function K for a thin resist, a negligible input-beam diameter and exposure of grating pattern of linewidth W, and period 2W. β_f is a Gaussian parameter which characterizes the spreading due to forward scattering. β_b is for spreading of the backscattered electrons [7.18]

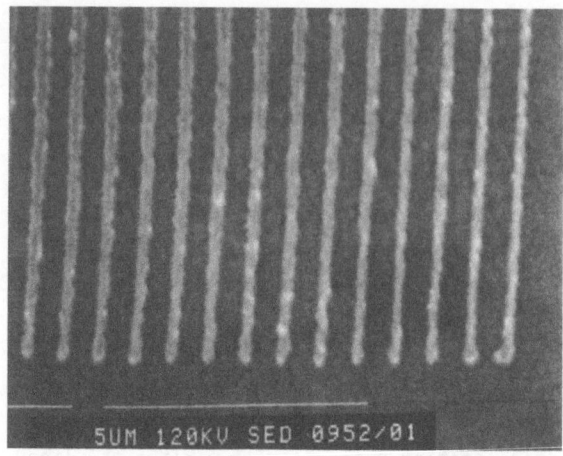

Fig.7.4 Scanning electron micrograph of 15 nm-thick, 20 nm-wide AuPd lines on GaAs produced by SEBL exposure of PMMA at 20 KeV, followed by development, deposition of the AuPd, and liftoff [7.20]

exercised. Thus, Fig.7.3 predicts that exposure in PMMA of 40 nm period gratings should be possible even on high atomic number substrates where $\eta_E \sim 0.9$. Recently, CRAIGHEAD et al. [7.20] demonstrated SEBL and liftoff of AuPd grating patterns of 70 nm period, 20 nm linewidth on GaAs, as shown in Fig.7.4. They used a 20 KeV electron beam of $\sim$ 2 nm linewidth to carry out the exposure. To compare their results with Fig.7.3, one should bear in mind that P > W/2. Since CRAIGHEAD et al. used a beam diameter much less than P/2, a somewhat higher K than predicted by Fig.7.3 is achieved. The results of CRAIGHEAD et al. indicate the feasibility of using SEBL to generate gratings with linewidths $\sim$10 nm on high or moderate-Z substrates. (To achieve 10-nm linewidth they used 120KeV.) BEAUMONT and co-workers demonstrated grating patterns of $\sim$10 nm linewidth on thin carbon substrates [7.21]. Zone plates should be somewhat more difficult to produce than gratings.

7.4 Holographic Lithography

Holographic lithography, illustrated in Fig.7.5, is widely used to expose grating patterns. Holographic generation of Fresnel zone patterns has been

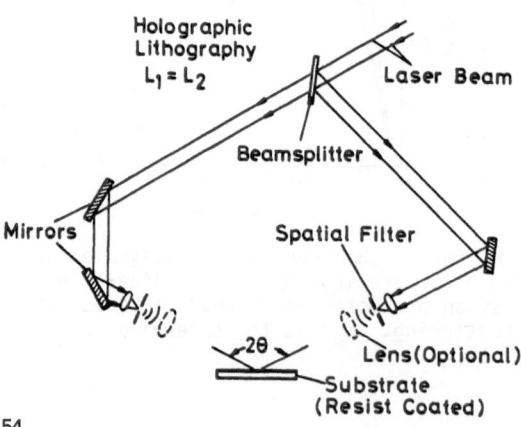

Fig.7.5. Schematic illustration of a configuration for holographic lithography of grating patterns. Note that the number of reflections in the two arms are equal

brought to a high state of development by SCHMAHL, RUDOLPH and co-workers [7.22, 23]. Articles no. 8-10 in this volume describe their techniques for correcting spherical aberration and their recent progress in achieving sub-100 nm zonewidths. Here we will limit our discussion to some new developments in grating-pattern exposure and resist-profile control.

Ordinarily, in holographic lithography, there is a reflection back from the substrate. This leads to an orthogonal standing wave with planes of maximum and minimum intensity parallel to the substrate surface. As a result, the profile of a developed resist line depends on the position during exposure of the resist relative to the orthogonal standing wave. On a reflecting surface, for example, the resist profile is rounded or "overcut" because intensity at the resist-substrate interface is a minimum. A dielectric film of proper thickness between the resist and the reflecting substrate can position the maximum intensity at the lower resist surface, leading to undercut profiles. We have demonstrated this, Fig.7.6, and also described two distinct but simple means for ensuring that the maximum is located at the interface [7.24,25]. Knowledge of the properties and thickness of the dielectric film are not required.

Figure 7.7 illustrates a simple but effective means of achieving vertical-sidewall profiles in holographic lithography [7.26]. It is similar in concept to the multilayer-resist schemes depicted in Fig.7.1: exposure takes place only in a top layer; the rest of the film is etched by directional O_2 reactive-ion etching (after deposition of a suitable mask material). Figure 7.8 shows the resulting resist profiles. This is currently the preferred method of holographic lithography in our laboratory. It is used, for

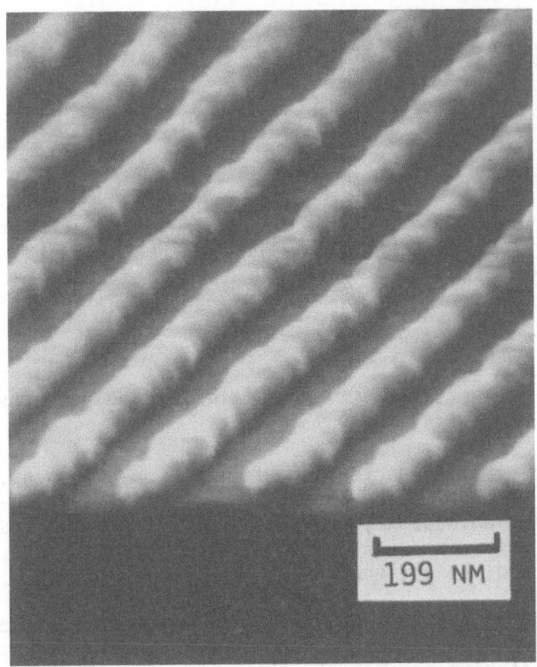

199 NM

Fig.7.6. Scanning electron micrograph of undercut profiles obtained in AZ 1350B type photoresist by holographic lithography when the maximum of the orthogonal standing wave is located at the resist-substrate interface [7.25]

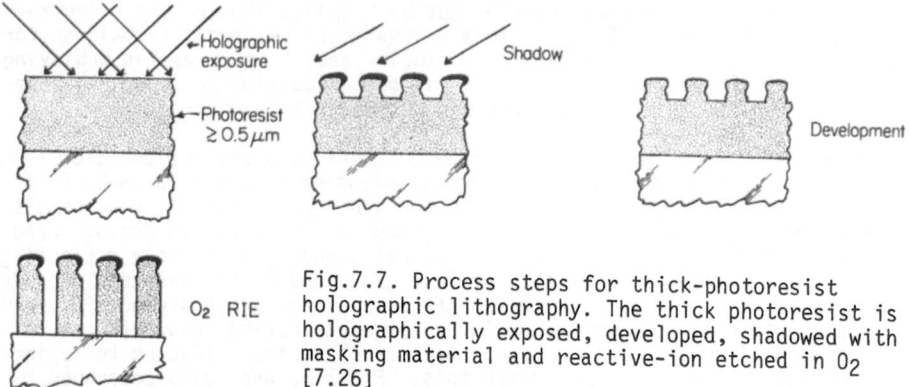

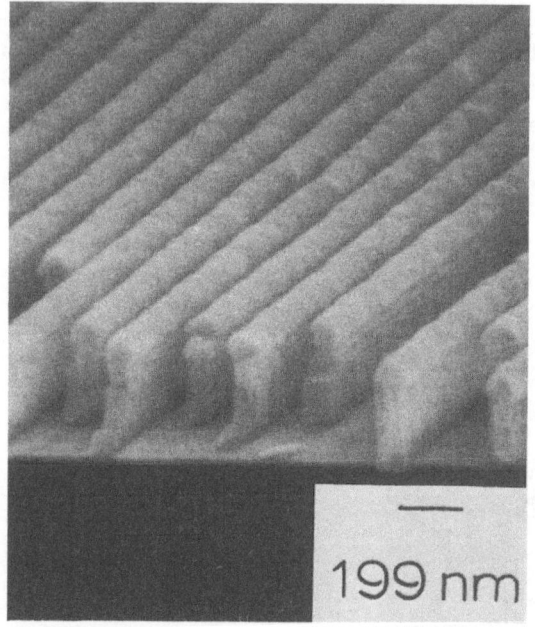

Fig.7.7. Process steps for thick-photoresist holographic lithography. The thick photoresist is holographically exposed, developed, shadowed with masking material and reactive-ion etched in O_2 [7.26]

Fig.7.8. Scanning electron micrograph of photoresist structure produced by the process illustrated in Fig.7.7

example, in making x-ray lithography masks of 0.2 μm-period grating patterns (see Section 7.7).

Figures 7.6-8 also illustrate a major shortcoming of using two-component, diazo/novolak type photoresists for sub-100 nm-linewidth lithography: granularity caused by component separation. One way of bypassing this problem is to use deep UV as the exposing radiation so that single-component resists, such as PMMA, can be used to record the pattern.

Grating patterns with spatial periods below 100 nm have been exposed using conventional holographic configurations and short wavelength lasers, [7.27] near UV lasers and indexmatching schemes, [7.28] and spatial-period

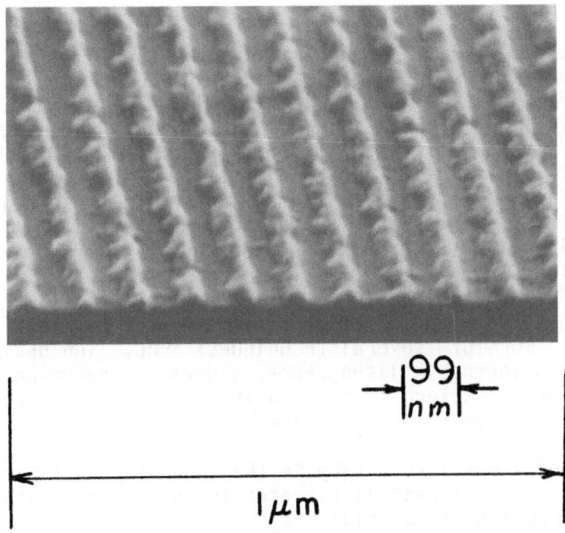

Fig.7.9   Scanning electron micrograph of a 99 nm-period grating pattern exposed in PMMA by deep-UV spatial frequency doubling using an ArF
laser. The parent grating was coated with a multilayer dielectric
mirror to suppress the zero order. Edge raggedness is due to defects in the grating and/or multilayer coating [7.33]

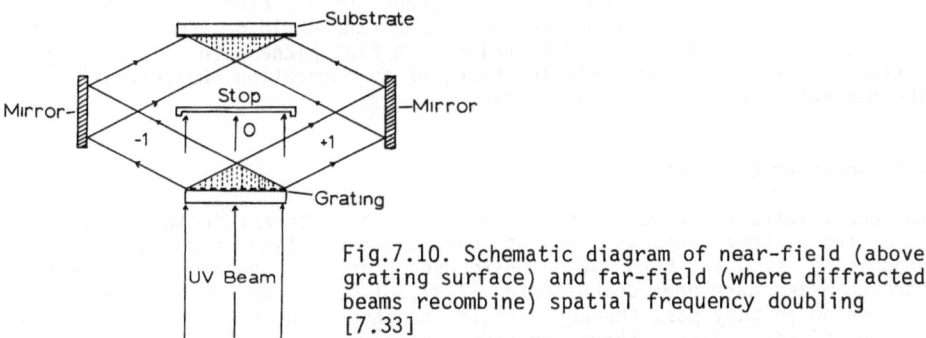

Fig.7.10. Schematic diagram of near-field (above
grating surface) and far-field (where diffracted
beams recombine) spatial frequency doubling
[7.33]

division [7.29-33]. The latter technique has been used with soft x-rays
(λ = 4.5 nm), as well as deep UV radiation from an ArF laser (193 nm). Recently, a multilayer dielectric coating over the parent grating was used to
attenuate the zero-order beam [7.33].

The result is shown in Fig.7.9. The line edges are quite rough, presumably due to defects in the mask and multilayer mirror.

Figure 7.10 illustrates a scheme for holographic lithography which would
be especially useful with deep UV (or x-ray) sources having relatively large
wavelength spreads (i.e., large $\Delta\lambda$) [7.33]. The spatial period exposed in
the far field would be one-half the period of the parent grating. Local defects in the parent grating would not degrade the far-field exposure pattern.
This method has not yet been developed.

7.5 X-Ray Lithography

X-ray lithography is a replication technique. Features created on the x-ray mask, by whatever means, can be replicated with a resolution of perhaps 15 nm. The instrinsic resolution (i.e., resolution dictated by photoelectron range) at λ = 4.5 nm is believed to be $\sim$5 nm. However, the need to have a finite mask-sample gap of at least the resist thickness plus the absorber thickness leads, inevitably, to diffraction, which limits the minimum spatial period that can be replicated to $\sim$30 nm (15 nm minimum "linewidth") [7.34]. Replication of a pair of lines with center-to-center spacing of 35 nm has been demonstrated [7.35]. Absorber lines were produced on the mask by a sidewall-shadowing technique.

One of the attractive features of x-ray lithography is that the mask pattern can be created by one or more of several techniques: SEBL, ion beam lithography, photolithography, holographic lithography, sidewall shadowing, or x-ray lithography. Once the mask pattern is created it can be replicated as many times as desired with relatively simple equipment.

Another attractive feature of x-ray lithography is the high-aspect-ratio structures that one can obtain. This is best illustrated in connection with the gold electroplating work described in Section 7.7.

A feature of x-ray lithography which may be especially useful in connection with Fresnel zone plate x-ray microscopy is an ability, at least in principle, to correct spherical aberration or other distortions. In the fabrication of an x-ray mask the membrane could be made to conform to some prefigured surface rather than a flat surface. Exposure of a pattern on the x-ray mask would be by holographic lithography or any other means, including x-ray replication. Thereafter, the mask membrane would be removed from the prefigured surface and allowed to relax to a flat plane. In this way, predistortion could be built into the mask, or an aberration characteristic of the generation method could be removed.

7.6 Ion-Beam Lithography

In recent years there has been a great deal of interest in the use of ion beams for lithography, as well as direct implantation and other forms of processing [7.36]. Both scanning-focused beam and parallel-ion exposure through masks have been demonstrated [7.37,38]. Ions undergo minimal lateral scattering as they pass through a resist and hence there is negligible proximity effect. Also, short exposure times are readily achieved. For sub-100-nm linewidth grating and Fresnel zone patterns, free-standing ion-beam stencil masks appear to be the most promising approach [7.38].

7.7 Electroplating

The efficiency of a diffraction grating or a Fresnel zone plate (i.e., the percentage of the incident radiation that is diffracted into a given order) depends on the x-ray attenuation and phase shift, and hence on the thickness and composition of the structure's lines. For homogeneous materials, optimum efficiency at a given wavelength is achieved for a particular choice of material composition and thickness [7.39]. (Multilayer materials [7.40] introduce new degrees of freedom in optimizing composition and thickness but, to date, have not been used in diffractive optical elements.) At x-ray energies greater than about 2 KeV requirements on material thickness pose rather formidable challenges for microfabrication technology, especially if high dispersion is also desired. For example, for a 0.25 nm x-ray wavelength the optimum 1st- order diffraction efficiency (~50%) of a grating is achieved with a gold thickness of 1.0 µm [7.39]. For gratings of 0.2 µm spatial period this implies an aspect ratio (i.e., height-to-width ratio) of about 10! Electroplating provides an effective means of producing grating and zone plate structures of a number of materials with minimum linewidths < 0.1 µm.

Reference [7.41] lists more than 20 electroplatable metals and alloys spanning the periodic chart from Al to Pb. Some of these, notably Al, require such high temperatures or harsh chemical baths that they are unsuitable for submicrometer electroplating into polymeric resist molds. Electroplating solutions and procedures are commercially available for more than a dozen metals and alloys. For some materials, such as gold, one can choose from more than 40 different baths. To be suitable for submicrometer electroplating, the bath and the procedure must provide low-stress material and controlled deposition rate. In our work with gold and silver, we have used SEL-REX BDT 510 and SEL-REX SILEREX II, respectively, [7.42,43].

The procedure used in microplating is: deposit a plating base, about 10-20 nm thick, of the material of interest, or an electrochemically compatible material; spin on a resist film with thickness less than or equal to the final material thickness required; create a mold in the resist film, either by lithographic exposure and development or by some two-stage process, such as depicted in Fig.7.1; electroplate; remove the resist mold, if desired. We have found that PMMA resist molds having height-to-width aspect ratios > 5 at linewidth ≲ 0.15 µm are distorted or destroyed by surface tension effects when they are immersed in a gold-plating solution. To achieve height-to-width ratios ≳ 10 we use x-ray lithography and transfer the substrate directly from the development bath to an alcohol rinse bath and then into the electroplating bath. In this way the mold is always under liquid. Dry etching methods (see Figs.7.1 and 7.8), although they are capable of

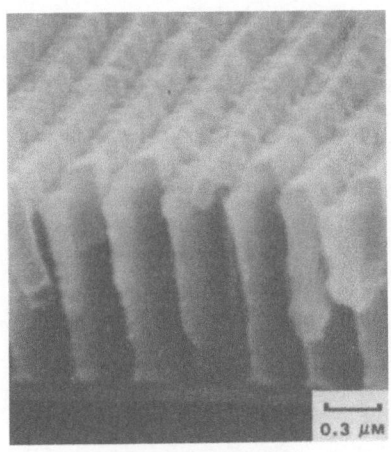 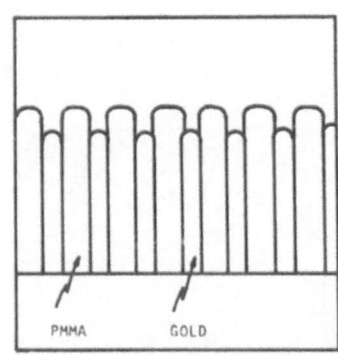

Fig.7.11 Scanning electron micrograph (left) and drawing (right) showing a grating pattern x-ray exposed (λ = 4.5 nm) in 1.2-µm thick PMMA resist, and electroplated gold filling the spaces between PMMA lines. The aspect ratio of the PMMA lines is 8. Grating spatial period is 0.3 µm

producing impressively large aspect ratios, would probably not be useful in preparing molds for high-aspect-ratio microplating because of the surface-tension problem. However, for some materials, it may be possible to bypass liquid electroplating using anisotropic dry etching directly [7.44].

Figure 7.11 illustrates both the high-aspect-ratio mold obtained in PMMA with C_K (λ = 4.5 nm) x-ray lithography, and the manner in which gold electroplating fills the spaces between PMMA lines. Figure 7.12 shows the structure of the gold grating after removal of the PMMA. (In this case, surface

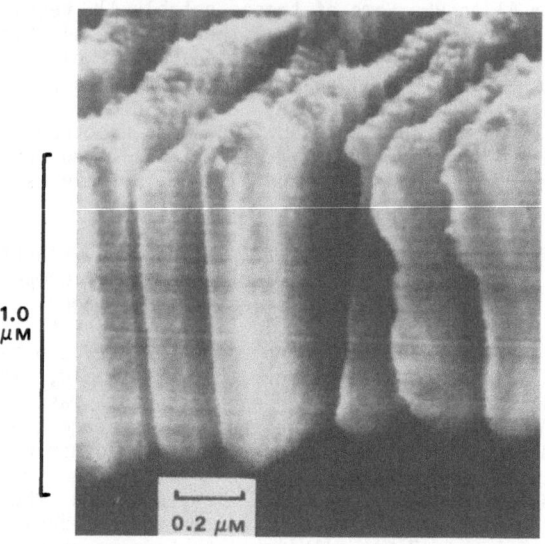

Fig.7.12 Scanning electron micrograph of 1-μm thick gold grating (0.3 μm period) after removal of the PMMA

Fig.7.13 Scanning electron micrograph of 0.2-μm period gold grating pattern. The cleaving of the sample for SEM viewing produced the irregular edge features

tension effects upon drying do not cause distortion.) Figure 7.13 shows an electroplated structure of 0.2 μm period. High-aspect-ratio submicrometer structures, such as shown in Figs.7.12 and 7.13, are likely to be used as high-dispersion, high-efficiency, diffractive optical elements for the next generation of laboratory and satellite-based spectrometers, microscopes and telescopes. The specific structures shown are part of a program to develop diffraction gratings for the AXAF x-ray astronomy telescope scheduled for launch in 1991. A gold thickness of 1.0 μm should enable the Fe^{+25} lines (∼7 KeV), which are of considerable astrophysical interest, to be analyzed.

Acknowledgements

The M.I.T. work reported here was supported by several sponsors, including the Joint Services Electronics Program, the National Science Foundation, the National Aeronautics and Space Administration. The authors are grateful to J. Carter for technical assistance.

References

7.1 H.I. Smith: Proc. IEEE 62, 1361 (1974)
7.2 D.C. Flanders, D.C. Shaver, A.M. Hawryluk and H.I. Smith: Annals of the New York Academy of Sciences 342, 203 (1980)
7.3 H.I. Smith: "Fabrication of Diffractive Optical Elements for X-Ray Diagnostics", in Low Energy X-Ray Diagnostics - 1981, pp. 223-224, Amer. Inst. Phys., New York, 1981. D.T. Atwood and B.L. Henke, editors
7.4 R.E. Howard and D.E. Prober: "Nanometer-Scale Fabrication Techniques", VLSI Electronics: Microstructure Science, Vol. 5, Chap. 4, Academic Press 1982
7.5 Scanned Image Microscopy, ed. E.A. Ash, Academic Press, 1980
7.6 M.J. Bowden: J. Electrochem. Soc. 128, 195C (1981)
7.7 M. Hatzakis: Solid State Technol. 24, 74 (1981)
7.8 M. Hatzakis, D. Hofer, T.H.P. Chang: J. Vac. Sci. Technol. 16, 1631 (1979)
7.9 B.J. Lin and T.H.P. Chang: J. Vac. Sci. Technol. 16, 1669 (1979)
7.10 J.M. Moran and D. Maydan: J. Vac. Sci. Technol. 16, 1620 (1979)
7.11 M. O´Toole, E. Liu and M. Chang: Proc. SPIE Conference on Semiconductor Microlithography VI, San Jose, CA., ed. J. Dey, V275, 128 (1981)
7.12 D.C. Shaver, D.C. Flanders, N.M. Ceglio and H.I. Smith: J. Vac. Sci. Technol. 16, 1626 (1979)
7.13 D.P. Kern, P.J. Houzego, P.J. Coane and T.H.P. Chang: J. Vac. Sci. Technol. B, Oct/Dec. 1983
7.14 H.I. Smith, S.R. Chinn and P.D. DeGraff: J. Vac. Sci. Technol. 12, 1262 (1975)
7.15 P.J. Coane, D.P. Kern, A.J. Speth and T.H.P. Chang: "An Electron Beam Microfabrication System for Lithography Below 1000A", Proceedings of 10th International Conference on Electron and Ion Beam Science and Technology, The Electrochemical Society, 83-2, 2, (1983)
7.16 M. Parikh: J. Appl. Phys. 50, 4371 (1979); M. Parikh: J. Appl. Phys. 50, 4378 (1979); M. Parikh: J. Appl. Phys. 50, 4383 (1979); M. Parikh and D.E. Schreiber: IBM J. Res. Dev. 24, 530 (1980)
7.17 A.N. Broers: 9th Int. Conf. on Electron and Ion Beam Science and Technology, ed. R. Bakish (The Electrochem. Soc., 1980), p. 396; and A.N. Broers: Proceedings of the International Conference on Microlithography, 1980, ed. R. Kramer (Delft University Press, The Netherlands, 1981), p. 9
7.18 R.J. Hawryluk: J. Vac. Sci. Technol. 19, 1 (1981)

7.19 T.H.P. Chang: J. Vac. Sci. Technol. 12, 1271 (1975)

7.20 H.G. Craighead, R.E. Howard, L.D. Jackel and P.M. Mankiewich: Appl. Phys. Lett. 42, 38 (1983)

7.21 S.P. Beaumont, P.G. Bower, T. Tamamura and C.D.W. Wilkinson: Appl. Phys. Lett. 38, 436 (1981)

7.22 G. Schmahl, D. Rudolph and B. Niemann: in Scanning Image Microscopy, ed. E.A. Ash, Academic Press, London 1980, p. 393

7.23 G. Schmahl, D. Rudolph and B. Niemann: "X-Ray Microscopy Using Fresnel Zone Plates". In Low Energy X-Ray Diagnostics - 1981, p. 225-227, Amer. Inst. Physics, New York, 1981, D.T. Atwood and B.L. Henke, eds.

7.24 N.N. Efremow, N.P. Economou, K. Bezjian, S.S. Dana and H.I. Smith: J. Vac. Sci. Technol. 19, 1234 (1981)

7.25 H.J. Lezec, E.H. Anderson and H.I. Smith: J. Vac. Sci. Technol. B, Oct/Dec (1983)

7.26 E.H. Anderson, C.M. Horwitz and H.I. Smith: Appl. Phys. Lett., 43, (1983)

7.27 G.C. Bjorklund, S.E. Harris and J.F. Young: Appl. Phys. Lett. 25, 451 (1974)

7.28 C.V. Shank and R.V. Schmidt: Appl. Phys. Lett. 23, 154 (1973)

7.29 D.C. Flanders, A.M. Hawryluk and H.I. Smith: J. Vac. Sci. Technol. 16, 1949 (1979)

7.30 A.M. Hawryluk, N.M. Ceglio, R.H. Price, J. Melngailis and H.I. Smith: J. Vac. Sci. Technol. 19, 897 (1981)

7.31 A.M. Hawryluk, H.I. Smith, R.M. Osgood and D.J. Ehrlich: Opt. Lett. 7, 402 (1982)

7.32 A.M. Hawryluk, Ph.D. Thesis, M.I.T. 1981; also republished as VLSI Memo 81-69, Dept. of Electrical Engineering & Computer Science, M.I.T.

7.33 A.M. Hawryluk, H.I. Smith and D.J. Ehrlich: J. Vac. Sci. Technol. B., Oct/Dec (1983)

7.34 D.C. Flanders: Microcircuit Engineering 81, International Conference on Microlithography Proc., p.22. Lausanne, Switzerland, Sept. 28-30, 1981. © 1981 Swiss Fed. Inst. Tech., Lausanne

7.35 D.C. Flanders: Appl. Phys. Lett. 36, 93 (1980)

7.36 W.L. Brown, T. Venkatejan and A. Wagner: Nuclear Instruments and Methods 191, 157-168 (1981)

7.37 J.L. Bartelt, et al: J. Vac. Sci. Technol. 19, 1166 (1981)

7.38 J.N. Randall, D.C. Flanders, N.P. Economou, J.P. Donnelly and E.I. Bromley: J. Vac. Sci. Technol. B, Oct/Dec (1983)

7.39 M.L. Schattenburg, Ph.D. Thesis, M.I.T., in preparation

7.40 T. Barbee, these proceedings

7.41 W.H. Safranek: "The Properties of Electrodeposited Metals and Alloys", American Elsevier, New York, 1974

7.42 Occidental Chemical Corp., SEL-REX Plating Systems, 75 River Road, Nutley, N.J. 07710

7.43 N.M. Ceglio, A.M. Hawryluk and M.L. Schattenburg: to be published, J. Vac. Sci. Technol., Oct/Dec 1983

7.44 G.A. Lincoln, M.W. Geis, S. Pang and N.N. Efremow: J. Vac. Sci. Technol., Oct/Dec (1983)

8. Zone Plates for X-Ray Microscopy

G. Schmahl, D. Rudolph, P. Guttmann, and O. Christ

Forschungsgruppe Röntgenmikroskopie, Universität Göttingen
Geismarlandstraße 11, D-3400 Göttingen, Fed. Rep. of Germany

8.1 Introduction

Zone plates are circular gratings with radially increasing line density. The imaging with zone plates of zone numbers n $\geq$ 100 obeys the same laws as imaging with thin refractive lenses. With r_1 = radius of the innermost zone (Fig.8.1), r_n = radius of the n-th zone, n = zone number and m = number of diffracted order, the focal length of a zone plate is approximately given by $f_m = r_1^2 \lambda^{-1} m^{-1}$. Because $f_m \propto 1/\lambda$, a zone plate has to be used with quasimonochromatic radiation with a bandwidth $\Delta\lambda$ given by $\lambda/\Delta\lambda \simeq n \circ m$. The width of the outermost zone is $dr_n = r_n/(2n)$. The smallest distance of two point sources which can be resolved with a zone plate is - according to the Rayleigh criterion - given by $\delta = 1{,}22\ dr_n/m$.

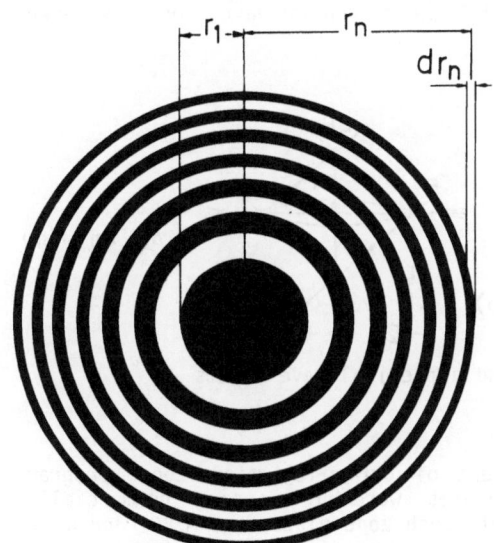

Fig.8.1 Principle of a zone plate

The maximum diffraction efficiency, i.e., the diffracted flux in a certain order divided by the incoming flux, is in the case of amplitude zone plates approximately 10% in the first, 2,6% in the second, and 1,2% in the third diffracted order. Higher diffraction efficiencies can be obtained by use of phase zone plates.

For x-ray microscopy experiments one needs two types of zone plates. First, the object has to be illuminated by a condenser. The condenser has to collect as much radiation as possible from the x-ray source. For our x-ray microscopy experiments at an electron storage ring this results in diameters of up to nine millimeters and rather high zone numbers of up to $3.8 \cdot 10^4$. Second, to image the object, a high resolution zone plate is necessary, i.e. a zone plate with an outermost zone width dr_n as small as possible. This zone plate should have only one hundred to some hundred zones in order not to restrict the usuable x-ray bandwidth to much. In consequence, high resolution zone plates have very small diameters in the region of about twenty microns to some hundred microns. We therefore call them micro-zone-plates.

Zone plates suited for x-ray microscopy can be constructed by different methods: for the construction of micro-zone-plates one can use electron-beam lithography techniques to write zone plate structures [8.1-3] or one can demagnify relative coarse zone plate structures using electron beam projection [8.4]. Another method, still under development, is to evaporate or sputter alternate layers of an opaque and an as transparent as possible material onto a thin wire and to slice the system to form zone plates (sputtered sliced zone plates) with a high aspect ratio [8.5].

Micro-zone-plates as well as condenser zone plates can be made using a holographic lithography technique. Using this technique zone plate patterns are exposed by superimposing two coherent uv or visible laser beams of wavelength λ_v on a photoresist-coated substrate [8.6-7]. The main problem in this case is to avoid aberrations caused by the fact that the zone plate structure is generated with uv or visible light and is used in the microscope with a wavelength which is about a factor of a hundred shorter. These aberrations can be taken into account by use of aspheric wave fronts as shown in Fig.8.2b.

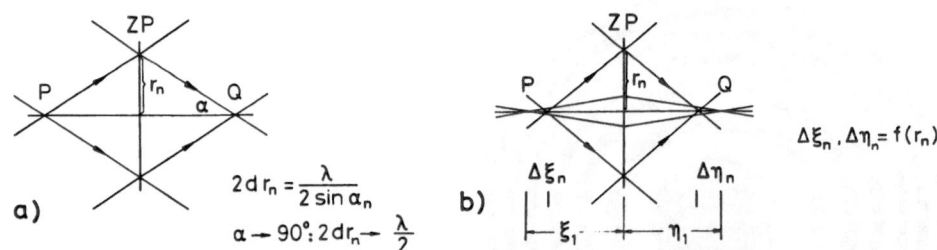

$$2 d r_n = \frac{\lambda}{2 \sin \alpha_n}$$

$$\alpha \to 90^\circ; 2 d r_n \to \frac{\lambda}{2}$$

$$\Delta \xi_n \cdot \Delta \eta_n = f(r_n)$$

Fig.8.2 Superposition of spherical and aspherical wave fronts

At the beginning of the development of our x-ray microscopy program micro-zone plate and condenser zone plates were built using commercially available optical elements. Examples of such zone plates are MZPO and KZP1 (compare Tables 8.2 and 8.3) which have comparatively large focal ratios f/D. For the construction of zone plates with improved focal ratios special optical arrangements are necessary.

The holographic lithography technique, the development of suited optical arrangements for the construction of micro-zone-plates and condenser zone plates as well as the properties of such zone plates are discussed in this paper and in the following two papers of this volume.

8.2 Micro-Zone-Plates

The procedure to construct zone plates with the holographic technique can be divided into the following steps:
1. A zone plate $r_n = f(\lambda_x, f_x, n, V)$ is calculated. λ_x is the wavelength used used in the x-ray microscope, f_x is the focal length, n is the number of zones and V is the x-ray magnification.
2. Aspherical wave fronts of wavelength λ_y are calculated, the superposition of which result in a zone plate interference fringe system according to 1 (compare Fig.8.2b).
3. An optical system is calculated and built which generates in good approximation the required two aspherical wave fronts.
4. A photoresist layer coated onto a substrate consisting of a glass plate, a polyimide layer and a gold layer is exposed with the zone plate interference structure generated by the superposition of the two wave fronts.
5. The zone plate interference structure is transferred into an amplitude transmission zone plate using the preparation steps: photoresist development, ion etching of gold, glueing of a support ring, etching of the glass plate with hydrofluoric acid. This yields a zone plate consisting of gold rings on a thin foil sufficient transparent to soft x rays.

As already discussed in [8.7] an optical solution for the above-described problem consists of an aplanatic lens system the principle of which is shown in Fig.8.3 on the basis of a two-element system. A spherical wave front converging to a point A is focused by the aplanatic lens system to point C. Points A and C - which are free of aberrations - are the aplanatic points of the system in a case where the radii R of the lenses and the coordinates (x,y) of the centers of radii R are chosen according to Table 8.1, which is based on the well-known Weierstrass construction (compare e.g., [8.8]). The convex surfaces of the lenses with the radii r, $r \cdot n^2$, $r \cdot n^4$, ... with n = refractive index of the lens material measured in respect to the surrounding air, produce the necessary refraction. The concave surfaces are chosen in such a way that no further refraction is introduced. This is reached by choosing the x, y coordinates of the centers of radii in the form $(r/n, 0)$, $(r \cdot n, 0)$, $(r n^3, 0)$....The radii of the concave surfaces can deviate from the values of Table 8.1 without introducing aberrations.

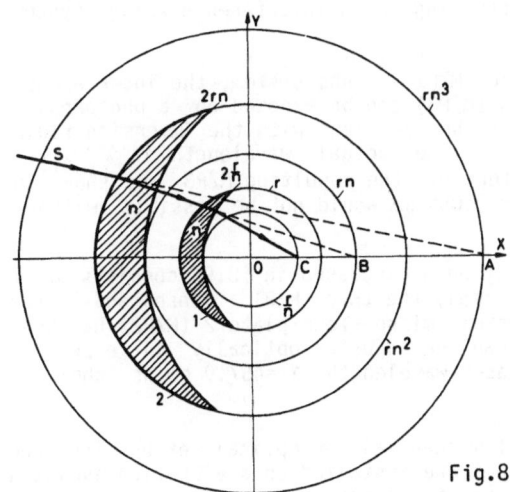

Fig.8.3 Aplanatic lens system

<u>Table 8.1</u> Construction scheme of a six-element aplanatic lens system.
The values in the upper rows correspond to the concave surfaces,
the values in the lower rows correspond to the convex surfaces.

lens number	radius R of lens surface	x,y coordinate of the center of radius R
1	$2 \cdot r/n$ r	$(r/n, 0)$ $(0, 0)$
2	$2 \cdot r \cdot n$ $r \cdot n^2$	$(r \cdot n, 0)$ $(0, 0)$
3	$2 \cdot r \cdot n^3$ $r \cdot n^4$	$(r \cdot n^3, 0)$ $(0, 0)$
4	$2 \cdot r \cdot n^5$ $r \cdot n^6$	$(r \cdot n^5, 0)$ $(0, 0)$
5	$2 \cdot r \cdot n^7$ $r \cdot n^8$	$(r \cdot n^7, 0)$ $(0, 0)$
6	$2 \cdot r \cdot n^9$ $r \cdot n^{10}$	$(r \cdot n^9, 0)$ $(0, 0)$

It should be noted that one can construct a more compact system with minimized air distances between the single lenses of an aplanatic lens system, resulting in changed values of R and (x, y) in comparision to Table 8.1.

It is the special characteristic of an aplanatic system that wavefronts focused in front of or behind the aplanatic point C (compare Fig.8.3) show aberrations as indicated in Fig.8.2b. The zone plate interference structure has to be recorded in a plane near point C.

Figure 8.4a shows an optical arrangement to generate the interference fringe system of three micro-zone-plates. A spherical wave front of 257 nm radiation is divided by a four-element partly transparent beamsplitter made from fused silica into two convergent spherical beams. The uv wavelength of 257 nm was obtained by second harmonic generation of the 514.5 nm Ar^+ laser line. As already discussed in [8.7] the beamsplitter can be combined with a three-element aplanatic system which focuses the two beams symmetrically to the aplanatic point of the system resulting in an interference fringe system of micro-zone-plate 1 (MZP1).

Adding two further lenses to the MZP1 - lens system, the interference fringe system of micro-zone-plate 3 (MZP3) can be exposed on a photoresist layer matched to the plane surface of the last lens with the immersion fluid glycerol in between. In this case the actual wavelength is $\lambda_{v,eff}$ = 257/1.5037 nm = 171 nm. The parameters of the resulting MZP3 are shown in Table 8.2. The small value of dr_n = 0.055 µm would not be possible without an immersion system.

Another solution which has already been discussed in [8.9] consists of a six-element aplanatic lens system, namely the three MZP1 - lenses and three additional lenses. With this system micro-zone-plate 2 (MZP2) has been built. The aplanatic lens systems can be adjusted optically in respect to the beamsplitter by use of the laser wavelength λ =457.9 nm as shown in Fig.8.4a.

Figure 8.4b shows a photograph of mechanical and optical set-ups for the generation of zone plates. The systems are installed on a vibration-isolated optical bench located in a temperature-stabilized room.

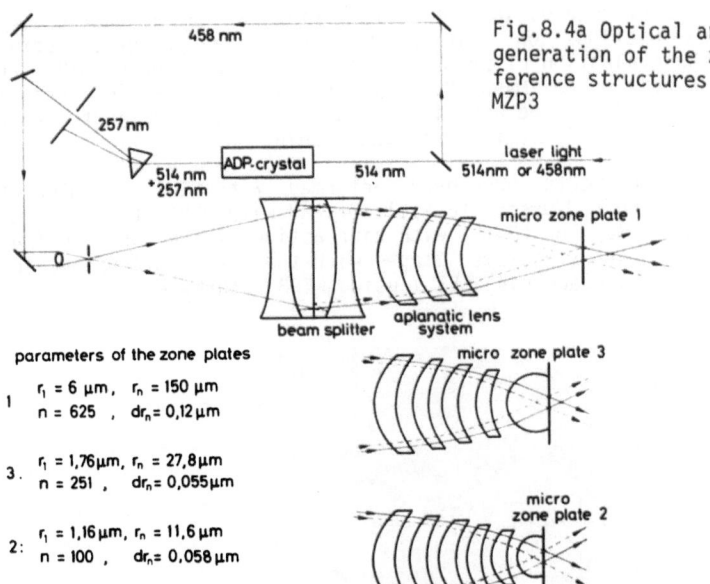

Fig.8.4a Optical arrangement for the generation of the zone plate interference structures of MZP1, MZP2 and MZP3

458 nm

257 nm

514 nm
+257 nm

ADP-crystal   514 nm   514nm or 458nm

laser light

micro zone plate 1

beam splitter   aplanatic lens system

parameters of the zone plates

1    $r_1 = 6\ \mu m$, $r_n = 150\ \mu m$
$n = 625$, $dr_n = 0{,}12\ \mu m$

micro zone plate 3

3 . $r_1 = 1{,}76\mu m$, $r_n = 27{,}8\mu m$
$n = 251$, $dr_n = 0{,}055\mu m$

micro zone plate 2

2:   $r_1 = 1{,}16\mu m$, $r_n = 11{,}6\mu m$
$n = 100$, $dr_n = 0{,}058\mu m$

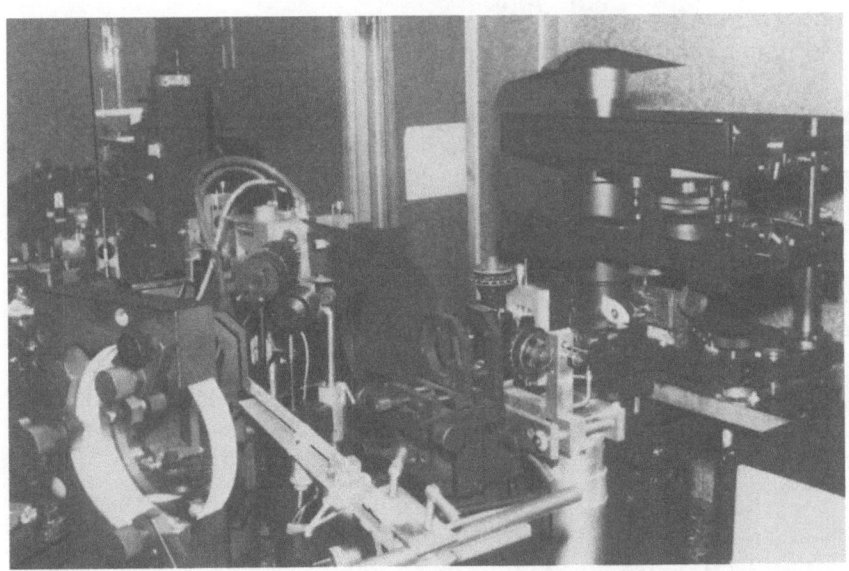

Fig.8.4b Mechanical and optical set-ups for zone plate generation. On the right side of the picture the vertical set-up of a beam splitter and an aplanatic lens system for micro-zone-plates is shown. In the background a set up for condenser zone plates can be seen

The calculations performed during the construction of an aplanatic lens system for MZP3, calculations concerning production tolerances for the optical systems, and the evaluation of zone plate imaging properties of MZP3 are discussed in more detail by GUTTMANN [8.10] (this volume, no. 9). One result is, for example, that MZP3 can be used in the rather large wavelength range 0.3 nm $\leq \lambda_X \leq$ 10.3 nm with diffraction limited resolution.

X-ray microscopy experiments have been performed with MZP1 and MZP2 at the electron storage ring ACO/Orsay [8.9,11,12] and with MZP3 at the electron storage ring BESSY/Berlin [8.13] with diffraction limited resolution. As an example, Fig.8.5 shows diatoms imaged with the x-ray microscope installed at the electron storage ring BESSY using MZP3 with λ_X=4.5 nm.

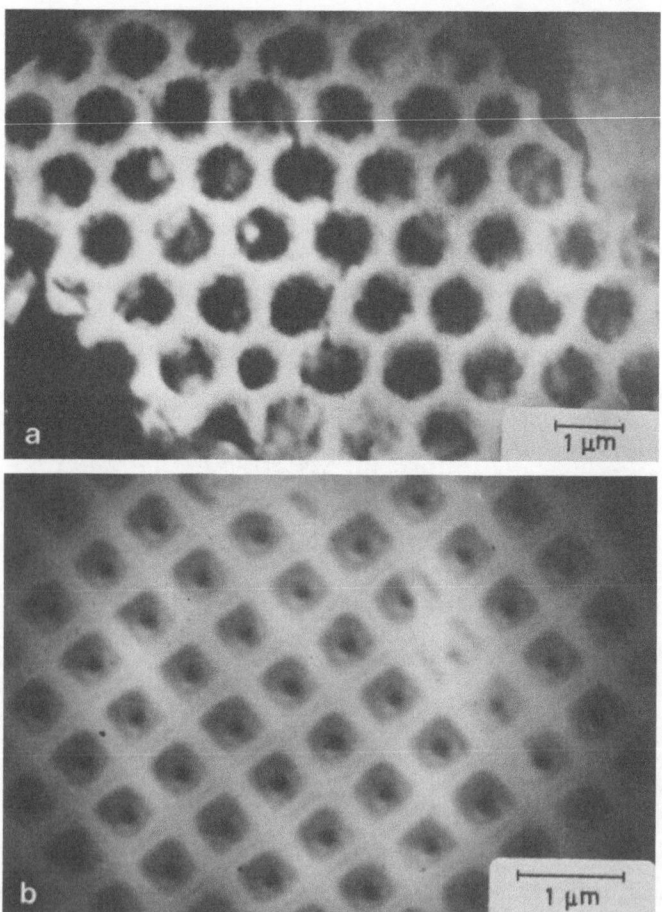

Fig.8.5a Diatom, λ_X = 4.5 nm, x-ray magnification 230x, exposure time 3^s, BESSY storage ring, METRO optic, electron current 130 mA, x-ray optics: KZP3 and MZP3

Fig.8.5b Diatom, λ_X = 4.5 nm, x-ray magnification 450x, exposure time 20^s, BESSY storage ring, METRO optic, electron current 100 mA, x-ray optics: KZP3 and MZP3

The smallest dr_n value which can be made using the above – described method is given by $dr_n = \lambda/(4\sin\alpha)$, compare Fig.8.2a. The smallest theoretical value, i.e., $\alpha \rightarrow 90^0$, is $dr_n = 0.043$ µm for $\lambda = 171$ nm. Ray-tracing calculations have shown that in practice α has to be smaller than 60^0 in the case of the optical system of Fig.8.4, resulting in $dr_n \simeq 0.05$ µm.

As already discussed, zone plates with a smaller zone width - and consequently better resolution - are under development using electron-beam lithography, resp. electron-beam projection techniques as well as the sputtered sliced zone plate technique. Another possibility is an x-ray interferometry technique as mentioned in [8.14] and shown in Fig.8.6.

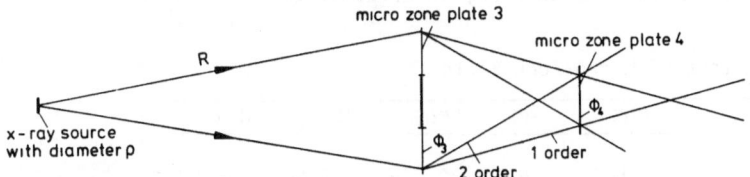

Fig.8.6 Schematic arrangement for the construction of MZP4

MZP3 can be illuminated with spatial coherent quasimonochromatic x radiation. The superposition of the first- and second-order radiation yields an interference figure which has the structure of a zone plate (MZP4). With a distance x-ray source - MZP3 of 200 mm and an x-ray wavelength of $\lambda_x = 2.36$ nm the resulting MZP4 has the following parameters: $r_1 = 0.587$ µm, $r_n = 9.68$ µm, $n = 270$ and $dr_n = 0.018$ µm. Figure 8.7 shows a ray-tracing made with MZP4 which illustrates that diffraction-limited resolution can be expected with this zone plate. In principle, this technique looks very simple. In practice, however, some difficulties have to be overcome. Because of the small dr_n value the zone plate interference fringe system must be recorded

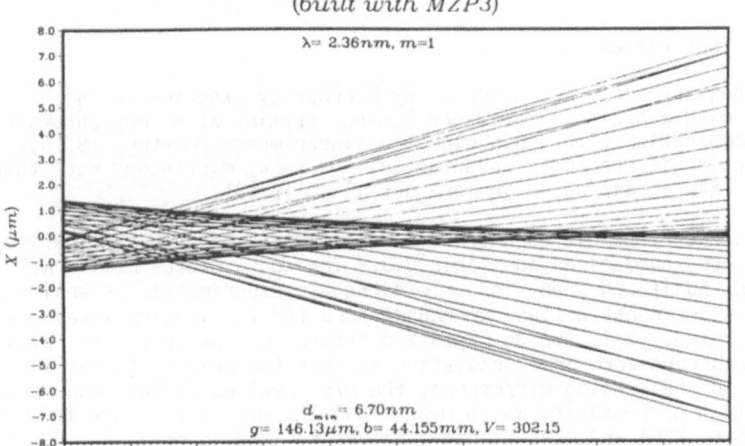

Fig.8.7 Ray-tracing, MZP4

in high-resolution x-ray resist which requires an x-ray source with a very high spectral brilliance. Calculations show that at the 0.8 GeV BESSY storage ring exposure times of the order of about 30 minutes would be necessary. (Much shorter exposure times are of course possible if one can use an x-ray undulator). Another experimental difficulty will be to transform the photoresist zone plate structure into an amplitude or phase zone plate with an adequate aspect ratio. This problem has to be solved, too, when using the electron-beam lithography technique.

In Table 8.2 the parameters of five micro - zone-plates are summarized. MZP0 has been built with commercially available optical components. MZP1 - MZP3 have been built using the optical arrangements of Fig.8.4a. MZP4 is under development using MZP3 as an x-ray interferometer.

Table 8.2 Parameters of Micro-Zone-Plates

	MZP0	MZP1	MZP2	MZP3	MZP4
radius of innermost zone r_1 [μm]	17,2	6,0	1,16	1,76	0,59
zone plate diameter $D = 2r_n$ [μm]	1000	300	23,2	55,6	19,4
zone number n	845	625	100	251	270
width of outermost zone dr_n [μm]	0,3	0,12	0,058	0,055	0,018
focal length $f_{4,5nm}$, 1. order [mm]	66	8	0,3	0,69	0,08
D/f, $\lambda = 4,5$ nm	1:66	1:27	1:13	1:12	1:4
focal length $f_{2,4nm}$, 1. order [mm]	123,7	15	0,56	1,3	0,15
D/f, $\lambda = 2,4$ nm	1:124	1:50	1:24	1:23	1:8

8.3 Condenser Zone Plates

Zone plate condensers are used in our x-ray microscopy experiments which act in combination with a grazing incidence laminar grating as a monochromator [8.15] or in combination with a pinhole as a linear monochromator [8.16,9]. The latter solution is much more advantageous for x-ray microscopy work than the first one. This is discussed in more detail in [8.13].

Several condenser zone plates have been developed up to now, the parameters of which are summarized in Table 8.3. The zone plates KZP1 - KZP3, which have been built and used in our microscopy experiments, as well as KZP6, which is under construction, are condensers for the imaging x-ray microscope [8.13]. They all have an apodized region in the center to avoid that in the microscope zero-order radiation reaches the object. To get short exposure times in the imaging microscope, the D/f values of the condenser have to be as high as possible, which means that the dr_n values have to be as small as possible. KZP4 and KZP5 are zone plates for the scanning x-ray microscope which is described by NIEMANN [8.17]. The optical arrangements for the generation of the zone plate interference fringe systems of KZP4 and KZP5 are discussed by THIEME [8.18].

Table 8.3 Parameters of condenser zone plates. KZP1-KZP3 and KZP6 are condensers for the imaging x-ray microscope, KZP4 and KZP5 are condensers for the scanning x-ray microscope

	KZP1	KZP2	KZP3	KZP6	KZP4	KZP5
radius of innermost zone r_1 [μm]	49	43,8	37	23,1	49,8	36,2
zone plate diameter $D = 2r_n$ [μm]	5000	6000	9000	9000	2500	2500
zone number n	2600	4700	15000	38000	630	1200
width of outermost zone dr_n [μm]	0,48	0,32	0,15	0,06	0,99	0,46
aberration limited diameter of the image of a point source d_a [μm]	20	12	3-5	3-5	1	0,5
focal length $f_{4,5nm}$, 1. order [mm] D/f, λ = 4,5 nm	511 1:103	426 1:71	304 1:34	118,6 1:13	551 1:220	290 1:116
focal length $f_{2,4 nm}$, 1. order [mm] D/f, λ = 2,4 nm	957 1:191	799 1:133	570 1:63	222,4 1:25	1033 1:413	544 1:218

Whereas the micro-zone-plates have been designed so that they image with diffraction-limited resolution, this is not necessary - and nearly impossible - for the condenser zone plates KZP1 - KZP3 and KZP6 for the imaging microscope. As a rule of thumb one can say that with increasing diameters D and increasing D/f of the zone plates it becomes more and more difficult to correct the aberrations. In consequence, with increasing D/f values from KZP1 to KZP6, as shown in Table 8.3, the optical arrangements for the generation of the zone plate structures become more sophisticated.

KZP1 has been built with commercially available optical elements and shows a rather bad resolution in spite of the low D/f. KZP2 has been built using a double hologram method [8.7]: two wave fronts of wavelengths λ_v = 476.5 nm and λ_v = 457.9 nm are stored successively in a photoresist phase hologram. Illuminating this hologram with one spherical wave front, two aspherical wave fronts are generated and superimposed, resulting in the zone plate fringe system of KZP2. A better condenser, namely KZP3, has been built by use of an optical arrangement as shown in Fig.8.8:

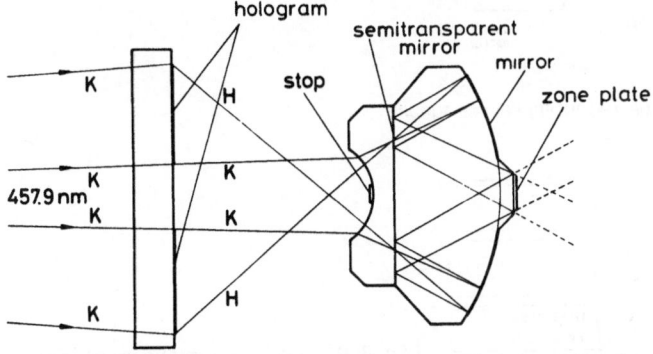

Fig.8.8 Optical arrangement for the generation of the zone plate interference structure of KZP3

The right part of the figure shows a quasiaplanatic lens-mirror system made of high refractive SF 59 glass. The aplanatic point of the spherical entrance surface is virtual. It is transformed into the plane where the zone plate is recorded by the reflections of the spherical and semitransparent mirrors. The system is fed by two spherical waves of $\lambda_V = 457.9$ nm. The first wave is the central part of the spherical wave K. The second wave H is generated from the outer part of the wave K by use of a ring-shaped phase hologram made in photoresist. The zone plate interference figure is recorded in a photoresist layer matched to the plane exit surface of the lens-mirror system by use of a thin glycerol layer.

For a further improved condenser a more sophisticated optical system has been designed and built. The optical system for the construction of KZP6 is shown in Fig.8.9. For the superposition of two aspherical wave fronts a photoresist phase hologram H is illuminated with 257.3 nm radiation as shown in Fig.8.9b. The zero-order radiation which transmits the hologram is reflected by a spherical mirror of 40 mm diameter and a semitransparent plane mirror, resulting in one aspherical wave front. The second aspherical wave front is the diffracted first-order radiation of the hologram H. The interference fringe system of this hologram has been designed so that the resulting first order wave front, when superimposed with the zero-order wave front - reflected by the spherical and semitransparent mirrors - results in the required KZP6 interference fringe system. Because of the large zone number $n = 3.8 \cdot 10^4$ and the rather small width of the outermost zone $dr_n = 0.06$ μm of KZP6, the aberration correction requires a special optical system for the construction of the hologram H. This system is shown in Fig.8.9a.

a) GENERATION OF HOLOGRAM H

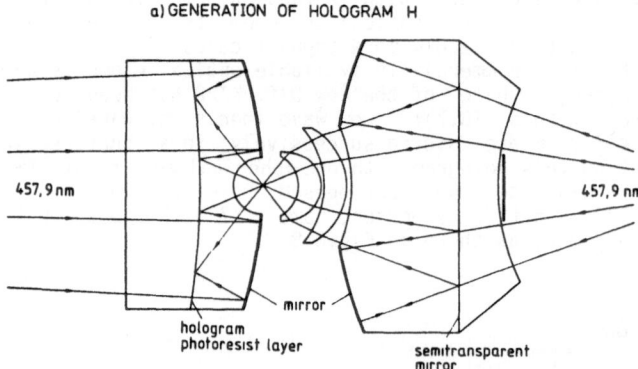

457,9 nm   457,9 nm

mirror

hologram
photoresist layer

semitransparent
mirror

b) GENERATION OF THE INTERFERENCE FIGURE

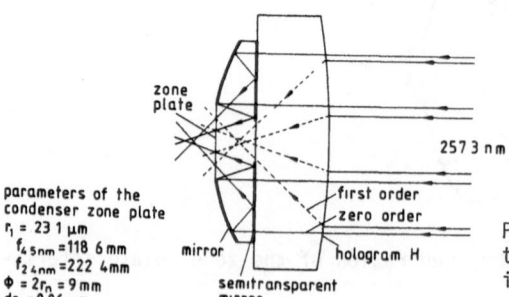

zone
plate

257 3 nm

parameters of the
condenser zone plate
$r_1 = 23.1$ μm
$f_{4.5nm} = 118.6$ mm
$f_{2.4nm} = 222.4$ mm
$\Phi = 2r_n = 9$ mm
$dr_n = 0.06$ μm

first order
zero order

mirror   hologram H

semitransparent
mirror

Fig.8.9 Optical arrangement for
the generation of the zone plate
interference structure of KZP6

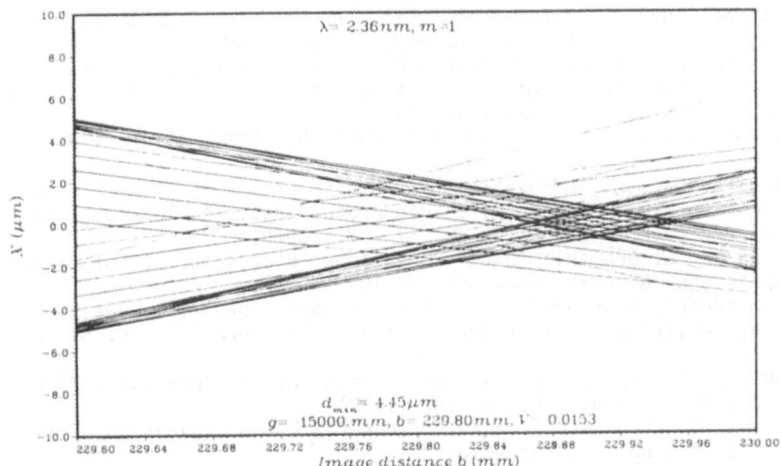

Fig.8.10 Ray-tracing, KZP6

The hologram is generated with λ_y = 457.9 nm and is recorded on a spherical surface. To match the fringe system of the hologram to the calculated fringe system the following parameters were varied during the calculations: curvature of the mirror in the left part of Fig.8.9a, the angles of the superimposed wave fronts in respect to the normal of the hologram surface and the curvature of the hologram. The calculations in the course of which the parameters of the optical system are varied to optimize KZP6 were performed using ray-tracing methods as discussed by GUTTMANN [8.10].

Figure 8.10 shows a ray-tracing for KZP6. In this ray-tracing a point source in a distance of 15 meter is imaged by KZP6 with the wavelength λ_x = 2.36 nm. The aberration-limited diameter of the image is 4.45 µm. KZP6 will be used in our x-ray microscopy experiments at BESSY. With a BESSY source diameter of,e.g., 500 µm and a demagnification of V = 0.0153, i.e. g = 15 m and b = 229.8 mm for λ_x = 2.36 nm, the image, when generated with a zone plate without any aberrations, would have a diameter of 7.7 µm. This value shows that the image diameter is not too much increased by the aberrations of the condenser. If it should be necessary in future to build a condenser with even smaller aberrations this has to be done with an optical arrangement which includes aspherical surfaces.

In Table 8.3 parameters of the condenser zone plates KZP1 - KZP3 and KZP6 for the imaging x-ray microscope and of KZP4 and KZP5 for the scanning microscope are summarized.

Acknowledgements

We are indebted to the Stiftung Volkswagenwerk for financial support. We are grateful fo Dr.Asmussen, FHI Berlin, for the Sundby diatoms, shown in Fig.8.5.

References

8.1 D.C. Shaver, D.C. Flanders, N.M. Ceglio and H.I. Smith: J. Vac. Sci. Technol. 16, 1626 (1979)
8.2 P.J. Coane, D.P. Kern, A.J. Speth and T.H.P. Chang: "An Electron Beam Microfabrication System for Lithography Below 1000A", Proceedings of

10th International Conference on Electron and Ion Beam Science and Technology, The Electrochemical Society, 83-2, 2 (1983)

8.3 A.G. Michette, M.T. Browne, P. Charalambous, R.E. Burge, P.J. Duke and M.J. Simpson: "Fabrication of Small linewidth Diffractive Optics for Use with Soft X-Rays", no. 12, this volume

8.4 H. Koops and J. Grob: "Submicron Lithography by Demagnifying Electron Beam Projection", no. 13, this volume

8.5 D. Rudolph, B. Niemann and G. Schmahl: "Status of the Sputtered Sliced Zone Plates for X-Ray Microscopy", SPIE Proceedings 316, 103-105 (1982)

8.6 G. Schmahl and D. Rudolph: Optik 29, 577-585 (1969)

8.7 D. Rudolph and G. Schmahl: "High Power Zone Plates for a Soft X-Ray Microscope", Ann. NY Acad. Sci. 342, 94-104 (1980)

8.8 Bergmann-Schaefer: Lehrbuch der Experimentalphysik Band III, Optik, Ed. H. Gobrecht, Walter de Gruyter Verlag Berlin, New York 1974, p. 57 and p. 106

8.9 G. Schmahl, D. Rudolph and B. Niemann: "Status of the Zone Plate Microscope", SPIE Proceedings 316, 100-102 (1982)

8.10 P. Guttmann: "Construction of a Micro-Zone-Plate and Evaluation of Imaging Properties", no. 9, this volume

8.11 G. Schmahl, D. Rudolph and B. Niemann: "X-Ray Microscopy of Biological Specimens with a Zone Plate Microscope", Ann. NY Acad. Sci. 342, 368-386 (1980)

8.12 B. Niemann, D. Rudolph and G. Schmahl: Nucl. Instr. and Meth. 208, 367-371 (1983)

8.13 D. Rudolph, B. Niemann, G. Schmahl and O. Christ: "The Göttingen X-Ray microscope and X-Ray Microscopy Experiments at the BESSY Storage Ring", no. 20, this volume

8.14 G. Schmahl, D. Rudolph and B. Niemann: "Imaging and Scanning Soft X-Ray Microscopy with Zone Plates", in Scanned Image Microscopy, ed. E.-A. Ash, Academic Press (1980)

8.15 B. Niemann, D. Rudolph and G. Schmahl: Applied Optics 15, 1883 (1976)

8.16 B. Niemann, D. Rudolph and G. Schmahl: Optics Communications 12, 160-163 (1974)

8.17 B. Niemann: "The Göttingen Scanning X-Ray Microscope", no. 22, this volume

8.18 J. Thieme: "Construction of Condenser Zone Plates for a Scanning X-Ray Microscope", no. 10, this volume

9. Construction of a Micro Zone Plate and Evaluation of Imaging Properties

P. Guttmann

Forschungsgruppe Röntgenmikroskopie, Universität Göttingen
Geismarlandstraße 11, D-3400 Göttingen, Fed. Rep. of Germany

Calculations are performed for an optical system to build a micro-zone-plate used as an imaging element for a high-resolution x-ray microscope. Ray-tracing methods are developed for this purpose. Calculations in the course of which the parameters of the optical system are varied are carried out to get production tolerances for the optical system. The imaging properties of the resulting micro-zone-plate are evaluated.

9.1 Introduction

The intention of this work was to build a micro-zone-plate which will be used as an imaging element for a high-resolution x-ray microscope.

The parameters of the desired micro-zone-plate MZP3 were chosen in such a way, that with the already existing micro-zone-plates MZP1 and MZP2 a set of micro-zone-plates with different focal lengths is available. Thus different magnifications are available for constant image distance.

The required focal length should be $f_x \simeq 1.3$ mm for the wavelength $\lambda_x = 2.36$ nm and the outer zone width should be $dr_n = 0.05$ µm, which will give the resolution of the zone plate as seen later on. The zone number n should be about 300 to satisfy the required monochromaticity $\lambda/\Delta\lambda \simeq 250$. The desired magnification will be $V \simeq 300$.

To build a micro-zone-plate with defined imaging properties for ultrasoft x-rays an optical arrangement (Fig.9.1) can be used with an aplanatic lens system [9.1]. Such a zone plate is called a holographically made zone plate because the construction of it with such an aplanatic lens system plus a beamsplitter (Fig.9.1) uses an analog to the Huygens-Fresnel principle: the superposition of two deformed spherical wave fronts produced with laser light.

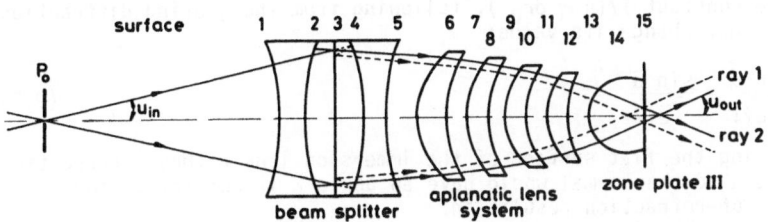

Fig.9.1 Optical arrangement to generate the interference figure of the micro-zone-plate

The resultant interference figure is recorded in a photoresist layer. The aplanatic lens system supplies in connection with the beamsplitter both deformed spherical wave fronts the "centres" of which lie on the right and the left side of one of the aplanatic points.

Because of the finite value of the optical free diameter of the beamsplitter and of the position of the point P_O a maximum angle of incidence of about 9° can be handled. If the aberrations in both construction wave fronts should be symmetrical, a ray is needed for the outer zone width $dr_n = 0.05$ µm of the micro-zone-plate leaving the aplanatic lens system with an angle of $u_{out} = 58.8^O$. This value results from the zone plate construction law:

$$\sin u_{out} = \frac{m \cdot \lambda_{v,eff}}{4 \cdot dr_n} \tag{9.1}$$

with $\lambda_{v,eff} = 171.08$ nm = effective construction wavelength, m = order of diffraction.

The angle u_{out} of a light beam which leaves the aplanatic lens system can be described by the approximation:

$$\sin u_{out} = n^k \cdot \sin u_{in} \tag{9.2}$$

with u_{in} = angle of incidence at the aplanatic lens system.

The aplanatic lens system is made from fused silica with the refractive index n = 1.50370 for $\lambda_v = 257.25$ nm. From (9.1) follows with $u_{in} = 9^O$ that the number of aplanatic lenses must be k = 5 for the desired angle $u_{out} = 58.8^O$.

An extension of an existing aplanatic lens system with 3 lenses to 5 lenses will be possible with the desired zone plate parameters [9.2].

The UV wavelength $\lambda_v = 257.25$ nm is too large for the construction of a micro-zone-plate with $dr_n = 0.05$ µm, because the smallest reachable value with this wavelength for the outer zone width is given by: $dr_n = \lambda_v/4 = 0.064$ µm.

To get a smaller zone width the last lens of the aplanatic system has to be built as an immersion lens. In that way as construction wavelength: $\lambda_{v,eff} = \lambda_v /n = 171.08$ nm can be used.

The used immersion liquid does not have exactly the same refractive index as fused silica - it is a little bit smaller - but the resulting refraction at the plan surface of the immersion lens does not distort the interference figure, if the constant $1/(4 \cdot dr_n)$, following from the grating diffraction equation, does not change its value:

$$\frac{1}{4 \cdot dr_n} = \frac{\sin \alpha}{\lambda_{eff}} = \frac{\sin \bar{\alpha}}{\bar{\lambda}_{eff}} \cdot \tag{9.3}$$

A beam leaving the flat surface of the immersion lens without refraction with the angle α to the normal would have an angle $\bar{\alpha}$ if refraction took place. The law of refraction results in:

$$\sin \bar{\alpha} = \sin \alpha \cdot \frac{n_{glass}}{n_{immersion}} \cdot \tag{9.4}$$

In spite of $\lambda_{eff} = \lambda_v/n_{glass}$ in the case of immersion

$$\overline{\lambda}_{eff} = \frac{\lambda_v}{n_{immersion}} = \lambda_{eff} \cdot \frac{n_{glass}}{n_{immersion}} \tag{9.5}$$

is used.

To prove the validity of (9.3) one must only substitute (9.4) and (9.5) in that equation. This result was verified in an experiment, too [9.3].

With the immersion lens
$$(dr_n)_{min} = \lambda_{v,eff}/4 = \lambda_v/(4\, n_{immersion}) = 0.0428\ \mu m$$
can be reached as smallest outermost zone width.

But the immersion lens itself limits the smallest reachable zone width to $(dr_n)_{min} = 0.0514\ \mu m$ because the ray which will produce this outermost zone must strike the immersion lens on the left side of the equator. So the construction optics is the limiting factor for the minimal outermost zone width dr_n and thereby for the resolution of the zone plate, too.

It can be shown [9.4] that according to the Rayleigh criterion the resolution of a zone plate is given by ($D = 2r_n$ = aperture of the objective, f_m = focal length for the m-th diffracted order, δ = resolution, λ = wavelength):

$$\delta = \frac{1.22 \cdot \lambda \cdot f_m}{2 \cdot r_n} \simeq \frac{dr_n}{m} \tag{9.6}$$

because the approximate width of the n-th zone is given by:

$$dr_n = \frac{r_n}{2 \cdot n} \ . \tag{9.7}$$

So it is shown that the width of the outermost zone is the determinative parameter for the resolution of a zone plate.

In case of the examined construction optics the magnitude of r_n is only limited by the immersion lens, because the geometry of this lens and of the path of rays in the aplanatic lens system permits only a maximum value $r_n \simeq 29\ \mu m$.

The optimization of the parameters of the construction optics for the micro-zone-plate as well as the evaluation of its properties was performed by ray-tracings. One can use ray-tracing methods for the evaluation of the properties of an optical system when the approximation condition for geometrical optics (wavelength neglibly small, $\lambda \to 0$ or dimensions of the optical system large compared to λ) is satisfied [9.5]. This condition is satisfied for ultrasoft x rays and for the UV light used for the construction of the zone plate.

9.2 Raytracing Method

The optical axis of the system should always describe the z axis of Cartesian coordinates orientated in the direction of the propagation of light. Owing to the radial symmetry of the optical system considered, for off-axis calculations it is only necessary to vary in x or y directions.

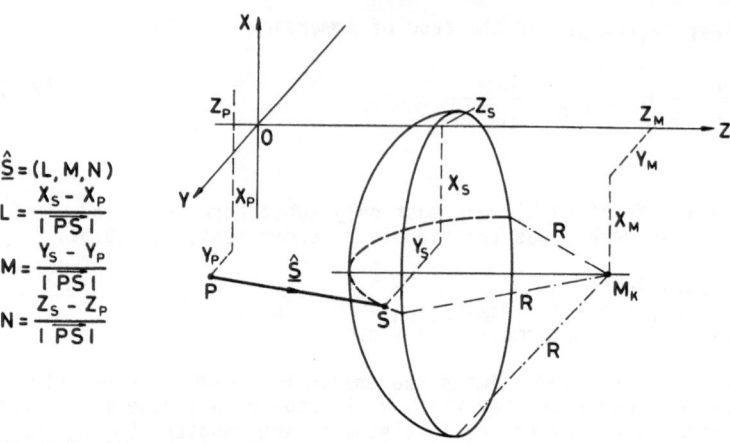

$\hat{S} = (L, M, N)$

$L = \dfrac{X_S - X_P}{|\overline{PS}|}$

$M = \dfrac{Y_S - Y_P}{|\overline{PS}|}$

$N = \dfrac{Z_S - Z_P}{|\overline{PS}|}$

Fig.9.2   Diagram for the derivation of the formulas to determine the intersection point of a light ray with a spherical surface

The radii of the spherical surfaces will get a positive sign if the centre
of the sphere lies on the left side of the surface and a negative sign in the
other case.

Light rays are always described by a point $P(x_p, y_p, z_p)$ through which they
go and by their direction cosines L, M, N to the axis of the coordinate system.

Contrary to the iteration methods described in the literature [9.6,7] the
determination of the intersection point $S(x_s, y_s, z_s)$ between a spherical surface and a light ray (see Fig.9.2) was carried out by using the following
exact formulas [9.2]:

$$x_s = \frac{L}{N} \cdot (z_s - z_p) + x_p \tag{9.8a}$$

$$y_s = \frac{M}{N} \cdot (z_s - z_p) + y_p \tag{9.8b}$$

$$z_s = N \cdot \{ A + \text{sign}(NR) \cdot \sqrt{A^2 - B} \} \tag{9.8c}$$

with

$$A = \frac{z_p}{N} - \{ L(x_p - x_M) + M(y_p - y_M) + N(z_p - z_M) \} \tag{9.8d}$$

$$B = \{ \frac{L}{N} z_p - (x_p - x_M) \}^2 + \{ \frac{M}{N} z_p - (y_p - y_M) \}^2 + z_M^2 - R^2 . \tag{9.8e}$$

Snell's law in vector form can be used in the calculations for the refraction and reflection of a light ray:

$$n' \cdot \hat{\underline{S}}' \times \hat{\underline{r}} = n \cdot \hat{\underline{S}} \times \hat{\underline{r}} \qquad \text{where (see Fig.9.3)} \tag{9.9}$$

$\hat{\underline{S}} = (L, M, N)$ = incident light ray in medium with refractive index n
$\hat{\underline{S}}' = (L', M', N')$ = refracted resp. reflected light ray in medium with
                   refractive index n'
$\hat{\underline{r}} = (\alpha, \beta, \gamma) = \text{sign}(N) \cdot \overline{M_k S}/R$ = normal of the surface at the intersection
                   point S of the light ray with the surface (M_k = centre of
                   sphere).

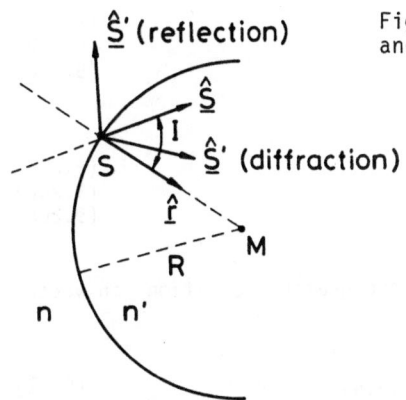

$\hat{\underline{S}}'$ (reflection)

$\hat{\underline{S}}$

I

$\hat{\underline{S}}'$ (diffraction)

$\hat{\underline{r}}$

S

M

R

n   n'

Fig.9.3 Notation for ray-tracings, reflection and diffraction at a spherical surface

The general solution of (9.9) is given by:

$$\hat{\underline{S}}' = T \cdot \hat{\underline{S}} + Q \cdot \hat{\underline{r}} \tag{9.10}$$

where $\quad T = \dfrac{n}{n'}$ (9.11)

and for refraction $\quad Q = +\sqrt{A_1^2 - B_1} - A_1$ (9.12)

resp. for reflection $\quad Q = -2 \cdot A_1 = -2 \cdot \cos I$

(9.13)

with

$$A_1 = T(L\alpha + M\beta + N\gamma) = T \cdot \cos I \tag{9.14a}$$

$$B_1 = T^2 - 1 \tag{9.14b}$$

(I = included angle between $\hat{\underline{S}}$ and $\hat{\underline{r}}$: $\cos I = \hat{\underline{r}} \cdot \hat{\underline{S}}$).

The value of $\cos I$ except for the normalization to the radius R of the sphere is given by the determination of the intersection point :

$$\cos I = \frac{\sqrt{A^2 - B}}{R} \quad . \tag{9.15}$$

The intersection point $S(x_S, x_S, z_S)$ of a light ray with a plane surface which is perpendicular to the z axis, described by $z = z_E$, will be determined by:

$$x_S = \frac{L}{N} \cdot (z_E - z_p) + x_p \tag{9.16a}$$

$$y_S = \frac{M}{N} \cdot (z_E - z_p) + y_p \tag{9.16b}$$

$$z_S = z_E \quad . \tag{9.16c}$$

The use of (9.10), (9.12), (9.13) in connection with:

$$\hat{\underline{r}} = \text{sign}(N) \cdot (0,0,1) \tag{9.17}$$

and $\quad A_1 = T \cdot N \cdot \text{sign}(N) = T \cdot |N|$ (9.18)

results in:
$$\begin{aligned} L' &= T \cdot L & (9.19a) \\ M' &= T \cdot M & (9.19b) \\ N' &= T \cdot N + Q \cdot \text{sign}(N) & (9.19c) \end{aligned}$$

with Q from (9.12) for the refraction resp.

$$\begin{aligned} L' &= L & (9.20a) \\ M' &= M & (9.20b) \\ N' &= -N & (9.20c) \end{aligned}$$
for reflection.

For ray-tracings through the zone plate the grating equation in vector form is used:

$$\underline{\hat{S}}' \times \underline{\hat{r}} = \underline{\hat{S}} \times \underline{\hat{r}} + \frac{m \cdot \lambda}{d} \cdot \underline{\hat{q}} \qquad \text{where (see Fig.9.4)} \qquad (9.21)$$

$\underline{\hat{S}} = (L,M,N) =$ incident light ray
$\underline{\hat{S}}' = (L',M',N') =$ diffracted light ray
$\underline{\hat{r}} =$ normal to the zone plate at intersection point P of the light ray with the zone plate
$\underline{\hat{p}} = (u,v,w) =$ unit vector vertical to the rulings, that means in radial direction $(w \equiv 0)$
$\underline{\hat{q}} = -\underline{\hat{p}} \times \underline{\hat{r}} =$ unit vector parallel to the rulings
$m =$ number of diffracted order
$d = 2 \cdot dr_n =$ grating spacing constant
$\lambda =$ wavelength.

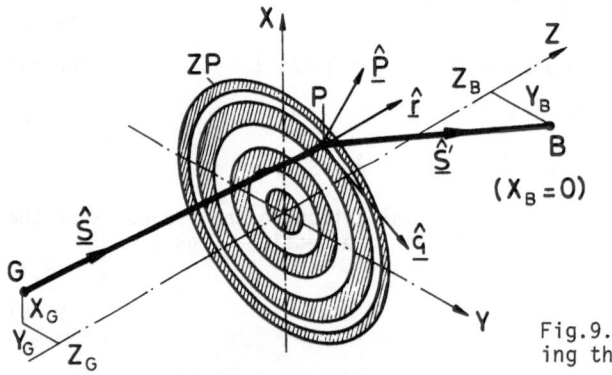

Fig.9.4 Notation for ray-tracing through a zone plate

In the meridional case (9.21) will become the usual grating equation:

$$\sin \alpha + \sin \beta = \frac{m \cdot \lambda}{d} \ . \qquad (9.22)$$

The general solution of (9.21) is given by :

$$\underline{\hat{S}}' = \underline{\hat{S}} - \Lambda \cdot \underline{\hat{p}} + Q \cdot \underline{\hat{r}} \qquad (9.23)$$

where

$$\Lambda = \frac{m \cdot \lambda}{2 \cdot dr_n} \qquad (9.24)$$

and (for transmission zone plates)

with
$$Q = +\sqrt{A_2^2 - B_2} - A_2 \tag{9.25}$$
$$A_2 = N \tag{9.26a}$$
$$B_2 = \Lambda^2 - 2\cdot\Lambda\cdot(Lu + Mv) . \tag{9.26b}$$

The aberrations should be divided symmetrically to both wave centres to have a symmetrical construction process, i.e. $\xi_n = \eta_n$ (see Fig.9.5). For this reason the place of formation of the zone plate is chosen so that $\xi_1 = \eta_1$ is valid, whereby ξ_1, η_1 are calculated with an input light ray in the beamsplitter of slope $tgu_{in} = 10^{-6}$.

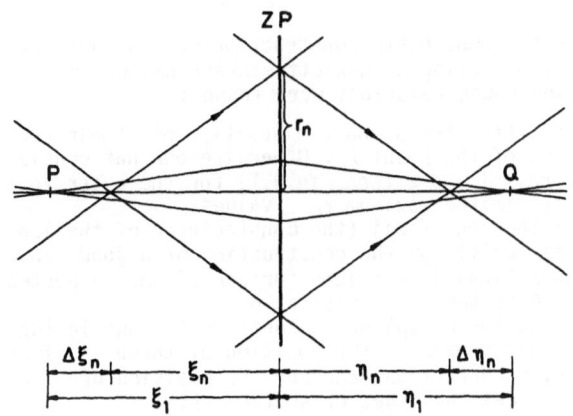

Fig.9.5 Diagram for the method to correct the spherical aberration of the zone plate

The aplanatic lens system does not work ideally, so incident light rays with greater slopes will not produce identical ξ_n and η_n ($\xi_n(tg\ u_{in}) \neq \eta_n(tg\ u_{in})$). The incident ray with $tg\ u_{in} = 10^{-6}$ will be used to determine the focal length:

$$f_v = (\frac{1}{\xi_1} + \frac{1}{\eta_1})^{-1} \tag{9.27}$$

of the zone plate for the construction wavelength λ_v, too. The first zone radius r_1 will then be given by:

$$r_1 = (\lambda_v \cdot f_v)^{1/2} . \tag{9.28}$$

Now the focal length for the desired x-ray wavelength λ_x is given by :

$$f_x = \frac{r_1^2}{\lambda_x} \tag{9.29}$$

and for a desired magnification V the object must be put in the distance :

$$g = \frac{V+1}{V}\cdot f_x \tag{9.30}$$

from the zone plate. For aberration free imaging one should have the image at the distance :

$$b = (V+1)\cdot f_x = V\cdot g . \tag{9.31}$$

If aberrations occur the distance b for the minimal confusion circle will have a different value, so will the magnification V.

To make a statement about the aberration-limited resolution d_{min} of the examined micro-zone-plate the diameter δ_{min} of the minimal confusion circle has to be determined:

$$d_{min} = \frac{\delta_{min}}{V} \, .$$

(9.32)

The micro-zone-plate will be a good one, if the aberration-limited resolution is substantially smaller than the diffraction-limited resolution: $d_{min} \ll dr_n$.

9.3 Raytracing Calculations

To find an optimum solution for the zone plate construction optics ray-tracings have been performed [9.2] by using a Hewlett-Packard HP41C pocket calculator. Thereby the following rough relations were found :

a) Variation of the radii of the spheres and the position of their centres will force into modification of the point P_o. Otherwise one has considerably different slopes of the rays 1 and 2 (see Fig.9.1) for the outer zone radii, which will result in very limited maximum r_n values.

b) the possible variation of the lens radii (the manufacturer of the lens can produce only discrete sphere radii) for the construction of a good zone plate is too rough. The most uncritical is the lens surface 13. As expected the radius of the lens surface 14 is the most critical one.

c) The displacement of the intrinsic aplanatic surfaces 12 and 14 (see Fig.9.1) is allowed in only a small region. The position of these surfaces can therefore be used as precision correction, whereby the position of surface 14 is considerably more critical than that of surface 12.

d) The displacement of surface 13 is allowed in a greater region. Its position can be used for the rough determination of the solution for the zone plate construction optics.

According to these statements the optimum solution was found by variation of the position of surface 13 and variation of the point P_o.

With the optimum solution for the optical system a micro-zone-plate with the following parameters can be constructed:
$r_1 = 1.756 \ \mu m$, $f_x = 1.307$ mm for $\lambda_x = 2.36$ nm,
$r_n = 27.8 \ \mu m$, $dr_n = 0.055 \ \mu m$, N = 251.

With an object distance of g = 1.311 mm an image distance of b = 437 mm is obtained for the minimal confusion circle, corresponding to a magnification of V = 333. The aberration-limited resolution is d_{min} = 3.6 nm. Figure 9.6 shows the ray-tracing for the micro-zone-plate.

Figure 9.7 shows the distribution of the aberrations to both spherical construction wave fronts of the micro-zone-plate (curves 2 and 3). Curve 1 shows the aberration curve of the nominal micro-zone-plate with $r_1 = 1.756 \ \mu m$ for symmetrical construction. This curve will be excellently reproduced if the arithmetic mean of curve 2 and 3 is built. Thus it is shown that the desired aberrations exist, but are not symmetrically distributed in both spherical wave fronts.

Opposite to the existing beamsplitter and aplanatic lens system only the 4th and 5th lenses are new, therefore only for these lenses are variation calculations carried out to get production tolerances within which a good micro-zone-plate is yielded. These calculations were performed [9.2] with the UNIVAC 1100 at the GWD Göttingen.

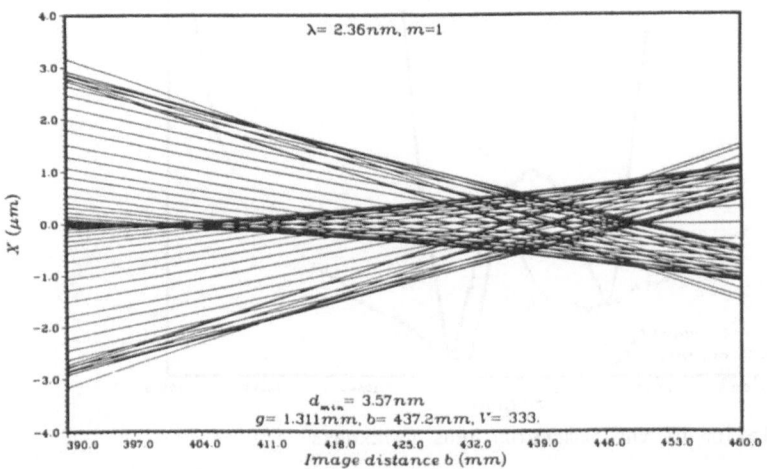

Fig.9.6 Ray-tracing through the micro-zone-plate (optimum case)

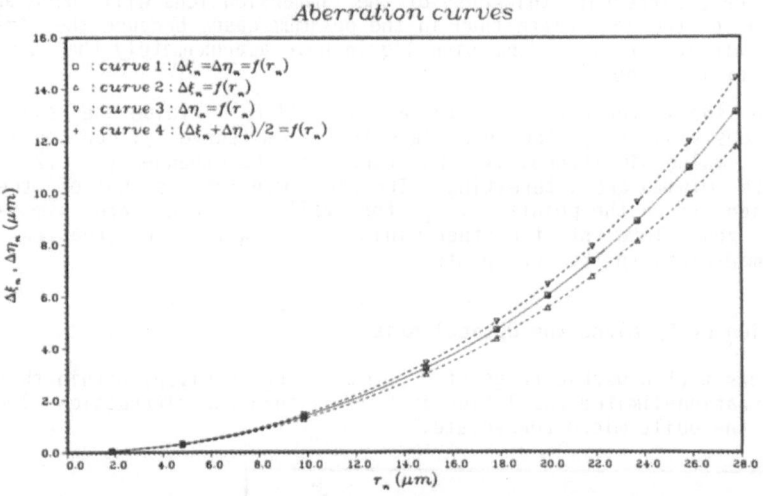

Fig.9.7 Distribution of the aberrations to both spherical construction
wave fronts of the micro-zone-plate

The starting point for the calculations is the optimum solution for the
optical system. If not otherwise stated, all curves for the aberration-limited resolution of the zone plate are valid for the x-ray wavelength
$\lambda_x = 2.36$ nm and the magnification $V = 300$.

9.3.1 Variation of the Immersion Lens Thickness

These calculations are essential because it is not known exactly where the
micro-zone-plate will be built. The thickness of the immersion liquid and

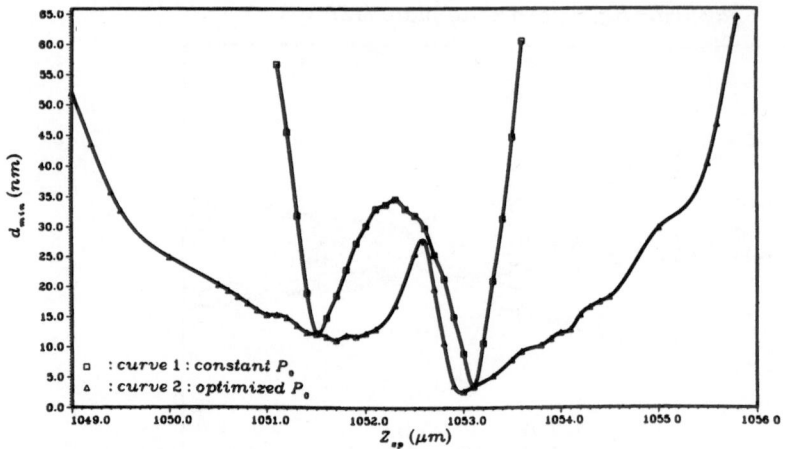

Fig.9.8 Variation of the immersion lens thickness

eventual dust particles will influence the position of the zone plate. It is the same whether the immersion liquid between the lens surface and the resist layer or a different thickness of the immersion lens will give another position for the zone plate than in the optimum case, because the immersion lens material and the immersion liquid have approximately the same refractive index (see above).

Figure 9.8 shows the result : for a constant P_0 value the usable range for the position ZZP of the zone plate is 2.4 μm (curve 1). For optimizing the P_0 value simultaneously this range will be expanded to 6.7 μm (curve 2). Both minima are interesting. The right minimum is that of the optimum solution, where the points ξ_1, η_1 for small r_n values are symmetrical to the zone plate. At the other minimum the ξ_n, η_n for great r_n values are symmetrical to the zone plate.

9.3.2 Variation of P_0 Along the Optical Axis

Figure 9.9 shows that a usable range of 0.36 mm for P_0 is given. Within this range the aberration-limited resolution is smaller than the diffraction-limited one for the built micro-zone-plate.

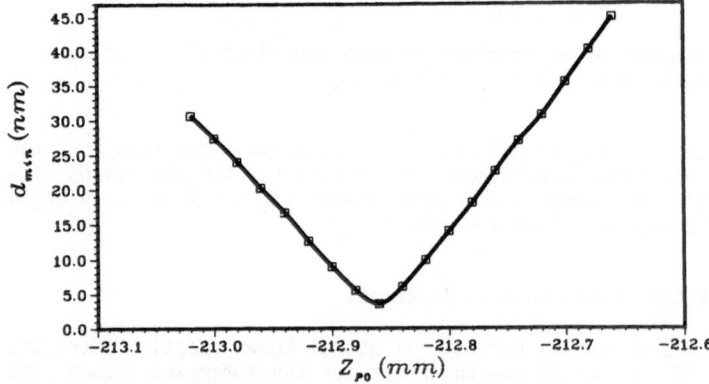

Fig.9.9 Variation of P_0 along the optical axis

9.3.3 Variation of the Thickness of Lens 4 of the Aplanatic System

In keeping all other parameters of the optical system constant, Fig.9.10 shows that lens 4 may be up to 3 µm thicker and up to 6 µm thinner than in the optimum case. Calculations have shown that a wrong thickness of a lens can at least partly be corrected by changing the suitable air distances.

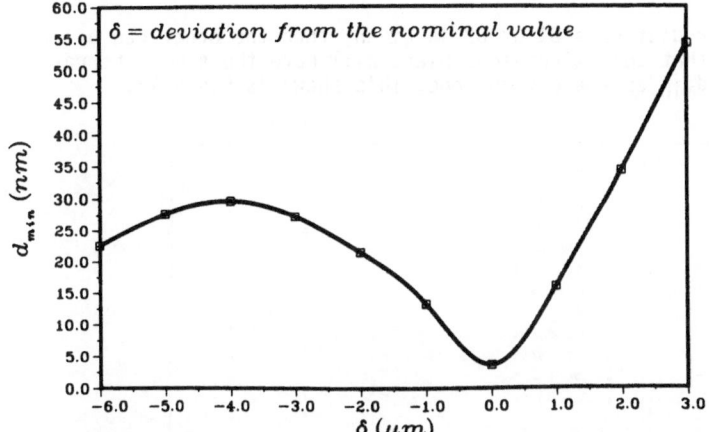

Fig.9.10 Variation of the thickness of lens 4 of the aplanatic system for constant P_O

9.3.4 Variation of the Air Distances Between the 3th and 4th and the 4th and 5th Lens

In Fig.9.11 is shown how accurate the air distances must be. The air distance D1 between lens 3 and 4 is substantially more uncritical than D2, the air distance between lens 4 and 5. The allowed range for D1 is very great : -22 µm $\leq \delta 1 \leq 14$ µm ($\delta 1$ = deviation from the desired value D1).

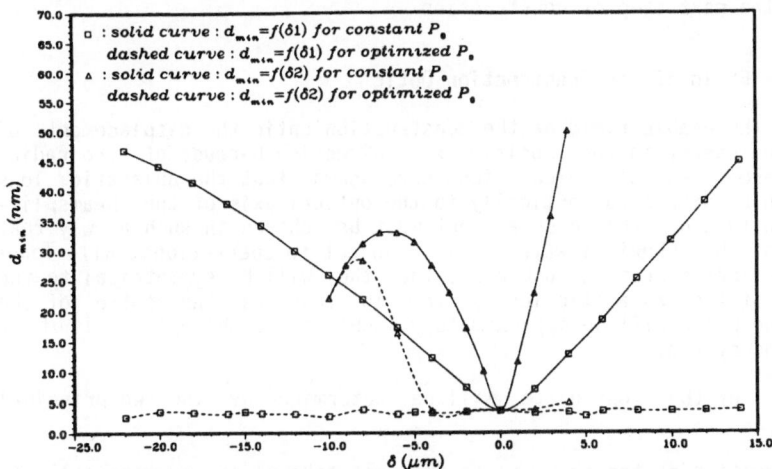

Fig.9.11 Variation of the air distance between the 3th and 4th resp. 4th and 5th lenses of the aplanatic system

If the value P_0 is optimized too, an approximate constant aberration lim-ited resolution of d_{min} = 3.6 nm (dashed curve) can be reached in the whole range.

The allowed range for D2 is smaller : -10 μm < δ2 < 4 μm. By optimizing P_0 an approximate constant aberration limited resolution of d_{min} = 3.6 nm in the range: -4 μm ≤ δ2 ≤ 4 μm can be reached, too.

If one of the air distances D1 or D2 is given, then the other can be cho-sen in such a way that the micro-zone-plate will have the same aberration-limited resolution d_{min} as the optimum one. This shown is Fig.9.12.

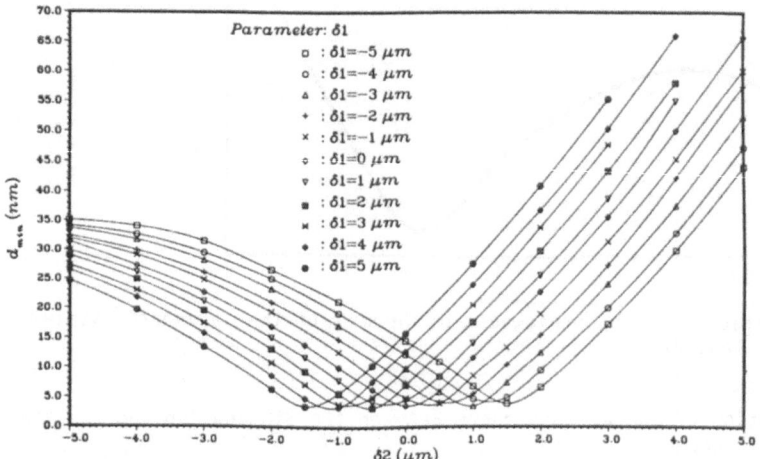

Fig.9.12 Variation of the air distances between the 3th and 4th and 4th and 5th lenses of the aplanatic system

Thus the already observed deviations from the desired values can be com-pensated in the next step of construction.

9.3.5 Usable Field of the Construction Optic

To determine the usable field of the construction optic the displacement of P_0 is only necessary in the positive x direction because of the radial symmetry. These off-axial calculations have shown that the aplanatic lens system has to be displaced vertically to the optical axis of the beamsplit-ter. The magnitude of this displacement must be chosen in such a way that the centres of the spherical wave fronts, subject to aberrations, will leave the aplanatic lens system in such a way that they will be symmetrical to the optical axis of the aplanatic lens system. In this case the centre of the resultant zone plate will be approximately identical to the optic axis of the aplanatic lens system.

The centre of the zone plate will be determined by the two principal rays.

The construction of the zone plate does not take place symmetrically to the z axis, but to the principal ray which forms the centre of the zone plate. Thus calculations have shown that the best confusion circles were obtained when the zone plate was used off-axis in imaging, that means, in the experiment to tilt the zone plate opposite to the optical axis.

The consideration of the ray-tracings through the zone plate has to be carried out with reference to the Gaussian principle ray. This is the usual procedure in optics.

The results of the off-axis calculations are shown in Fig.9.13. The solid curve shows the aberration-limited resolution in x direction as a function of the x coordinate of P_0, whereas the dashed curve shows the aberration-limited resolution in y direction. As can be seen, P_0 can be up to 0.75 mm beside the optical axis without the aberration-limited resolution being worse than the diffraction-limited one.

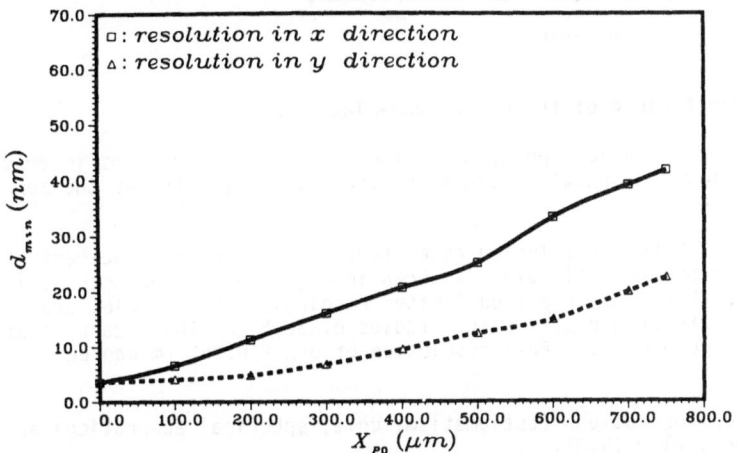

Fig.9.13 Usable field of the construction optics

9.4 Imaging Properties of the Micro-Zone-Plate

In the following sections the imaging properties of the micro-zone-plate, built with the optimum optical system, are evaluated. The interesting properties are the depth of field and the object field which can be imaged with full resolution.

9.4.1 Depth of Field

Figure 9.14 shows the aberration-limited resolution d_{min} for constant image distance as a function of the deviation δg from the desired object distance g. The usable range is $\delta g = \pm 1.3$ μm.

The theoretical depth of focus is given by:
$$\Delta f = \frac{\lambda_x \cdot f_x^2}{4 \cdot r_n^2} \quad . \qquad (9.33)$$

With the zone plate parameters mentioned above (9.33) will give $\Delta f = 1.30$ μm and thus are in agreement with the ray-tracings $\delta g = 1.31$ μm.

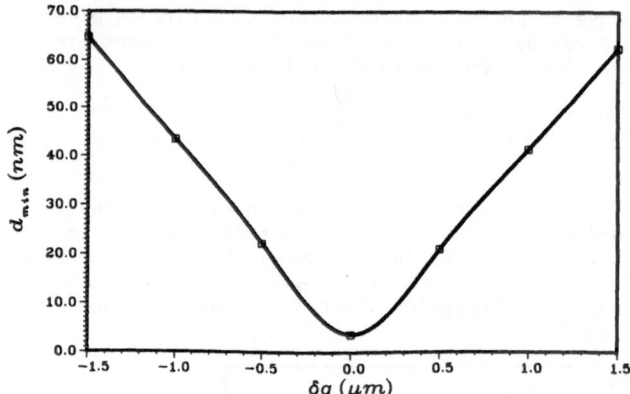

Fig.9.14 Depth of field of the micro-zone-plate

9.4.2 Usable Object Field of the Micro-Zone-Plate

The displacement of the object point G for the off-axis ray-tracings is only necessary in positive x direction because of the radial symmetry of the zone plate.

The aberration-limited resolution as a function of the displacement of the object point from the optic axis is shown in Fig.9.15. The object field which is imaged with full diffraction limited resolution is a circle around the optic axis of the zone plate with a radius of 32.5 μm. This means that 1047 x 1047 image points with full resolution of $dr_n = 0.055$ μm can be imaged.

The third-order aberrations (astigmatism, coma, spherical aberration) are calculated theoretically [9.8].

$$\frac{\Delta y'}{x_0} = - \frac{3}{2} r_n \cos \phi \{ \frac{y_0'^2}{x_0'^3} - \frac{y^2}{x^3} \} + \frac{1}{2} r_n^2 \{1 + 2\cos^2\phi\} \cdot \{ \frac{y_0'}{x_0'^3} - \frac{y}{x^3} \}$$

$$- \frac{1}{2} r_n^3 \cos \phi \{ \frac{1}{x_0'^3} - \frac{1}{x^3} - \frac{\Theta}{f^3} \}.(9.34)$$

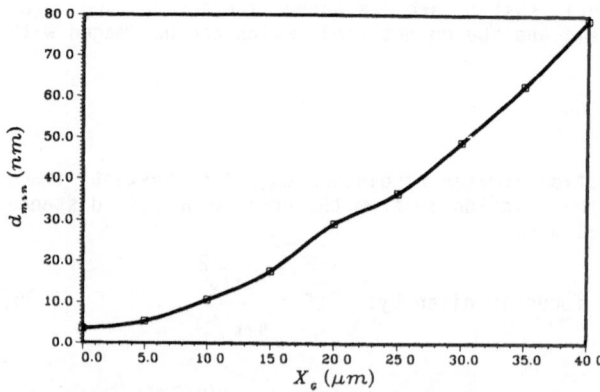

Fig.9.15 Usable object field of the micro-zone-plate

The spherical aberration can be corrected for a given x-ray wavelength and a given magnification V when building the zone plate, so only the first and second terms of (9.34) are relevant.

In substituting the suitable parameters in (9.34) the maximal deviation of the object point from the optic axis for imaging with full resolution is given by y = 28.1 μm.

9.4.3 Other Properties of the Micro-Zone-Plate

Calculations have shown that the use of the micro-zone-plate for other than the desired wavelength is possible, because in the range from λ_x = 0.3 nm up to λ_x = 10.3 nm the aberration–limited resolution will be smaller than the diffraction–limited one. But the object distance and image distance must be changed with the wavelength.

Using the micro-zone-plate in the third diffraction order should give theoretically a diffraction-limited resolution of dr_n/m = 18 nm. But the ray-tracing shows that the aberration–limited resolution is poorer: d_{min} = 24.3 nm. This MZP3 is not well corrected for this diffraction order.

For use in a scanning x-ray microscope [9.9] the described micro-zone-plate demagnifies the image of the synchrotron source which has already been demagnified by a condenser zone plate. The desired demagnification with the micro-zone-plate is V = 0.01. As shown by Fig.9.16, the minimum circle of confusion for the image of a point source will have a diameter of 3.6 nm in this case. This has no influence on the desired size of 50 nm for the demagnified image of the synchrotron source.

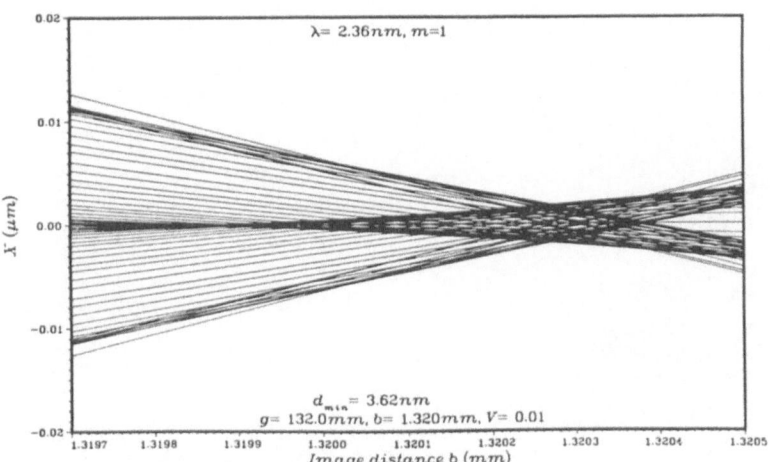

Fig.9.16 Ray-tracing through the micro-zone-plate used in the scanning x-ray microscope with V = 0.01

Acknowledgements

This work was supported by the Stiftung Volkswagenwerk. The ray-tracings were performed with the UNIVAC 1100 of the Gesellschaft für wissenschaftliche Datenverarbeitung Göttingen (GWDG).

References

9.1 G. Schmahl, D. Rudolph, P. Guttmann, and O. Christ
"Zone Plates for X-Ray Microscopy", this volume, no. 8

9.2 P. Guttmann: Theoretische Untersuchungen über hochauflösende Zonen-platten als abbildende Systeme für weiche Röntgenstrahlung, Diplom-arbeit, Universität Göttingen, 1982

9.3 G. Schmahl: private communication

9.4 D. Rudolph, G. Schmahl: "High Power Zone Plates for a Soft X-Ray Microscope", Ann. N. Y. Acad. Sci. 342, 94 (1980)

9.5 M. Born, E. Wolf: Principles of Optics, Sixth Edition, Pergamon Press, Oxford, 1980

9.6 W.T. Welford: Aberrations of the Symmetrical Optical System, Academic Press, London, 1974

9.7 G.H. Spencer, M.V.R.K. Murty: J. Opt. Soc. Am. 52, 672 (1962)

9.8 D. Rudolph: Holographische Zonenplatten als abbildende Systeme für die Röntgenastronomie, Forschungsbericht W74-07 des Bundesministeriums für wissenschaftliche Forschung, Sept. 1974

9.9 B. Niemann: "The Göttingen Scanning X-Ray Microscope", this volume, no. 22

10. Construction of Condenser Zone Plates for a Scanning X-Ray Microscope

J. Thieme

Forschungsgruppe Röntgenmikroskopie, Universität Göttingen
Geismarlandstraße 11, D-3400 Göttingen, Fed. Rep. of Germany

10.1 Introduction

A scanning x-ray microscope consists of an optical system, which images the source of x-ray light into a very small scan spot. An object can be scanned through it and because all optical elements that absorb radiation are between the source and the object, the radiation impact on the specimen is minimized.

Using zone plates as optical elements, monochromatic light is needed, because the focal length f is proportional to the inverse of the wavelength λ. A condenser zone plate together with a diaphragm as a linear monochromator and a micro-zone-plate as a high-resolution element yield together the optical system (see Fig.10.1).

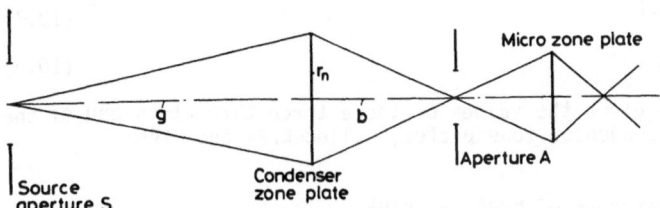

Fig.10.1 An optical system for a scanning x-ray microscope

The monochromaticity of the linear monochromator has to be

$$\lambda/\Delta\lambda \simeq nm/2 \tag{10.1}$$

where m is the number of the diffracted order and n is the number of zones of the micro-zone-plate used.

The geometrical dimension of the x-ray source at the BESSY storage ring and the aperture S near the source point with a diameter of $d_S = 100$ μm lead to a useful beam divergence of $\Theta = 0.2$ mrad [10.1]. An object distance of g = 12 m between aperture S and the condenser zone plate means that the diameter of the condenser zone plate should be

$$D = 2r_n = 2.4 \text{ mm} . \tag{10.2}$$

The source emits polychromatic radiation, which is focused by the condenser zone plate. If there is a diaphragm A in the image plane, the monochromaticity of the radiation going through that diaphragm is given by

$$\lambda / \Delta \lambda \simeq D/2d_A \tag{10.3}$$

where D is the diameter of the condenser zone plate and d_A is the diameter of the diaphragm A. A micro-zone-plate behind the diaphragm will therefore be supplied with quasimonochromatic radiation.

10.2 Calculation of the Zone Plate Parameters

It is intended in this work to construct two condenser zone plates, which shall be used in combination with two already existing micro-zone-plates, called MZP2 and MZP3. The required monochromaticities, calculated by (10.1), are $\lambda / \Delta \lambda = 100$ for MZP2 and $\lambda / \Delta \lambda = 250$ for MZP3 for first-order radiation. The diameter d_A of diaphragm A can be calculated by (10.3). With the magnification

$$V = d_s/d_A \tag{10.4}$$

one can get the focal length

$$f = g/(V+1). \tag{10.5}$$

The radius r_1 of the innermost zone, the width dr_n of the outermost zone and the total zone number n can be obtained by the following three equations [10.2]:

$$r_1 = \sqrt{f \circ \lambda} \tag{10.6}$$

$$n = r_n^2 / r_1^2 \tag{10.7}$$

$$dr_n = r_n / 2n . \tag{10.8}$$

The following table gives the values of these three parameters and of the radius r_n for the two condenser zone plates, called KZP4 and KZP5:

Table 10.1   The parameters of KZP4 and KZP5

	KZP4	KZP5
r_1 [μm]	49.8	36.15
r_n [mm]	1.25	1.25
dr_n [μm]	0.992	0.458
n	630	1196

KZP4 is suited for the wavelength $\lambda_x = 2.36$ nm in combination with MZP2 and for the wavelength $\lambda_x = 4.5$ nm with MZP3. KZP5 is suited for the wavelength $\lambda_x = 2.36$ nm in combination with MZP3.

10.3 Construction of an Optical System to Build the Condenser Zone Plate

Zone plates can be built by the superposition of a convergent and a divergent beam of laser light. If a lens system focuses two wave fronts of laser light into two focal points on the optical axis, the required superposition is possible in between these two points. The needed laser light comes from an argon laser with the wavelength $\lambda = 457.936$ nm. Because this wavelength

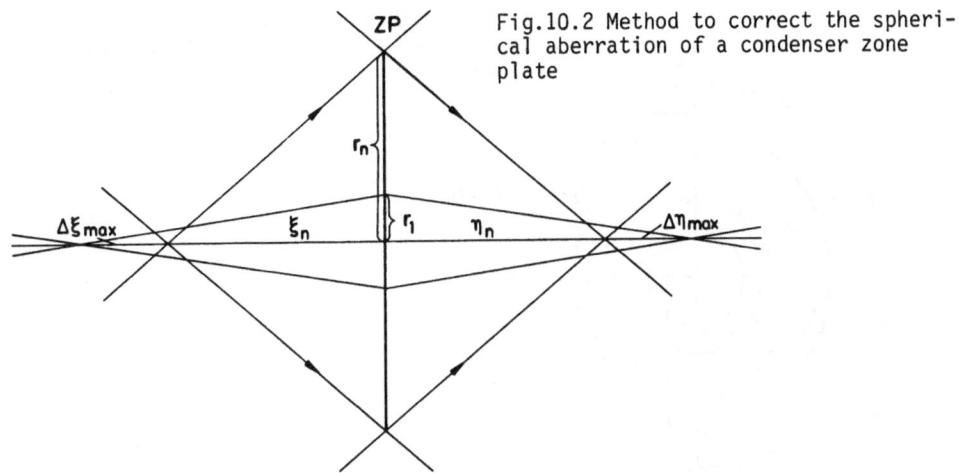

Fig.10.2 Method to correct the spherical aberration of a condenser zone plate

differs from the x-ray wavelength used, spherical aberration occurs. The way to correct it is to use wave fronts with spherical aberrations $\Delta\xi_n$ and Δn_n for the construction (see Fig.10.2). It is necessary that the aberration $\Delta\xi_n$ has an oppositive sign compared to the aberration Δn_n with the zone plate as a plane of symmetry. The values of $\Delta\xi_n$ and Δn_n are equal in the case of symmetry, so only the ξ_n and $\Delta\xi_n$ have to be calculated.

The lens system that satisfies these conditions is a lens system calculated by the aplanatic theory of Weierstraß [10.3]. The interference fringe system of the zone plates is near the innermost aplanatic point of the system. The outer refracting surfaces of the lenses are calculated by that theory, the inner surfaces are used to obtain good zone plate parameters. Because the aberrations $\Delta\xi_n$ are very small compared with ξ_n, lenses with long focal lengths are needed. According to the aplanatic theory, the radii of the outer surfaces of the lenses increase with the square of the refraction index n. If the radius of the convex surface of the first lens is r_1, the radii of the convex surfaces of the following lenses are given by

$$r_2 = n^2 r_1, \ r_3 = n^2 r_2 = n^4 r_1, \ \ldots . \tag{10.9}$$

In this case the air distances between the lenses will become very long. To improve the system there is the following possibility (see Fig. 10.3).

In the frame of reference I there are two spherical surfaces SS_1 and SS_2, arranged by the aplanatic theory. The radius of SS_1 is r_1, the radius of SS_2 is $r_2 = n^2 r_1$. The point A is the inner aplanatic point of SS_1, the point B is the outer aplanatic point of SS_1 and the inner aplanatic point of SS_2. The distance between SS_1 and SS_2 is

$$d = r_1 n^2 - r_1 = r_1(n^2-1), \tag{10.10}$$

the distance from the origin to the point B is r_2/n.

Rays, which shall be focused by SS_1 in A, have to be convergent in B after refraction by SS_2. One axis of the frame of reference I will be displaced parallel by the distance D into the reference system II. That means that the distance between the origin and B is shortened by D. In the new reference system II the point B is still the outer aplanatic point of SS_1.

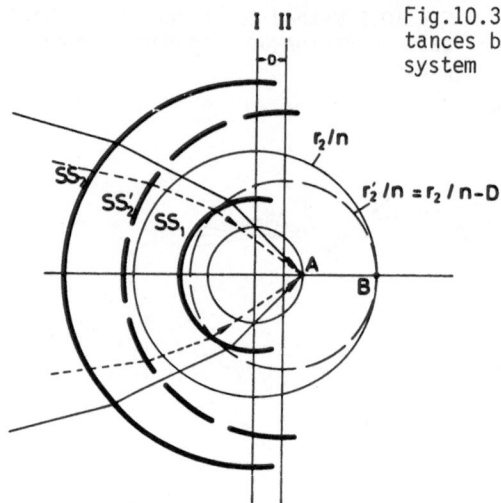

Fig.10.3 Method to minimize the air distances between the lenses of an aplanatic system

Therefore, there has to be a surface SS_2, for which B is the inner aplanatic point. The distance from the origin to B is

$$r_2'/n = r_2/n - D \tag{11.11}$$

and it follows the radius r_2' of SS_2'

$$r_2' = r_2 - nD = n(r_1n - D). \tag{11.12}$$

Rays, which are refracted by SS_2 into the point of convergence B, will be focussed by SS_1 into the point A. This is exactly according to the aplanatic theory of Weierstraß. The distance between SS_1 and SS_2 is

$$d' = r_2' - r_1 - D = r_1n^2 - Dn - r_1 - D$$

$$= r_1(n^2 - 1) - D(n+1) . \tag{11.13}$$

The comparison of (10.10) and (10.13) show that d' is shorter than d by $D(n+1)$. The radius r_2' of SS_2' is shorter than the radius r_2 by nD. Thereby an aplanatic lens system with two lenses has been minimized. This procedure can be expanded to systems with several lenses without violation of the aplanatic theory. The concave surfaces of the lenses are used to correct aberrations to get optimum zone plate parameters.

The values of ξ_n and $\Delta\xi_n$ for KZP4 are greater than the corresponding values for KZP5. Therefore, one lens has been added to the optical system for KZP4. The parameters of this lens have been calculated so that ξ_n, $\Delta\xi_n$ for KZP5 have been reached.

As a result, a two-lens optical system has been calculated for the construction of KZP4 with one additional lens for the construction of KZP5.

To supply the lens system with the needed convergent laser beams, a beam splitter is calculated, which splits one divergent into two convergent beams as required. Figure 10.4 shows the beamsplitter and the optical rays, originating from a point source. The beamsplitter consists of two spherical mir-

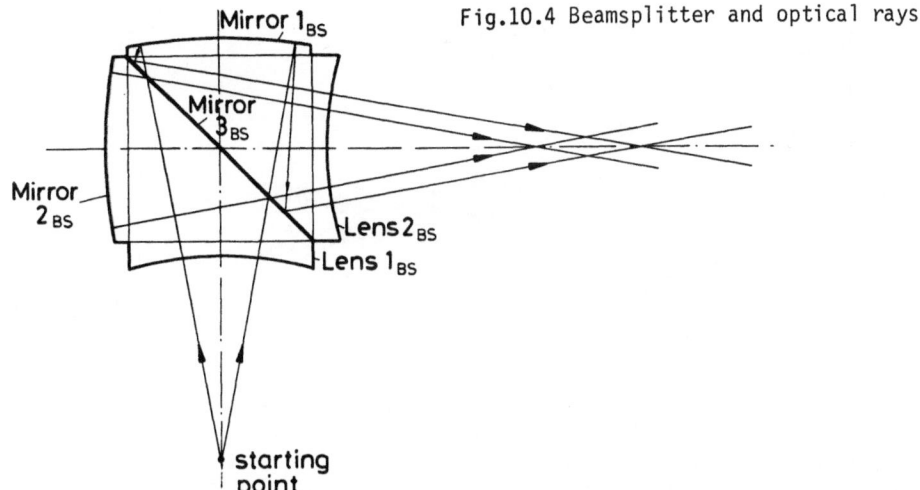

Fig.10.4 Beamsplitter and optical rays

rors, one semitransparent mirror and two lenses. The semitransparent mirror is the dividing element, which splits one beam into two. The two spherical mirrors focus the beams as required by the lens system. Lens 1_{BS} and lens 2_{BS} are necessary to minimize aberrations.

If the right aberrations for the construction of the zone plate shall appear, the two focusses of mirror 1_{BS} and mirror 2_{BS} have to be near the outer aplanatic point of the total lens system. It is not possible to calculate a beamsplitter for two exact foci. There will be aberrations. Since the inner focus is very sensitive to displacements and aberrations, it will be imaged in the relation 1:1 by mirror 2_{BS}. The radii of lens 1_{BS} and lens 2_{BS} are calculated for that and therefore the outer focus will have aberrations, but these are too small to show effects. The complete optical system, with which the interference fringe system of KPZ4 and KPZ5 can be constructed, is shown in Fig. 10.5. For KPZ5 the optical system is an immersion system. For KPZ4 the optical system is used without the lens 1_{LS}.

The optimization of the parameters of the construction optics for the two condenser zone plates as well as the evaluation of the properties of the zone plates were performed by ray-tracing calculations. Fig.10.6 shows a ray-tracing through KZP4 with $\lambda_x = 4.5$ nm, resulting in a circle of confusion with a smallest diameter of less than 40 nm. The same holds for KZP5.

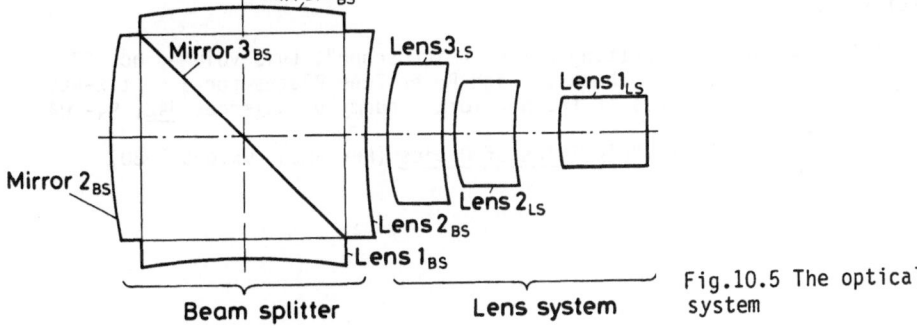

Fig.10.5 The optical system

95

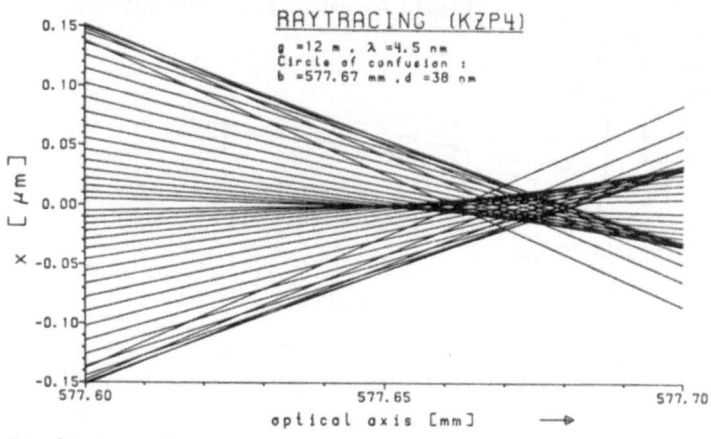

Fig.10.6 Ray tracing through KZP4

This value is small in comparison to the diffraction-limited resolution given by the width of the outermost zone, which is 992 nm for KZP4 and 523 nm for KZP5.

10.4 Variation of Parameters of the Optical Systems

The calculations were performed with ideal optical systems. In practice, however, optical systems can only be built with certain tolerances. The radii of the spherical surfaces can be manufactured in very good agreement with the theoretical values. The thickness of the optical elements, however, as well as,e.g.,the position of the semitransparent mirror 3_{BS} can vary. To investigate the influence of the variations of optical parameters, ray-tracing calculations with changed parameters of the system were performed, using the diameter of the circle of confusion - as discussed above - as a measure for the quality of the zone plate.

The calculations show that variations in the thickness of the optical elements can be compensated by a variation of the air distances between the lenses, and that a misalignment of mirror 3_{BS} by,e.g.,$\pm$ 2' can be compensated by a readjustment of lens 1_{BS} and mirror 1_{BS} without influencing the imaging quality of the zone plates built with these optical arrangements.

References

10.1 B. Niemann: "The Göttingen X-Ray Microscope", this volume, no. 22
10.2 D. Rudolph and G. Schmahl: "High Power Zone Plates for a soft X-Ray Microscope", Annals of the New York Academy of Sciences 342, 94-104 (1980)
10.3 M. Born, E. Wolf: Principles of Optics (Pergamon, Oxford 1980)

11. Recent Advances in X-Ray Optics

N. M. Ceglio

Lawrence Livermore National Laboratory, University of California, P.O.Box 5508
Livermore, CA 94550, USA

Recent developments in x-ray optics are reviewed. Specific advances in coded
aperture imaging, zone plate lens fabrication, time-and space-resolved spectroscopy, and CCD x-ray detection are discussed.

11.1 Introduction

The rapid development of x-ray optics and the important role played by microfabrication technology has been an active topic of review in recent years
[11.1]. This paper is intended as an update to the comprehensive body of
review literature of x-ray optics. The past 12 - 24 months have produced significant technological contributions, which are advancing x-ray optical capabilities along a broad front. Specific advances reported here are in the
areas of coded aperture imaging, x-ray lens development, time-and space-resolved x-ray spectroscopy, and CCD x-ray detection. This work has involved
LLNL personnel as well as the collaborative efforts of researchers from other institutions: IBM's Thomas J. Watson Research Center; Cornell University's National Research and Resource Facility for Submicron Structures; MIT's
Center for Space Research; MIT's Submicron Structures Laboratory; the University of Rochester's Laboratory for Laser Energetics; and Purdue University's Department of Mechanical Engineering.

Recent theoretical work [11.2] in coded aperture imaging has extended our
understanding of the basic principles of Zone Plate Coded Imaging (ZPCI). An
eigenfunction analysis of continuous source distributions has provided new
insight to linear as well as nonlinear effects in the microscopy of small
laboratory x-ray sources. In addition, the development of new techniques for
the fabrication of ultrathick zone plate coded apertures has extended high
resolution coded imaging to shorter wavelength radiation [11.3] (e. g., 100
keV x rays).

Significant new work has been done in the fabrication of advanced zone
plate lensing elements for x rays. Sophisticated scanning electron-beam lithographic (SEBL) techniques [11.4] have been applied to the production of
Fresnel zone plate objective lenses with minimum linewidths as small as
1500A. SEBL techniques have also been applied to the production of "large"
diameter ($\sim$1mm) Fresnel zone plate condenser lenses for use in x-ray microscopy [11.5]. In addition, a detailed design strategy for x-ray phase lenses
(i. e., Fresnel phase plates) has been carried out, and the first x-ray phase
lens (for use at Al $K_\alpha \simeq 1.5$ keV) has been fabricated [11.6].

New capabilities in time-and space-resolved x-ray spectroscopy have emerged from the coupling of high-precision x-ray transmission gratings [11.7]
with x-ray streak cameras [11.8] and grazing incidence reflection (GIR)

x-ray microscopes [11.9]. These new instruments have been applied to the investigation of laser-produced laboratory plasmas with striking results [11.10].

A collaborative effort [11.11] is underway for the development and fabrication of a CCD camera for direct x-ray recording. This instrument will provide a capability for real-time, spatially resolved x-ray detection with high sensitivity and low noise. It offers significant advantages for x-ray microscopy and spectroscopy at low signal levels.

In the following sections each of these topics is discussed in greater detail.

11.2 Coded Imaging

Coded imaging, first proposed by Mertz in the 1960's, is a two-step imaging technique [11.12]. In the first step, source information is recorded (encoded) by simple geometrical shadowcasting through a coded aperture. In the second step, image reconstruction (decoding) is achieved using a numerical or optical procedure matched to the coded aperture design. When the coded aperture design is a Fresnel zone plate, image decoding may be achieved by coherent optical reconstruction. In this case the entire imaging process can be viewed as a form of Fresnel transform microscopy, the spatial analog of Fourier transform spectroscopy. This point of view is based on the appreciation that the process of shadow casting through a Fresnel zone plate may be mathematically represented as a Fresnel transform. The subsequent optical reconstruction of the shadow graph (Fresnel diffraction) may also be represented as a Fresnel transform. This perspective has particular value, since the eigenfunctions of the Fresnel transform are well known [11.13]. In particular, the Gaussian-Laguerre polynomials in polar coordinates form a complete, orthogonal set that retain their functional form when Fresnel transformed. Therefore, when a source is represented as a superposition of Gaussian-Laguerre polynomials, all the integrals in the coded image recording and reconstruction process can be analytically evaluated, and a generalized expression for the reconstructed image in any order can easily be found.

A number of useful insights have emerged from the eigenfunction analysis of continuous source distributions. These are summarized below and discussed in greater detail in reference [11.2]:

(i)   In order to achieve ZPCI with good S/N and negligible background effects, it is necessary that the characteristic large scale dimension of the source be less than the diameter of the central zone of the coded aperture.

(ii) For continuous sources of finite size, tomographic resolution worsens with increased order number. In particular, the tomographic resolution element increases linearly with order number.

(iii) In the image reconstruction process the focussed intensity into a given image order varies inversely as the fourth power of the order number.

In addition to contributions to the linear theory, the eigenfunction analysis has been used to simulate numerically nonlinear effects in ZPCI cases of practical interest. The results of several simulations are shown in Figs. 11.1-4 . Each figure shows: (a) a radial lineout of the calculated shadow-

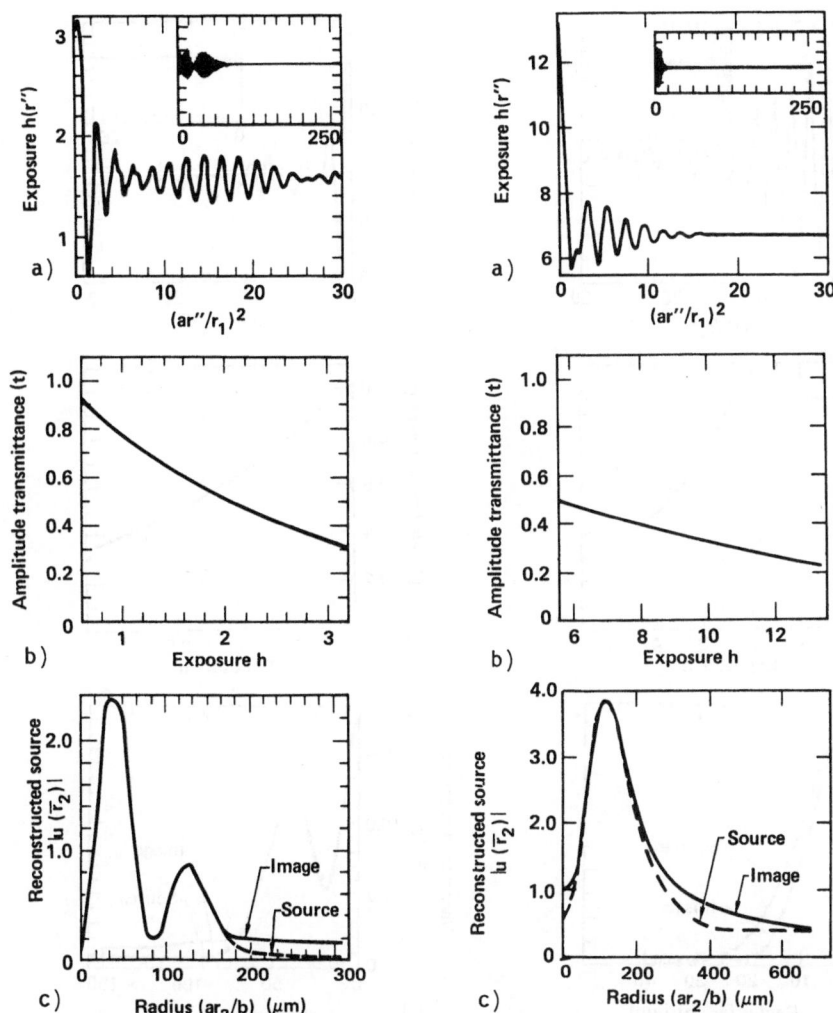

Fig.11.1,2 Simulation of the effects of nonlinear shadow-graph recording on the construction of continuous x-ray source distributions. In Fig.11.1 the coded aperture has 250 zones and the ZPCI Fresnel number is 0.074. In Fig.11.2 the coded aperture has 250 zones and the ZPCI Fresnel number 0.342

graph exposure; (b) the nonlinear relationship used to map exposure to amplitude transmittance of the coherently illuminated coded image; and (c) a comparison between the scaled first-order image (i. e., reconstructed source) and the original source. With the exception of Fig.11.3, imaging errors introduced by nonlinear effects do not influence image integrity. Source widths and amplitude ratios are not altered by more than a few percent. The largest imaging error is in Fig.11.3, where the low level exposure is severely clipped by threshold effects in the shadow graph recording film. In this case the nonlinearity results in a 15 % error in the apparent FWHM of the source. These and other simulations demonstrate that so long as severe film

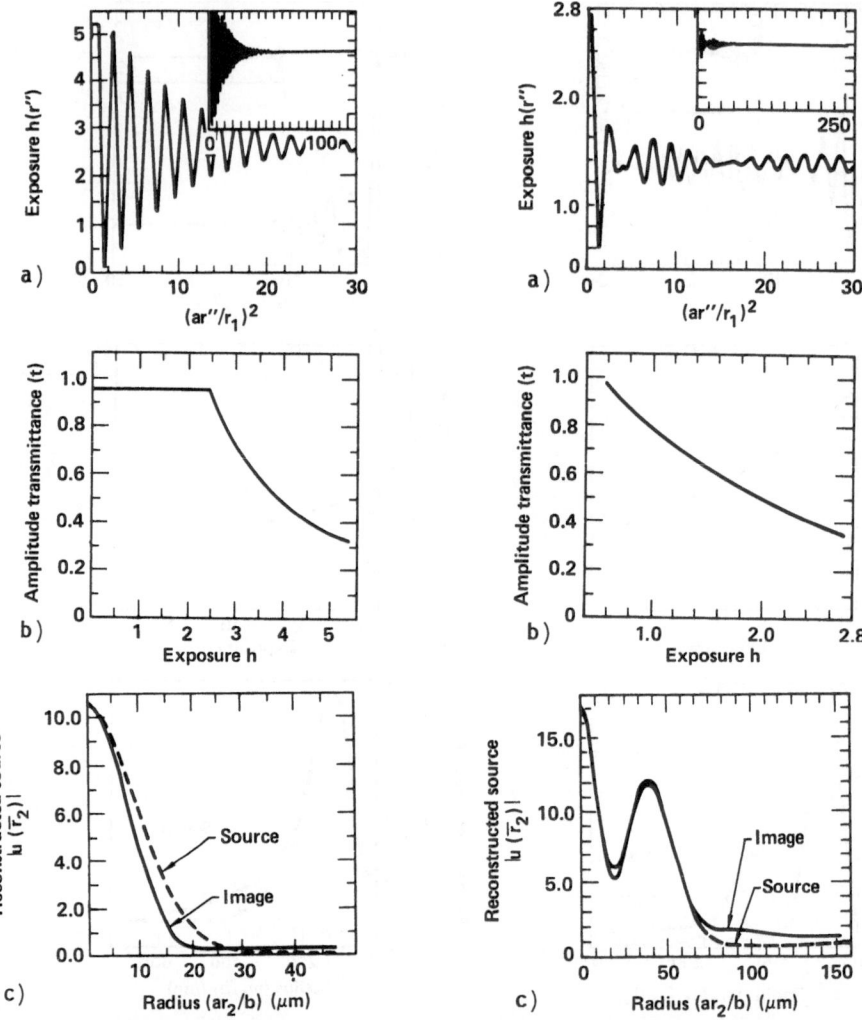

Fig.11.3,4 Simulation of the effects of nonlinear shadow graph recording. In Fig.11.3 the coded aperture has 100 zones and the ZPCI Fresnel number is 0.028. In Fig.11.4 the coded aperture has 250 zones and the ZPCI Fresnel number is 0.045

nonlinearities are avoided, the linear reconstruction terms dominate, and the optically reconstructed first-order image is an accurate representation of the source.

Zone Plate Coded Imaging has been used extensively for the microscopy of laser-driven fusion targets [11.14]. Progress in this field has led to requirements for imaging capabilities at shorter x-ray wavelengths. This necessitates the fabrication of thick, micro-Fresnel zone plates sufficiently opaque to produce high contrast shadow graphs suitable for optical reconstruction. Such zone plate coded apertures have been fabricated using reactive ion etch techniques, and used to image 100 keV x rays with a higher

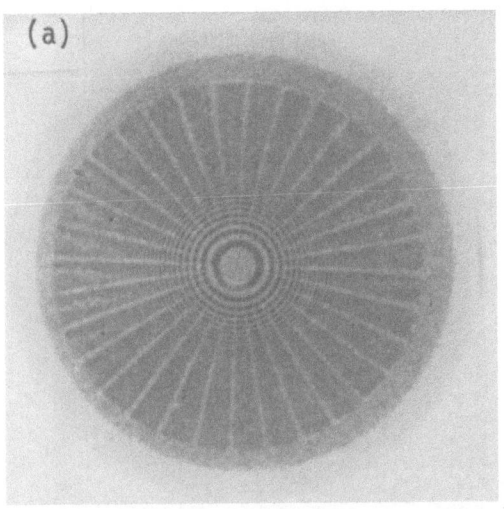

(a)

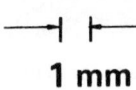

1 mm

100 keV x-ray
7th order

(b)

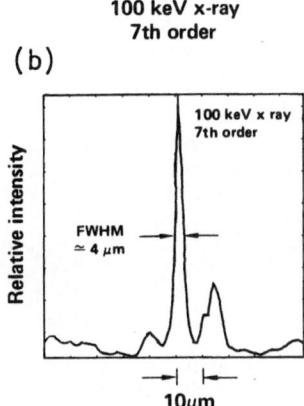

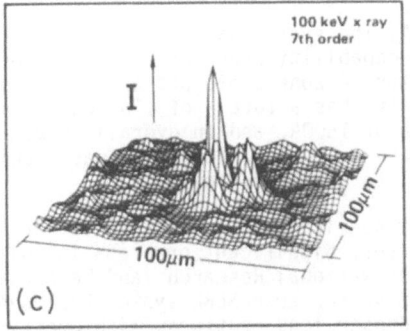

Fig.11.5 Image data from the 100 keV
x-ray point response test.
(a) Coded image;
(b) and (c) Reconstructed image data
in seventh order

order resolution ∿ 4μm [11.3]. The coded aperture used in these experiments
was a free-standing, 75 μm thick, gold zone plate with a total of 100 zones,
and 30 μm minimum zone width. Data from the 100 kev test experiments are
presented in Fig.11.5. Shown are: (a) the coded image produced by the 100
keV 'point' source; (b) a 1-D linear plot of image intensity along a line
through the center of the 100 keV point image; and (c) a 3-D representation
of the point response data.

11.3 Advanced X-Ray Lensing Elements

Among the areas of primary interest in x-ray lens development are improvements in spatial resolution, radiation collection efficiency, and x-ray
diffraction efficiency. Improved spatial resolution requires the production
of narrow linewidth zone plate objective lenses. Significant progress in
this area has been made at IBM's Thomas J. Watson Research Center [11.4]. An
advanced SEBL system has been used to generate Fresnel zone plate master

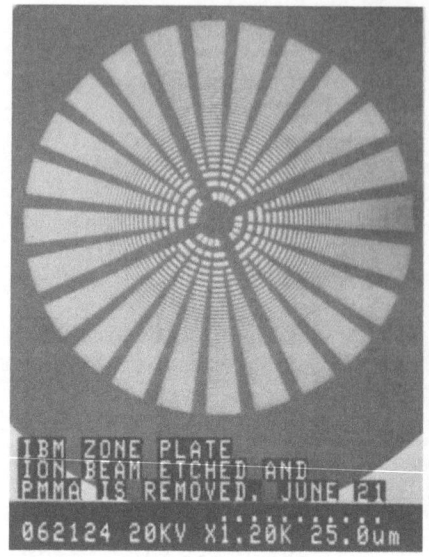

Fig.11.6 SEBL generated zone plate objective lens pattern. Minimum zone
width is 1500Å; overall pattern diameter is 60 μm

patterns with linewidths as small as 1500Å. The system has 16 bit D/A converters and a polar coordinate vector scan capability provided by a 'hard-
wired' polar rectangular coordinate converter. A zone plate pattern generated by this system is shown in Fig.11.6. It has a total of 100 zones, 24
radial support struts, a minimum zone width of 1500Å, and an overall diameter of 60 μm. This pattern has been made into an x-ray lithography mask, and
replicated to produce free-standing zone plate objective lenses.

Improved radiation collection efficiency requires the production of large
diameter, multi-zone, Fresnel condenser lenses. Significant progress in this
area has been made at Cornell University's National Research and Resource
Facility for Submicron Structures [11.5]. A vector scan SEBL system has been
used to generate Fresnel zone plate condenser lens patterns with overall
diameter approaching 1mm. Shown in Fig.11.7 is a condenser lens master pattern of 450 zones with ∿4500Å minimum zone width, and 0.8mm overall diameter. The edge structure of the outermost zones in Fig.11.7 arises from the
rectangular coordinate format of the vector scan system and the field quan-
tization limitations of the 13 bit D/A converters.

Increased x-ray diffraction efficiency can be achieved in Fresnel structures by the judicious use of phase effects [11.15,16]. In particular, if
the material and thickness of the solid zones are chosen to produce negligible absorption and a phase shift of π radians, a first-order diffraction efficiency as high as 40 % can, in principle, be achieved. For a material having refractive index, $n = 1 - \delta + i\beta$, the appropriate thickness for the
phase shifting zones is $\pi = \lambda_x/2\delta$. In passing through a phase shifting zone
the local x-ray intensity is reduced by the factor $e^{-2\pi\eta}$, where $\eta = \beta/\delta$ is
the attenuation parameter upon which lens efficiency solely depends. η and π
are the critical parameters in phase lens design. η determines lens efficiency and must be much less than unity for significant phase enhancement of
efficiency. π is restricted by practical fabrication limitations on high

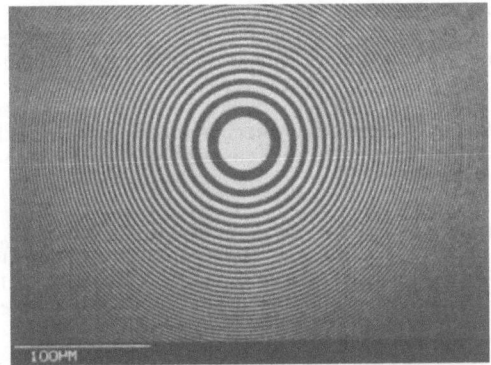

Fig.11.7 SEBL generated zone plate condenser lens pattern. Minimum zone width is approximately 4500A; overall pattern diameter is 0.8mm

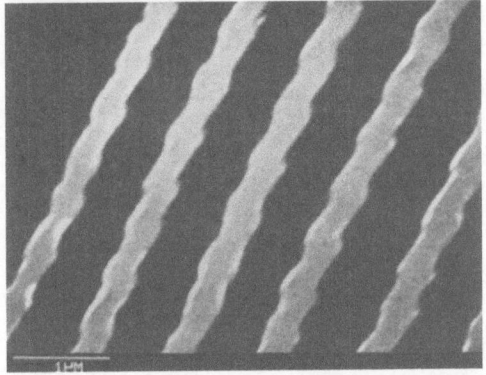

Fig.11.8 Phase lens design scheme for silver. The energy bands for efficient operation extend from 110 - 180 eV and from 1.4 - 2.7 keV

aspect-ratio ($\pi / \Delta r$) microstructures. Figure 11.8 illustrates a simple scheme for phase lens design. Shown are overlaid curves of π and η versus x-ray energy for a silver phase lens. The curves are overlaid such that chosen upper limit values for π and η coincide. In this case, we required $\pi < 1$ μm (a submicron Δr is presumed), and $\eta < 0.175$ (leading to a first-order diffraction efficiency > 25%). Only at those x-ray energies for which both η and π lie below the common upper-limit line can a high-resolution, 'reasonably efficient', x-ray lens be fabricated.

A high-resolution Fresnel phase plate designed for operation at the AL K_α line ($E_x \approx 1.5$ keV) has been fabricated [11.6]. The phase lens, shown in Fig.11.9, is made of silver approximately 0.55 μm thick. It has a total of 100 zones with a minimum zone width of 3200A [11.17]. Its calculated operating characteristics at 1.5 keV are as follows:

Attenuation Parameter	: η	≈ 0.16
Energy Fraction in First-Order Focus	: ε_1	≈ 0.26
Energy Fraction in Zeroth Order	: ε_0	≈ 0.04
Energy Fraction Absorbed	: ε_{ab}	≈ 0.32
Energy Fraction in Other Orders	: $\Sigma \varepsilon_j$	≈ 0.38
Spatial Resolution	: δ	≈ 0.4 μm
f/number	: $f^\#$	≈ 387

Fig.11.9 SEM micrographs of the silver phase lens. The structure has 100 zones, a minimum zone width of 3200A, and is 5500A thick

11.4 Time- and Space-Resolved X-Ray Spectroscopy

X-ray transmission gratings with submicron spatial period are attractive dispersion elements for time- and space -resolved spectroscopy. Because of their simple geometry and broad spectral range, they are easily coupled with instruments of high temporal or spatial resolution. In particular, a free-standing x-ray transmission grating with 3000A period has been coupled with a soft x-ray streak camera as illustrated in Fig.11.10. This time-resolved x-ray spectrometer [11.10] measures continuous x-ray spectra with excellent temporal resolution ($\Delta t \approx 20$psec) and moderate spectral resolution ($\Delta\lambda \approx 1$A)

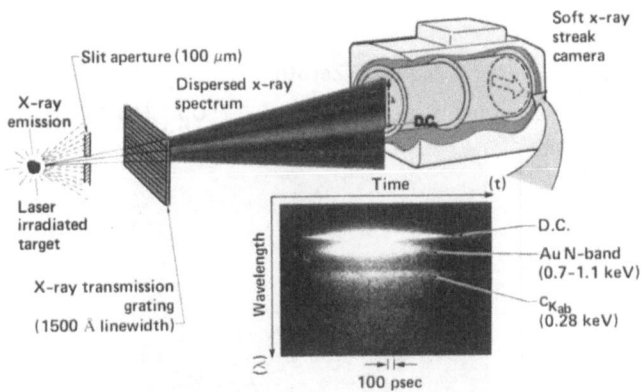

Fig.11.10 A time-resolved x-ray transmission grating spectrometer

over a broad spectral range (2A - 120A) with high sensitivity and large information recording capacity. It has been used to investigate the temporal
variation of soft x-ray spectra from laser irradiated targets. A typical
data set from these experiments [11.10] is presented in Fig.11.11. Shown
are: (a) the direct photographic streak record (x-ray wavelength vs time);
(b) the unfolded x-ray source spectra at two different times (accounting for
instrument's response function); and (c) the time history of target emission
in spectral bands centered at 150 ev and 500 ev, as well as that of the total, undiffracted emission.

In addition to time-resolved spectra, x-ray transmission gratings have
been used to provide spatially resolved spectra from laser-driven fusion
targets [11.10]. As illustrated in Fig.11.12, an x-ray transmission grating
has been coupled with a GIR x-ray microscope to produce an imaging spectrometer of broad spectral range (~ 0.3 - 4 keV), (~ 3 μm, and moderate spectral dispersion $d\theta/d\lambda \sim 3.3 \times 10^{-4}$ rad/A.

This versatile diagnostic capabiity allows spatial localization of spectrally distinct emissions from laser compressed targets. Figure 11.12 shows,
for example, the spatially localized emission from the compressed argon fill
gas of a laser fusion target [11.18] (shot #7668). The zeroth-order microscope image shows that without spectral dispersion the argon feature would
be lost amidst the strong, low-energy emission from the surrounding glass
shell. Such spectrally resolved image data allow us to address issues of
implosion dynamics, shell stability, and mix in laser fusion experiments.

11.5 X-Ray CCD Camera

A balanced development program in x-ray optics must not overlook the significant advantages that can accrue from improvements in x-ray detection. Advances in microelectronic technology have, in recent years, had important
impact on solid-state x-ray detectors. In particular, newly developed charge-coupled devices (CCD's) offer low noise electronic x-ray detection with
high sensitivity, good spatial resolution, real time read out, and extreme
positional stability [11.19]. Such devices provide a significant advantage
over x-ray film with its combined problems of quality control, high grain
noise, limited sensitivity, and slow data turn around.

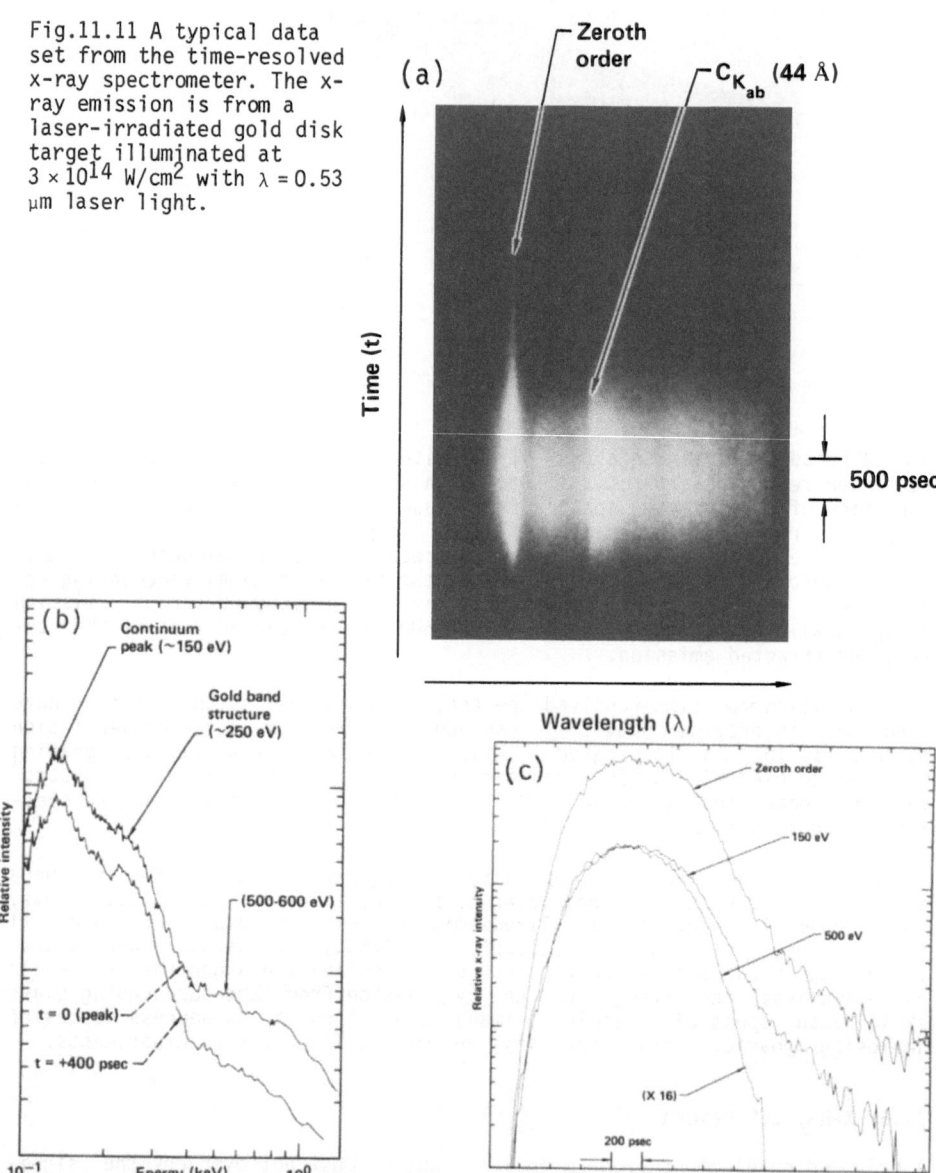

Fig.11.11 A typical data set from the time-resolved x-ray spectrometer. The x-ray emission is from a laser-irradiated gold disk target illuminated at 3×10^{14} W/cm^2 with $\lambda = 0.53$ μm laser light.

(a)

Zeroth order

$C_{K_{ab}}$ (44 Å)

Time (t)

500 psec

Wavelength (λ)

(b)

Continuum peak (~150 eV)

Gold band structure (~250 eV)

(500–600 eV)

t = 0 (peak)

t = +400 psec

Relative intensity

10^{-1} Energy (keV) 10^0

(c)

Zeroth order

150 eV

500 eV

(X 16)

Relative x-ray intensity

200 psec

Time (psec)

We are involved in a collaborative effort [11.11] for the development and implementation of a CCD camera for direct x-ray imaging. Design specifications for the camera are as follows. It will initially operate in an energy range from 1.5 keV to 15 keV. (A near term goal will be to extend this range to include subkilovolt x rays). The camera will utilize a front illuminated, virtual phase CCD made by Texas Instruments, Inc. The CCD has a 390 x 584 pixel array. Each pixel is 22μm square. The overall active detector area is 8.6mm x 12.8mm. The CCD will operate in a temperature range from -60°C to -100°C, and have an equivalent noise figure ~10-20 electrons rms. (By com-

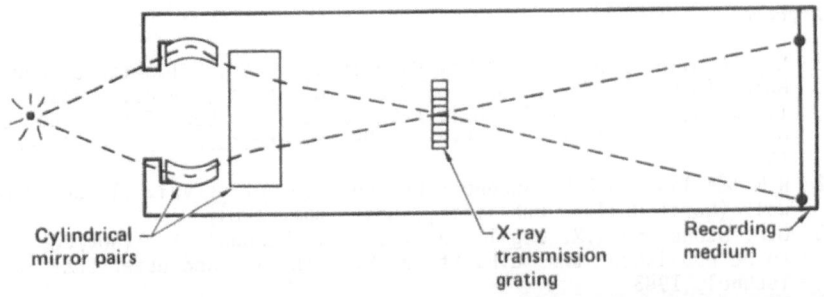

Cylindrical
mirror pairs

X-ray
transmission
grating

Recording
medium

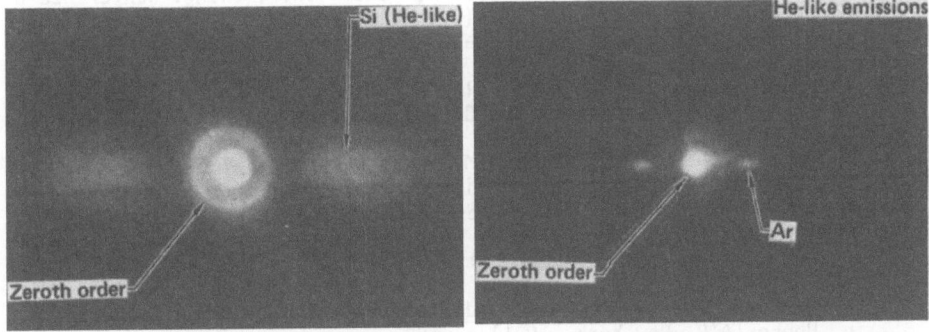

DT filled glass microsphere (#7529)

Argon filled glass microsphere (#7668)

Fig.11.12 A schematic diagram of the x-ray imaging spectrometer. Typical
data are shown for laser-irradiated, gas-filled, microsphere,
laser fusion targets

parison, a 1.5 keV x-ray deposits over 400 electrons in the CCD). The camera
will be capable of single photon detection, and have a capability for nondispersive energy discrimination. It will significantly enhance our ability
for x-ray microscopy and spectroscopy using low brightness laboratory sources.

Acknowledgement

The development and implementation of new x-ray optical capabilities is a
multidisciplinary effort involving a variety of diverse technologies and
disciplines in an interlocking collaboration. I am delighted to recognize
the skilled contributions of colleagues and collaborators at institutions
throughout the U.S. At IBM's T.J. Watson Research Center: D.P. Kern, T.H.-
P. Chang, and P.J. Coane; at MIT's Submicron Structures Laboratory: A.-
M. Hawryluk, M. Schattenburg, H.I. Smith, and J. Carter; at MIT's Center for
Space Research: G.R. Ricker, J. Doty, J. Vellarga, and G. Luppino; at Cornell University's NRRFSS: E. Wolf, R. Tiberio, B. Whitehead, and D. Costello; at University of Rochester's Laboratory for Laser Energetics: M. Richardson, D. Villeneuve and A. Letzring; at Purdue University: D.W. Sweeney;
and finally at LLNL: G. Stone, G. Howe, R.L. Kauffman and H. Medecki.

References

11.1 N.M. Ceglio: Low Energy X-Ray Diagnostics: Eds. D.T. Attwood and B. L. Henke (AIP, New York, 1981) p. 210; T.W. Barbee, ibid., p. 131; Henry I. Smith, ibid, p. 223; D.T. Attwood, et al. OSA Topical Conference on Laser Techniques for Extreme UV Spectroscopy, Boulder, Colorado 1982

11.2 N.M. Ceglio and D.W. Sweeney: Progress in Optics Vol. 21, Ch. 4 Ed. E. Wolf (North Holland Publishing, Amsterdam, 1984)

11.3 G.F. Stone and N.M. Ceglio: J. Vac. Sci. Technol 21, (Nov/Dec - 1983) to be published; and G.F. Stone, M.S. Thesis, Rochester Institute of Technol, 1983

11.4 P.J. Coane, D.P. Kern, A.J. Speth, and T.H.P. Chang: Conference on Microcircuit Engineering, 1982 (Grenoble, France); D.P. Kern, P. Houzego, P.J. Coane, and T.H.P. Chang, J. Vac Sci Technol 21 (Nov/Dec - 1983) to be published

11.5 The SEBL work on the condenser lenses was done by R. Tiberio, B. Whitehead, and D. Costello of the Cornell University NRRFSS. The work is as yet unpublished

11.6 N.M. Cegio, A.M. Hawryluk, and M. Schattenburg: J. Vac. Sci. Technol. 21, (Nov/Dec - 1983) to be published

11.7 A.M. Hawryluk: et al. J. Vac. Sci.Technol. 19, 897 (1981)

11.8 C.F. McConagy and L.W. Coleman: Appl. Phys. Lett. 25 (268) 1974; D.T. Attwood, et al., Phys. Rev. Lett. 37, 499 (1976); 38, 282 (1977); and G.L. Straddling, et al.: Bull. Am. Phys. Soc. 23, 880 (1978)

11.9 R.H. Price: Low Energy X-Ray diagnostics, Eds. D.T. Attwood and B. L. Henke (AIP. New York, 1981) p. 189

11.10 N.M. Ceglio, A.M. Hawryluk, and R.H. Price: Applied Optics 21, 3953 (1982); and N.M. Ceglio, R.L. Kauffman, A.M. Hawryluk, and H. Medecki: Applied Optics 22, 318 (1983)

11.11 The primary collaborators for design and development of the x-ray CCD camera are George Ricker, John Doty, John Vellarga, and Gerrald Luppino of the MIT Center for Space Research

11.12 L. Mertz and N.O. Young: Proc. Int'l. Conf.on Opt. Instr. p. 305: (Chapman and Hall, London 1961); L. Mertz: Transformations in Optics (John Wiley and Sons, New York, 1965)

11.13 A.E. Siegmann: An Introduction to Lasers and Masers p. 304 (McGraw Hill, New York, 1971)

11.14 N.M. Ceglio and L.W. Coleman: Phys. Rev. Lett. 39, 20 (1977); N.M. Ceglio and J.T. Larsen: Phys. Rev. Lett. 44, 579 (1980); N.M. Ceglio, D.T. Attwood, and J.T. Larsen: Phys. Rev. A, 25, 2351 (1982)

11.15 J. Kirz: J. Opt. Soc. Amer. 64, 301 (1974)

11.16 N.M. Ceglio and H.I. Smith: Proc. VIII Int'l. Conf. x-ray Optics and Microanalysis (Boston, 1977)

11.17 The SEBL master pattern shown in Fig.11.9 was originally generated at Lincoln Laboratory; see D.C. Shaver, et al.: J. Vac. Sci. Technol. 16, 1626 (1979)

11.18 The experimental data shown in Fig.11.12 was generated on the Omega laser facility at the University of Rochester Laboratory for Laser Energetics in collaboration with M. Richardson, D. Villeneuve, and S. Letzring

11.19 G.R. Ricker, et al.: Proc. S.P.I.E. 290, 190 (1981); R. A. Stern, et al.: Rev. Sci. Instrum. 54, 198 (1983); P. Burstein, et al.: Ann. N. Y. Acad. Sci. 342, 252 (1980)

12. Fabrication of Small Linewidth Diffractive Optics for Use with Soft X-Rays

A. G. Michette, M. T. Browne, P. Charalambous, R. E. Burge, and
M. J. Simpson
Physics Department, Queen Elizabeth College, University of London
Campden Hill Road, London W8 7AH, UK

P. J. Duke
Daresbury Laboratory, Daresbury, Warrington WA4 4AD, UK

12.1 Introduction

Recently there has been much interest in the development and use of soft x-ray microscopy, as other contributions to this volume show. In the United Kingdom, a soft x-ray microscopy program is being carried out on the Synchrotron Radiation Source at the Daresbury Laboratory [12.1]. Although it is intended ultimately to develop a scanning microscope, the initial stages of the program have concentrated on the construction of high-resolution optical components for use in a prototype imaging microscope. For this purpose a lithographic technique using a Vacuum Generators HB5 scanning transmission electron microscope (STEM) is being used. This technique is capable of drawing accurate patterns, with linewidths much less than 50 nm, over small areas. It is therefore suitable for the manufacture of soft x-ray diffractive optics, in particular zone plates. This is discussed in this paper, as are the effects of possible manufacturing inaccuracies on the zone plate imaging properties. Deliberate modifications to the Fresnel zone plate are also considered.

12.2 Beam Writing

A major problem in the use of a STEM for the study of biological or other material is image degradation caused by deposition of a layer of contamination on the specimen surface. Although very high vacuums ($\leq 10^{-7}$ torr) are used, residual gas analyses of the STEM column atmosphere show the presence of, e.g., water vapor, nitrogen, oxygen, and various heavy hydrocarbon molecules. These hydrocarbons, from pumping oils, are the source of the contamination; they form a layer on the specimen surface and are polymerized by the electron beam. Thus they can no longer diffuse and a build-up of the polymer occurs; leaving the beam in one position causes the formation of a "contamination cone". These cones are formed on both surfaces since many electrons pass through the specimen. Much work has been done to try to reduce or eliminate contamination [12.2], while beam writing is an attempt to utilize it [12.3].

For the drawing of patterns suitable for use as soft x-ray diffractive optical components, the specimen is replaced by a thin ($\sim$ 10-20 nm) carbon film on a standard electron microscope aperture. The carbon is prepared on mica and is then floated onto water; the aperture is moved upwards through this to collect the film. To remove excess water the film and aperture are baked at $\sim$ 200 °C for a few minutes. Contamination patterns are drawn by a microprocessor controlled scan of the electron beam across the film [12.4].

In order to be able to draw patterns accurately and relatively quickly several problems had to be solved. Limitations of the electron optics prevent the beam from being scanned accurately over areas bigger than a few microns by a few microns ("primary fields"). For larger patterns the specimen stage must be mechanically scanned and the primary fields joined together ("patched") using appropriate registration marks. The quality of the patterns is closely related to both the focussing of the beam and the length of time for which any particular point (pixel) of the film is irradiated. A defocussed beam would create poorly defined lines, while fixed pixel times would result in patterns of uneven thickness due to variations in the contamination rate. To solve the former problem, focus-finding scan lines are periodically generated. During the course of these the microprocessor minimizes the time required for the line spots to reach a certain thickness by changing the focus in 20 nm steps, the minimum time per pixel corresponding to optimum focus. The microprocessor detects when each spot (of a focus-finding or a pattern line) has reached the required thickness by constantly monitoring the difference between the STEM dark field signal (elastically scattered electrons) and bright field signal (inelastically scattered and unscattered electrons). Since the bright field signal decreases with spot thickness while the dark field signal increases, at a particular thickness the two signals are equal. This condition is independent of beam current variations and the rate of contamination. By individually setting the starting level of each signal the contamination thickness can be controlled.

During the generation of a pattern the hydrocarbons are used up, leading to a decrease in the contamination rate unless they are replenished. This has been accomplished by constructing a special specimen holder incorporating a small oil reservoir. The oil and the film are together maintained at $\sim$ 100 $^{\circ}$C which leads to a high hydrocarbon vapor pressure ($\sim$ 10^{-4} torr) in the vicinity of the film. The high temperature also increases the mobility of the hydrocarbon molecules. Thus molecules immobilized by polymerisation are constantly replaced (through adsorption) leading to sustained periods of high contamination rate.

12.2.1 Manufacture of Linear Diffraction Grating

Although the (initial) aim of developing the contamination writing technique was to generate high resolution zone plate patterns, it was decided first of all to manufacture a linear diffraction grating, in order to perfect the method and to test for distortions. The nominal size of this grating was $\sim$ 11 μm x 30 μm with a pitch of $\sim$ 110 nm ($\sim$ 1.8×10^{4} lines/mm). The primary field was chosen to be a 5 μm x 5 μm area, addressable with an accuracy of 2.5 nm. A row of 16 registration lines, each $\sim$ 1.3 μm long, was drawn on the first primary field, each line being accurately registered with respect to the first. The last line was then brought to the position previously occupied by the first, using the mechanical scanning stage. By registering with this new first line, 15 more lines were drawn. This process was repeated until the row of lines was $\sim$ 30 μm long. The exact position of each line was then determined by registering with the first line of the relevant primary field and the lines were extended by about 5 μm on either side. Two more lines were drawn in each of the gaps between the lines. Thicker support lines were then drawn perpendicular to the grating structure. After baking at $\sim$ 500 $^{\circ}$C for a few minutes to remove excess oil, gold was evaporated onto the film (at an angle to prevent the gaps being filled) - the resulting grating is shown in Fig.2.1. By viewing the grating in a conventional transmission electron microscope and comparing it with a standard grid, and by analysis of a diffraction pattern (see below), the pitch was determined to be $\sim$ 125 nm, $\sim$ 10% larger than intended. This was found to be due to poor calibration of the STEM which has subsequently been corrected, along with other effects causing the distortions visible in Fig.12.1, such as uneven motion of the mechanical stage giving occasionally misplaced lines.

Fig.12.1 STEM generated linear diffraction grating

$$\overset{\bullet}{C_2} \quad \overset{''}{O_2} \overset{'' \, ' \, \bullet}{C_1 O_1 Si_2} \, \overset{\bullet \, ' \, ''}{Si_1} \quad \overset{''}{} \qquad \bullet$$
$$Si_3$$

Fig.12.2 Diffraction pattern obtained using a STEM grating

Diffraction patterns have been obtained from the grating, an example being shown in Fig.12.1. A rotating anode x-ray generator with a carbon coated anode was used. Lines due to carbon K_α, oxygen K_α (presumably from the carbon paint stabiliser) and silicon K_α (from vacuum grease) can be seen.

12.2.2 Manufacture of Zone Plates

The first zone plates manufactured on the STEM have 100 zones, an absorbing center zone, an outer diameter of $\sim$ 30 µm and an outermost zone width of $\sim$ 75 nm. This is well within the capabiltities of the technique, and subsequent zone plates will have much finer zones and therefore higher resolutions. The zone plates were drawn on carbon films initially coated, for stability, with a thin layer of contamination, giving a total substrate thickness of $\sim$ 20 nm. The patterns were generated in 24 equal angular sectors each containing 3 primary fields, the sectors being aligned by registration

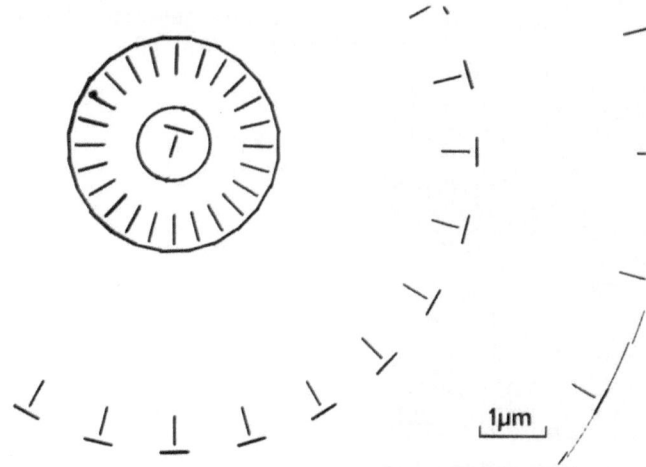

Fig.12.3 Registration marks used to define a zone plate pattern

with an accurately drawn inner circle. Each alternate zone was drawn as a series of ∿ 20 nm wide arcs, each registered with a mark drawn at the inside of the primary field. Fig.12.3 shows the registration marks, drawn before starting the zone plate pattern.

As the pattern was drawn, the registration marks were obscured (and possibly corrupted). To minimize the effect of this, the sectors were drawn from the outside in, the outer primary field containing 44 zones, the middle 36 zones and the inner 20 zones. A visual check on the positioning of the zones was made every third of an outer primary field, every half of a middle

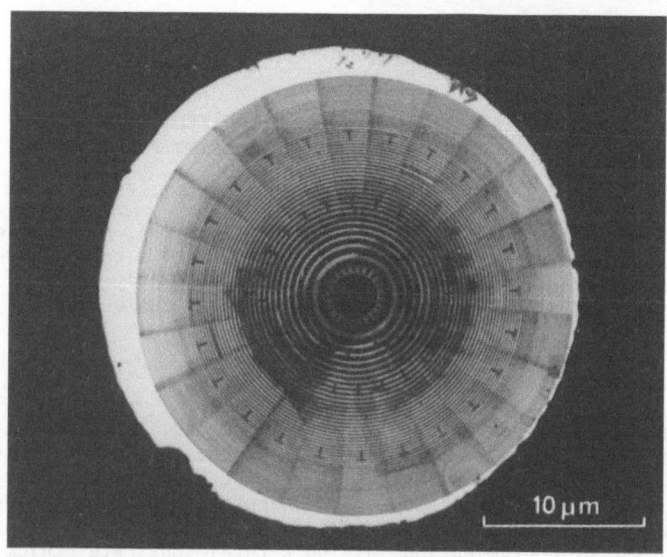

Fig.12.4 A zone plate manufactured on the STEM

primary field, and every inner primary field, thus ensuring that the zones were positioned with an accuracy of no worse than 5 nm. After each sector the stage was rotated through 15° and the process repeated, so that all the sectors were drawn with the beam in the same region, minimizing x-y distortion. Each sector covered an angle of 16°, so that there were slight overlaps at the ends of the arcs ensuring good joins. After all the sectors had been drawn, the central zone was completed by drawing a series of registered circles and finally filling in over the central registration mark - this last operation was unregistered but any drift within it was insignificant. An example of such a zone plate is shown in Fig.12.4 - the contamination thickness is ~ 0.15 μm and the pattern is uncoated. Possibilities for coating the patterns with heavy metal layers are being investigated.

12.3 Imaging Properties of STEM Zone Plates - The Effect of Manufacturing Inaccuracies

Manufacturing inaccuracies lead to a reduction in the efficiency and affect other imaging properties of zone plates. Although any manufacturing technique is subject to such inaccuracies, the discussion here is limited to those likely in the STEM process. A more general treatment is given in [12.5]. Results are presented for a) partial transmission through the absorbing zones and partial absorption in the transmitting zones; and b) displacements of whole or part zones and whole or part sectors. The calculations for b) were based on the Fresnel-Kirchhoff diffraction integral for the image amplitude at an on-axis point (see [12.5] for more detail). The effects considered do not move the major peaks of the diffraction pattern off axis and hence how closely the imaging properties of an aberrated zone plate approach the ideal may be estimated using the Strehl ratio [12.6]. This is the ratio of the peak intensity of the aberrated spread function (at a focus) to that of the equivalent perfect function; a value of 0.8 (the Strehl limit) is generally accepted to be tolerable. Shifting or distorting zone boundaries alters the focal length of a zone plate; where necessary, refocussing was carried out before calculating the Strehl ratio.

12.3.1 Incomplete Absorption or Transmission

Incomplete absorption in the absorbing zones does not distort the diffraction pattern but the focussed intensity is reduced and the background (unfocussed) radiation level is increased, leading to reduction in contrast. Alternatively, if the radiation transmitted by the absorbing zones undergoes a phase change, the focussed intensity could be enhanced. This is difficult to quantify because of the lack of good measurements of soft x-ray optical constants.

Partial absorption in the transmitting zones is equivalent to inserting an attenuator in the beam. Image quality is not affected but less of the available radiation is focussed.

Uncoated STEM zone plates used of a wavelength of ~ 4 nm absorb $\sim 75\%$ of the radiation incident of the absorbing zones (for a thickness of carbon of ~ 0.15 μm) and transmit $\sim 85\%$ of the radiation incident on the transmitting zones (for a carbon thickness of ~ 20 nm).

12.3.2 Displaced Zones and Sectors

If the displacements of zones are characterized in terms of zone boundary shifts, then for the case where complete boundaries are displaced, they may be left unshifted (0), moved inwards (-), or moved outwards (+). Expressing the boundary shifts in pairs, e.g.,(+,-) meaning that the odd numbered boundaries are moved outwards and the even ones inwards, allows the situations shown in Fig.12.5 to be identified. These fall into 4 groups, as indicated: (i) all zones left unchanged, (ii) every other boundary moved inwards or outwards resulting in the absorbing zones being too narrow and the transmitting zones too wide or vice versa, with displaced zone centers, (iii) alternate zone boundaries moved in opposite directions, giving zones too wide or too narrow with undisplaced zone centers, and (iv) all zones moved inwards or outwards.

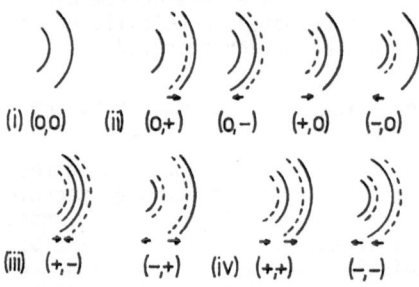

(i) (0,0)   (ii) (0,+)   (0,-)   (+,0)   (-,0)

(iii) (+,-)   (-,+)   (iv) (+,+)   (-,-)

Fig.12.5 Dashed lines, correct position. Full lines, actual positions of zone boundaries

The Strehl ratios for each of the groups are plotted against ε, the displacement of the zone boundaries as a fraction of the outermost zone width, in Fig.12.6. If the on-axis intensity in the first order is not to fall below the Strehl limit, then ε < 0.21 for the worst case, (iii). Less stringent limits on ε are given by combinations of shifts involving more than two zones, e.g.,(+,-,+), (-,+,0,+). STEM zone plates will tend not to have systematic combinations of shifts; the envelope obtained from 100 separate computations in each of which the zone boundaries were displaced in random directions with random values of ε is also shown in Fig.12.6. This gives the requirement ε < 0.3. Higher order foci can in principle give better resolutions, but the Strehl limit is reached for smaller values of ε, as shown in Fig.12.6 for the third order.

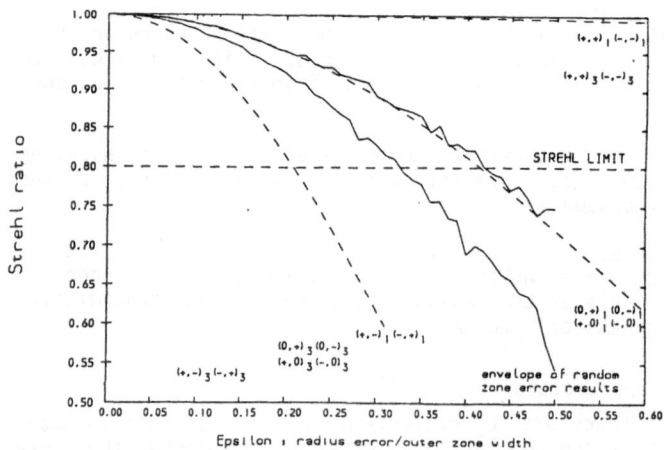

Fig.12.6 Strehl ratios for displaced zones

114

The STEM zone plates have boundary positions correct to within ∿ 5 nm with an outermost zone width of ∿ 75 nm, i.e.,ε ∿ 0.07 giving a Strehl ratio of ∿0.99 for first-order imaging.

Misalignments of the zones between the sectors in which the STEM zone plates are drawn could result in discontinuities (steps) in the zone boundaries. Possible configurations are shown in Fig.12.7. Calculations of the effects of these were carried out assuming that the configurations shown were repeated every two sectors; groups of more than two sectors give less stringent limits.

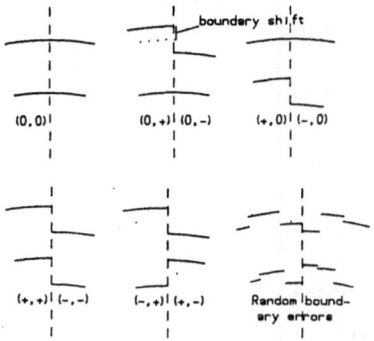

Fig.12.7 Stepped zone boundaries

The Strehl ratios for stepped zone boundaries are shown in Fig.12.8 for the worst case, ε < 0.21. For random step positions and sizes (see Fig.12.7 - this could happen if, for example, the STEM stage drifted between registrations), the envelope shown in Fig.12.8 is obtained, giving ε ≲ 0.35. Note that for a step size of ∿ 5 nm with an outermost zone width of ∿ 75 nm, the Strehl ratio is ∿ 0.99.

Since each sector of a STEM zone plate is drawn as three primary fields, perhaps the most likely combination of displaced zones is that a whole (or part) field is incorrectly positioned. This might happen if, for example, an

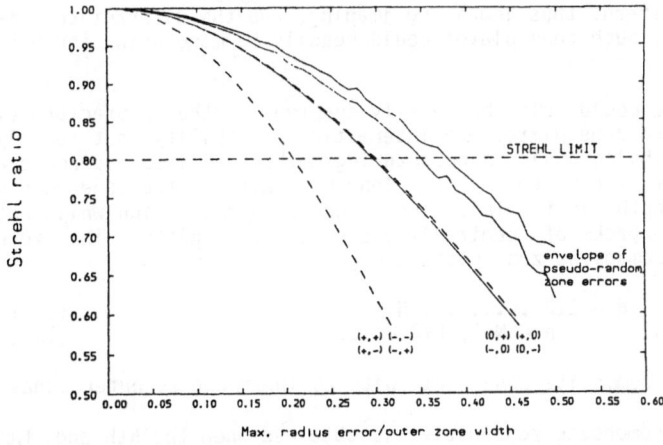

Fig.12.8 Strehl ratios for stepped zone boundaries

imperfection in the film obscures a registration mark. If just one field of the whole zone plate is displaced, the resulting Strehl ratios are as shown in Fig.12.9. Displacement of an outer field (i.e., of an outer part of the zone plate) has a larger effect than displacement of a middle or inner field, due to the larger area. Figure 12.9 also shows the effect of displacing more than one outer field; since the field positions are monitored visually during the manufacturing process, the effects of outer and middle field displacements are likely to be less than shown. Also, any gross errors can be detected and, if necessary, the process restarted. For field displacements of ∿ 5 nm, the Strehl ratio remains well above the Strehl limit.

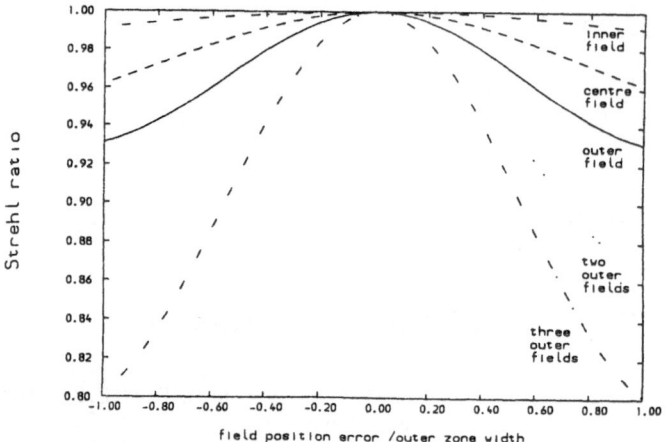

Fig.12.9 Strehl ratios for displaced primary fields

12.4 Modified Fresnel Zone Plates

The resolution of a zone plate is governed by the width of the outermost zone, which in turn is limited by the manufacturing process. One method of obtaining higher resolutions, by obstructing the central zones, has been discussed previously [12.7]. This also gives more energy in the outer rings of the diffraction pattern, thus impairing imaging, and the diffraction efficiency is decreased. Such zone plates could readily be made using the STEM process.

The STEM technique could also be easily adapted to the manufacture of other types of modified zone plate. One interesting possibility, not readily accessible to, e.g., holographic manufacturing methods, is to surround a central zone plate of first-order focal length f with extra zones whose third-order focal length is f. This gives increased resolution while not suffering from the drawbacks of centrally obstructed zone plates. The zone boundaries of such a composite zone plate are given by

$$r_n^2 = n \lambda f \qquad n = 1,2 \ldots\ldots\ldots N \qquad (12.1)$$
$$r_n^2 = n \lambda f \, (3-2N/n) \qquad n = N+1, N+2 \ldots\ldots\ldots M \qquad (12.2)$$

and Fig.12.10 shows a composite zone plate with 20 inner and 20 outer zones.

The best possible composite zone plate is obtained when the Nth and Mth zones both have width equal to the finest line possible in the manufacturing

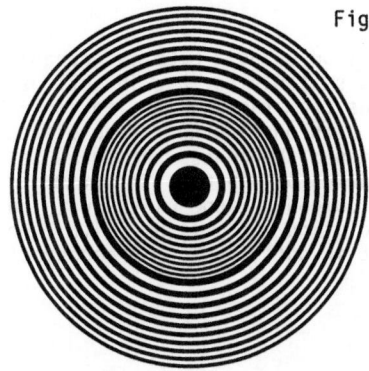

Fig.12.10 A composite zone plate

method. This gives the maximum value of M, 11/3 N. Figure 12.11 shows the axial intensity distributions at f (for plane wave illumination) of composite zone plates with 100 inner zones and various total numbers of zones. The on-axis intensity increases rapidly as M increases, due to the larger effective aperture, while the width of the central peak decreases, i.e., the resolution improves; for M = 11/3 N, the resolution is 0.46 times that of the central zone plate alone. The insets to Fig.12.11 show that the amount of energy in the outer rings decreases as M increases.

The principal disadvantage in using a composite zone plate is the increase in background radiation due to the zeroth and first orders of the outer zones. Also, the depth of focus is smaller (due to the third-order contribution from the outer zones) - this could lead to focussing difficulties in x-ray microscopy. For the same reason, when used with broad band radiation, the accepted bandwidth $(\Delta\lambda/\lambda)$ is narrower, and so the relative gain in focussed intensity is decreased [12.8].

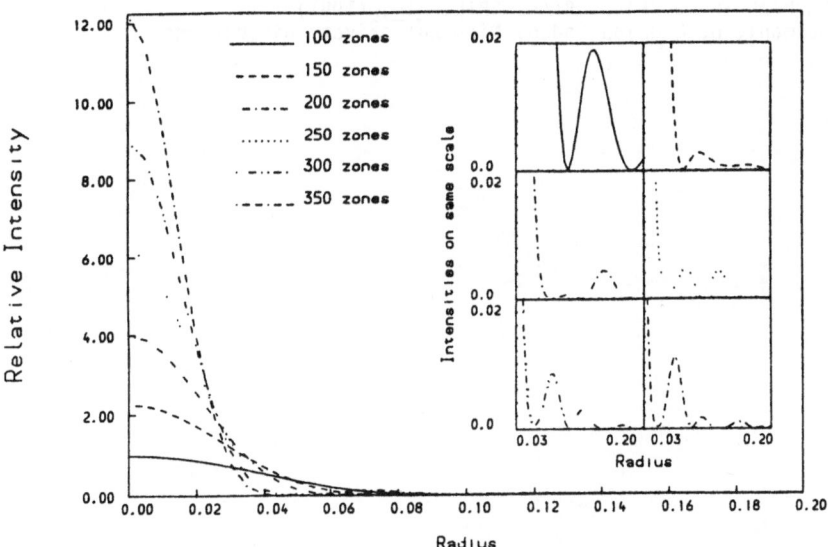

Fig.12.11 Radial intensity distributions of composite zone plates

12.5 Conclusions

It has been demonstrated that a lithographic method using a STEM is capable of drawing high-resolution zone plate patterns suitable for use in soft x-ray microscopy. The success of the method arises from the ability to accurately register different parts of the pattern with one another. Zone plates with minimum linewidths of $\sim$ 75 nm have been made; the zones were drawn as series of arcs each $\sim$ 20 nm wide. Hence zone plates of minimum linewidth of $\sim$ 20 nm (or even less) should be possible; a composite zone plate with an outermost zone of this size would have a resolution of $\sim$ 11 nm, i.e.,the order of the radiation damage resolution limit in soft x-ray microscopy of living biological material.

Acknowledgements

We are grateful to the Science and Engineering Research Council (SERC) for financial support. MJS is supported by an SERC grant.

References

12.1 P.J. Duke et al., this volume, no.24
12.2 See, e.g.,L. Reiner and M. Wechter: Ultramicroscopy 3, 169, (1978)
     V. Harada, T. Tomita and T. Watabe: Scanning Electron Microscopy II, 103, (1979)
12.3 A.N. Broers, J. Cuomo, J. Harper, W. Molzen, R. Laibowitz and M. Pomerants: Ninth International Congress on Electron Microscopy, Toronto, Vol. II, 343 (1978)
12.4 M.T. Browne, P. Charalambous and R.E. Burge: Inst. Phys. Conf. Ser. 61, 43, (1982)
12.5 M.J. Simpson and A.G. Michette: to be published in Optica Acta (1983)
12.6 W.T. Welford: in "Aberrations of the Symmetrical Optical System", p. 206, Academic Press (1974)
12.7 H. Rarback and J. Kirz: Proc. SPIE, 316, (1981)
12.8 G. Schmahl, D. Rudolph and B. Niemann: private discussions

13. Submicron Lithography by Demagnifying Electron-Beam Projection

H. W. P. Koops and J. Grob

Institut für Angewandte Physik, Technische Hochschule Darmstadt
Hochschulstraße 2, D-6100 Darmstadt, Fed. Rep. of Germany

Demagnifying electron-beam projection is a technique well suited to generate submicron structures for various applications. A two- or three-lens projection system demagnifies a selfsupporting master stencil into a registration plane. The stencil is fabricated by optical or electron-beam lithography in conjunction with standard planar etching and plating technologies. The design rules for reducing image projection lens systems are discussed. To produce nanometer structures from micrometer master structures large reduction factors are to be used. For inspection of the recorded nanometer pattern a conventional transmission electron miscroscope (CTEM) is an adequate tool. A single field condenser objective lens offers the advantage to control the diminishing and the magnifying objective lens field by one lens excitation.

We show that a CTEM equipped with a single field condenser objective lens of 2 mm focal length and a two-lens illumination system is capable to render a variable demagnification factor of 1:20 to 1:110 at an accelerating voltage of 36 kV. Line gratings and grids having a periodicity in the range of 400 nm to 80 nm are fabricated using a micromesh of 8 μm period as a master stencil and selfsupporting collodion films as a recording medium. The use of foil masks and the contrast mechanism in the demagnifying projection system is discussed. We show that foil masks render sufficient scattering absorption contrast to be employed as an original in projection systems.

13.1 Introduction

Submicron lithography is required to produce finest structures for physical and technical applications; e.g. diffraction gratings and zone plates for x-ray optics, structured gates for semiconductor investigations, surface relief patterns for graphoepitaxy, structures for scaled-down integrated circuits, for superconducting circuits, and molecular device research. Structures having submicrometer dimensions are commonly fabricated using scanning electron beam writers with computer control having a scanning electron microscope incorporated as a monitoring instrument. High brightness electron guns like lanthanumhexaboride or field emission sources have the required small spot size with a high current density necessary to record submicron structures in a reasonable time. The number of lines per frame is limited by deflection aberrations. Typically 10.000 lines per frame are obtained [13.1]. Distortion is hard to measure and to control. The data transfer rate limits the recording velocity. One-level resist techniques render submicron structures only on thin film recording foils because of the proximity effect [13.2]. If two-level resist techniques are used even on bulk samples, submicron structures can be obtained applying reactive ion etch techniques [13.3,1].

Demagnifying electron-beam projection systems are a tool to transfer the structure of a selfsupporting master stencil demagnified onto a recording medium. Until now a maximum number of 37 000 lines per frame has been obtained by LISCHKE et al. [13.5] who used a machine of reasonable size. Independent of the resolution obtained the maximum number of lines per frame is defined for a system by the refractive power of the two or three demagnifying lenses. This determines the size of the aberration field curvature D and isotropic and anisotropic field astigmatism C and c. In geometric combination with the diffraction aberration they define an optimum imaging aperture to a given resolution or field diameter [13.4]. The imaging optic is controlled to be free of isotropic and anisotropic distortion E and e. Telescopic systems are also achromatic lens systems [13.5]. The axial aperture aberrations like spherical aberration C_3, isotropic and anisotropic coma F and f, and axial chromatic aberration can be neglected since the imaging aperture is smaller than 10^{-4} rad. The optimum aperture results from the electron gun crossover diameter being adequately demagnified with a condenser lens. A field lens produces a parallel and homogeneous illumination of the stencil. Imaging the original to the registration plane increases the illumination aperture α_0 by the reduction factor V. Nevertheless the depth of the image field is larger than in scanning electron microscopes because of the small imaging aperture. To observe the registration plane and the recording process on thin films a CTEM is fitted to the demagnifying optics. A magnification of 5 000 diameters is sufficient to monitor submicron structures. Figure 13.1 gives a schematic representation of a demagnifying electron projection system.

Similar optics have been developed to fabricate micron structures for application in the semiconductor technology [13.5-8]. They offer more than 10 000 lines per field having a final resolution of 0.1 μm. Submicron structures as optical elements for x rays have been fabricated with demagnifying projection systems in the fifties and sixties [13.9-11]. Table 13.1 presents the development of projection systems.

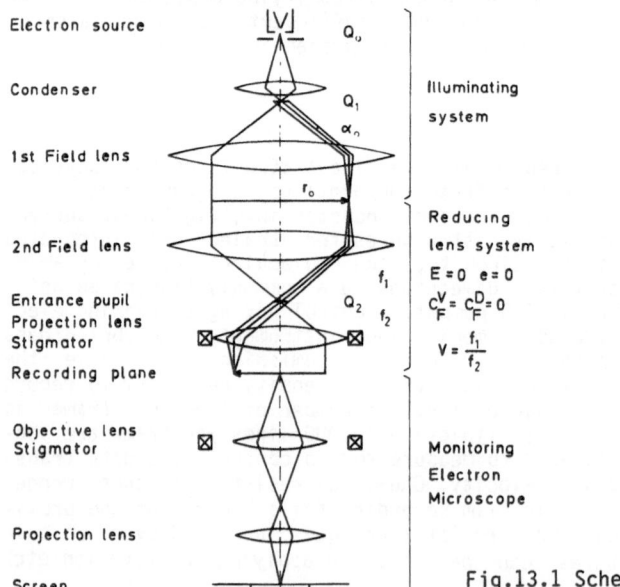

Fig.13.1 Schematic of a reducing image projection system

Since synchroton and other high brightness x-ray sources became applicable during the last years demagnifying electron-beam projection systems again become interesting tools to fabricate high-resolution structures having submicron dimensions.

This paper gives a review of our investigations on demagnifying projection systems aimed at submicron lithography. A new designed system is described which renders gratings of 80 nm grid constant. It offers the possibility of a variable reduction factor having low distortion.

The mask stencil problem in general requires a second recording step to clear away a supporting structure. To overcome this problem we suggest the application of foil replica masks in the demagnifying lens system. Contrast is enhanced using a space frequency filter at the entrance pupil of the lens system. This diaphragm produces scattering absorption contrast as it is known from the imaging process in the CTEM. The preparation of foil replica masks is described. These foil masks can withstand higher power densities than bulk stencils.

Table 13.1 Development of reducing image projection systems [13.5-14]

Year	Author	Scale	Reso-lution	Lines per frame	Applica-tion
1944	M. v. Ardenne	1:500 ÷ 1:10^4 (one demag. lens or two demagnifying lenses)			project
1959	H. Boersch et al.	1:17	3 μm		grids
1963	K.H. Löffler	1:17	0.1 μm	100	microgrids
1964	F.H. Plomp et al.	1:40	0.1 μm	10,000	zone plates
1965	K.H. v. Grote et al.	1:10 ÷ 1:20	0.1 μm	2,500	zone plates
1969	H. Koops et al.	1:15 ÷ 1:80	0.1 μm	10,000	gratings
1972	H. Koops	1:125	0.05 μm	6,000	gratings
1975	M. Heritage	1:10	0.25 μm	12,000	semicond.
1977	R. Speidel et al.	1:5 (photocathode)	1 μm	6,000	semicond.
1978	B. Lischke et al.	1:4	0.10 μm	37,000	semicond.
1979	T. Asai et al.	1:4	0.50 μm	10,000	semicond.
1983	J. Grob et al.	1:20 ÷ 1:110 one or two demagnifying lenses	0.04 μm		gratings

13.2 Design of Reducing Image Projection Systems with Submicron Resolution

Considering the two - lens optics of a reducing image projection system with respect to resolution and number of lines per field, eleven axial and field aberrations of third order have to be considered in combination with the diffraction aberration [13.4]. Because of their different dependency on the imaging aperture α_0 and the mask radius r_0, resolution and image field are limited by the chromatic field aberrations of magnification C_M^V and image rotation C_R^V which blur the image according to the relative width of the electron energy, the relative stability of the lens current I and the high tension supply U:

$$\Delta E/E = [(\Delta eU/eU)^2 + (\Delta U/U)^2 + (2\Delta I/I)^2]^{1/2} \quad . \tag{13.1}$$

Here E and e are the energy and the charge of the electrons respectively. In addition image curvature and the field astigmatism introduce aberrations on the border of the image field. Their size depends on the aperture used. This is optimized to an adequate contribution of the diffraction aberration error $d_d = 1.22 \, \lambda/\alpha_0$. λ is the wavelength of the electrons. Quadratic superposition of the aberration disks renders the resolution limit d_a:

$$d_a^2 = [2(D+C+(C^2+c^2)^{1/2})r_0^2\alpha]^2 + [1.22\lambda/\alpha]^2 + [2 \, C_F^D \, r_0 \, \Delta E/E]^2 \quad . \tag{13.2}$$

As shown earlier, reducing image projection systems can be controlled for distortion-free imaging conditions [13.4]. This is also a condition where a low chromatic aberration of magnification is obtained. Therefore we neglect C_F^V.

C_F^D is directly related to the image rotation ψ in radians [13.15] by $C_F^D = \psi/2$. The strong excited single field condenser objective lens (SCOL) rotates the image to the highest extent possible for an objective lens, producing $C_F^D = 1$. The limitation in resolution and field size is given in Fig.13.2.

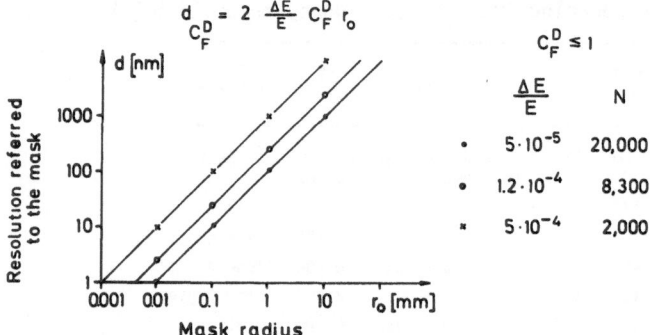

Fig.13.2 Resolution and field size are limited by the chromatic aberration of image rotation C_F^D and the relative stabilities of the electron energy and the power supplies $\Delta E/E$. N is the number of lines per field

For submicron resolution at a low reduction factor a small relative width of the electron energy is required. This is obtained with common electron guns at accelerating voltages > 40 kV. For projection systems therefore the maximum number of lines per frame results to N > 2 000.

The chromatic aberration of image rotation vanishes if the pole piece and field dimensions of the lenses of the telescopic system are scaled with a factor $V = f_1/f_2$. Simultaneously their excitations must be equal and contrary. In this case resolution and line number depend only on the size of image curvature, field astigmatisms, and diffraction aberration. They define an optimum aperture

$$\alpha_0 = (0.61\lambda/[D+C+(C^2+c^2)^{1/2}])^{1/2} / r_0 \tag{13.3}$$

which increases with decreasing mask diameter. Petzvals theorem connects refractive power, image curvature, and isotropic field astigmatism: $D-C = 2/f$. Telescopic systems have $C < D/2$ [13.16]. Approximately $D < 4/f$ results, which is in good agreement with all earlier calculations [13.4,5.16].

Because of technical reasons the focal length of the reducing lens should not be smaller than 2 mm. This produces $D = 2$ mm^{-1}. Therefore the optimum aperture is approximately given by

$$\alpha_0 = 6.83 \cdot 10^{-4}/[(UD^2)^{1/4} \cdot r_0] \text{ [rad]},\qquad(13.4)$$

with U in [volts], D in [mm^{-1}] and r_0 in [mm]. Here the approximation $c < C/2$ gathered from calculations was introduced, e.g. $U = 40{,}000$ V, $D = 2$ mm^{-1}, $r_0 = 1$ mm gives $\alpha_0 = 3.4 \cdot 10^{-5}$ rad.

The most rigorous design of a reducing image projection system by LISCHKE et al. [13.5] reached $D = 0.04$ mm^{-1} which widens the optimum aperture by 10 with respect to this estimation. Consequently the optimum aperture ranges from $3 \cdot 10^{-4}/r_0$ to $3 \cdot 10^{-5}/r_0$. Even with high diminishing factors the imaging aperture is still small compared to the aperture used in scanning beam writers. The depth of the image field $T = d_a/2\alpha_0 V$ is at least $T = 1{,}000\, d_a$. This means with respect to submicron structure fabrication the axial position of the recording film is not a very critical parameter.

The maximum number of lines per field is $N = 2r_0/d_0$, with d_0 being the fabricated line referred to the mask of radius r_0, $d_0^2 = d_p^2 + d_a^2$. d_p is the mask opening and d_a the contribution of the aberrations. Introducing $A = d_0^2 / d_a^2$ or $A = 1 + (d_p^2/d_a^2) > 1$, we obtain $N_{max} = 2r_0/(d_a^2 A)^{1/2}$, and with the optimum aperture

$$N_{max} = [A(1.22\lambda(D+C+(C^2+c^2)^{1/2})+((\Delta E/E)C_F^D)^2)]^{-1/2} .\qquad(13.5)$$

The aproximations given above result in

$$N = (2.4 \cdot 10^{-6}AD/U)^{1/2} +(C_F^D \cdot \Delta E/E)^2)^{-1/2} .\qquad(13.6)$$

For $U = 40$ kV, $\Delta E/E = 5 \cdot 10^{-5}$, $C_F^D = 1$, $D = 2$ mm^{-1}, and $A = 4$ we obtain $N = 3{,}234$ lines per frame, which is sufficient for an experimental system designed to explore the fabrication of submicron structures. These formulae show that image curvature and accelerating voltage are the critical parameters. In general more than 1 000 lines per field are reached without difficulty.

13.3 The Image Projection System with Variable Reduction Factor

To investigate the submicron capabilities of reducing image projection systems we developed a lens system which allows to vary the reduction factor and therefore just one master of a special structure is necessary. This provisionally solves the master stencil problem. To generate gratings having a different grating constant we used a copper grid with 8.5 μm grid constant supplied by EMI Ltd. as master stencil. For the lens system we modified the Philips electron microscope EM 301 column. Following the ideas of previous work [13.12],we found that the objective lens of the microscope can be excited to render at $U = 40$ kV accelerating voltage a single field condenser objective lens. A focal length of 2 mm is produced at 4 590 Ampere turns with a pole piece of gapwidth $S = 6.21$ mm, and bore radius $R = 2.07$ mm. This lens has a spherical aberration $C_3 = 1.33$ mm, and an axial chromatic aberration $C_F^A = 1.46$ mm.

Introducing the mask stencil into the midplane of the second condenser lens via the condenser aperture holder we reached a fixed reduction factor of 63 $\pm$ 2% according to the geometry of the microscope and the focal length of the condenser field of the SCOL. It projects the demagnified image to the middle plane of the strong lens. The pole piece surfaces have been positioned under the constraint that the specimen table plane coincides with the field maximum of the SCOL. This desgin is similar to the arrangement of the "twin lens" in the Philips EM 400 microscope. We therefore have the original specimen airlock employed to change our exposed samples. The stereo tilt drive incorporated in the high-resolution table allows to position off-axial parts of the recording foil to the image field. Focusing by mechanical shift is observed with the monitoring microscope having a smaller depth of field than the reducing lens system because of the scattering of the electrons in the recording foil. Exposures were placed next to the position already used for focusing by shifting the table. The illumination aperture required for submicron resolution at a mask radius of 1 mm is α_0 = 3.4 · 10^{-5}. It is controlled diminishing the crossover of the W-hairpin cathode with the first condenser lens. Distortion is compensated by exciting the second condenser lens until the image monitored with the objective field and the projection lens is free of distortion.

If the stencil is inserted in the middle plane of the first condenser lens via the condenser aperture holder and the second condenser is switched off again,a fixed reduction factor of 110 is obtained. Distortion is compensated by exciting the first condenser lens to the appropriate amount. In this case the illumination aperture of 3.6 · 10^{-5} cannot be reached with a conventional hairpin filament. A tipped cathode,however, having a crossover diameter of 10 μm, supplies an illumination aperture of 2.5 · 10^{-5} rad and is therefore employed [13.17].

To get a variable reduction factor the mask is situated in the middle plane of the first condenser lens and both condenser lenses and the SCOL are employed. Reduction factors ranging from 1:20 to 1:150 are obtained,exciting the first condenser lens to form a diminished source image in front of the second condenser lens. This in turn forms a source image at the entrance pupil of the SCOL and also an image of the mask in front of the entrance

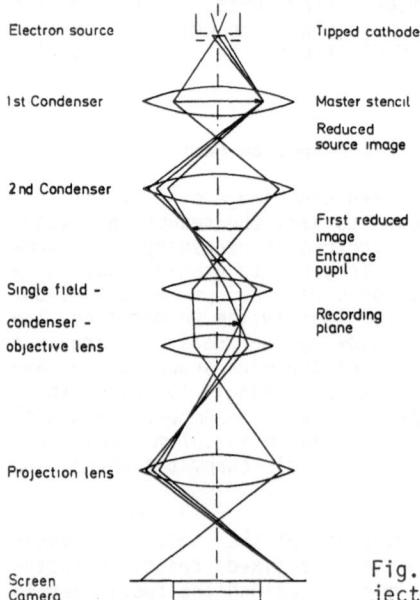

Electron source — Tipped cathode

1st Condenser — Master stencil
Reduced source image

2nd Condenser — First reduced image
Entrance pupil

Single field - condenser - objective lens — Recording plane

Projection lens

Screen Camera

Fig.13.3 Schematic of the reducing image projection system with variable reduction factor

pupil. This weak or strong reduced mask image is projected at a reduced
scale by the condenser field of the SCOL onto the recording plane. Distortion is controlled by a proper excitation of the first condenser lens. The reduction factor is set with the excitation of the second condenser lens. Figure 13.3 gives the schematic of the reducing image projection system having
a variable diminishing factor.

13.4 The Recording Process and Experimental Results

Reducing image projection systems having a monitoring CTEM require transparent recording foils. These foils can be used with 40 kV electrons if
their thickness is less than 50 nm. Liquid developers have a surface tension
which in general is destructive to films and foils. To avoid them we applied
collodion foils as selfsupporting films of 100 nm thickness mounted on a
brass ring with 3 mm outer and 1.3 mm inner diameter. To anneal the films
5 nm of carbon is evaporated from the rear. Collodion has a sensitivity of
0.005 C/cm^2 for 40 keV electrons. We apply this dose during a recording time
of 10 sec. The irradiated part of the foil is reduced to 25% of the initial
film thickness by the electrons. This results in a surface relief which is
developed by a shadowing evaporation with gold at an incident angle of
10 degrees to the film surface. Inspection is done in a CTEM or in a light
microscope if polarizing effects are to be measured. Recordings with 23 diameters and 110 diameters reduction are shown in Fig.13.4 in CTEM magnification. The left-hand side presents recordings of the microgrid shadowed with
gold. During exposure the image of the multibeam mask can be shifted parallel to a grid direction to record line gratings. The shift is introduced to
the image using the lower coils of the wobbler deflection element situated
near the entrance pupil of the SCOL. The corresponding line gratings are
given on the right-hand side of Fig.13.4. At the moment the system is capa-

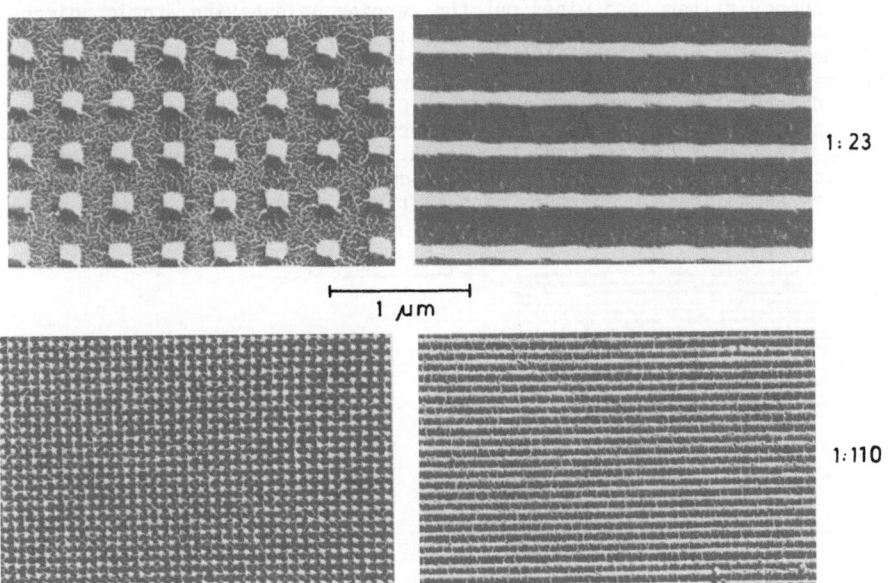

1 μm

1:23

1:110

Fig.13.4 CTEM magnification of grid patterns and line gratings recorded in
collodion film fabricated with the reducing image projection system
having a variable reduction factor, top 1:23, bottom 1:110

ble of printing 200 (1:63) respectively 80 periods (1:110) caused by a field
limiting aperture in the first condenser lens and by the period of the actual master. A master stencil with micron structures covering an area of
2 mm in diameter allows to record the above estimated number of 2 000 lines
per frame. The mask stencil problem is,however, still present. To expose
annular structures various methods have been developed, e.g. complementary
masks [13.18]. They can be fabricated with high resolution using e-beam lithography and by applying various reactive ion and chemical etch steps. Here
dual exposure is necessary. This technique is difficult to be applied in the
reducing image projection system.

13.5 The Reducing Image Projection System with Foil Mask and Space Frequency Filter

To overcome the stencil problem we suggest the application of selfsupporting
foil masters in a reducing image projection system having a contrast aperture stop in the entrance pupil. A foil mask is prepared as a replica of a
surface relief using standard transmission electron microscopy preparation
techniques [13.19]. From photoresist line patterns recorded by laser interferometry,carbon grating replicas were fabricated. They are shadowed with
platinum at an angle of 10 degrees to the surface. The evaporation is done
at + 90 and -90 degrees to the line direction. These replicas can be prepared covering a diameter of 5 mm to serve as an original. Contrast is formed
in the reducing image projector like in a CTEM as scattering absorption contrast. Electrons passing the original are scattered in the heavy metal lines
of the replica but can penetrate the carbon foil with less interaction. The
diffracted waves however are focused by the field lens to form a source image at the entrance pupil. Therefore a diffraction pattern is produced at
this plane. Now an aperture can stop all higher orders of the scattered
waves and admits the passage of the zero order of the beam. It acts as a
space frequency filter and wipes out the carrier grid having atomic dimensions. Therefore no more additional recording steps have to be applied to
cancel the carrier grid. The stencil problem is solved and all geometries
can be imaged. We measured the contrast K produced by scattering absorption
with the diaphragm by scanning a magnified CTEM image of the replica over a
single-channel electron counter. The contrast approaches K = 1 if an aperture of $\alpha_0 = 7 \cdot 10^{-3}$ rad is used. Figure 13.5 shows a CTEM magnification of
the replica and a table of the contrast produced by different apertures. For
experiments an adjustable aperture stop must be placed in the entrance pupil

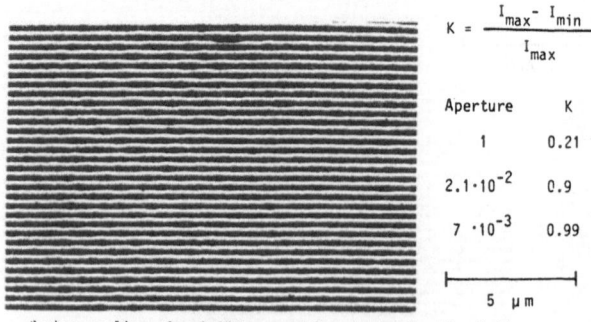

$$K = \frac{I_{max} - I_{min}}{I_{max}}$$

Aperture	K
1	0.21
$2.1 \cdot 10^{-2}$	0.9
$7 \cdot 10^{-3}$	0.99

5 μm

Carbon replica of a 0.452 μm grating shadowed with platinum

Fig.13.5 CTEM micrograph of the grating replica obtained by a standard
C-replica technique shadowed with platinum. An imaging aperture
$\alpha_0 = 7 \cdot 10^{-3}$ produces scattering absorption contrast K = 0.99

plane. This is not possible in the EM 301 system because of technical reasons. We will be able to perform these experiments using the microprojection system developed by LISCHKE et al. [13.5]. This instrument allows to place a scattering contrast aperture stop at the entrance pupil of the projection system and is therefore suited to investigate this question.

13.6 Conclusions

The reducing image projection systems with a variable diminishing factor offers a large flexibility in fabricating line gratings for x-ray spectroscopy. Being capable to reduce micron master structures opens the possiblity to produce high-resolution zone plates and other structures at a varied scale required for experiments. This makes it easy to investigate effects depending on the size of the structures. There is the possibility to reach the estimated maximum number of lines per frame with the EM 301 system. Fields with even higher numbers of lines per frame will be recorded using the Siemens microprojector. The flexibility of the projector will be increased modifying the focal length of the projection lens with an appropriate increase of the accelerating voltage to incorporate the variable scale image projection mode in the system.

Acknowledgments

The authors are indepted to T. Tschudi for encouragement and strong interest in the experiments. We appreciate the assistance of our colleagues J.P. Kotthaus and D. Heitmann from the Institut für Angewandte Physik, Hamburg, providing the gratings fabricated by using laser interferometry and G. Jourdan and P. Daab from our institute for preparing grating replicas.

We thank the Siemens AG, München, especially I. Dietrich, E. Fuchs, and B. Lischke, for the gift of the complete 1:4 microprojection system to our institute which widens our possibilities for submicron lithography. The work was sponsored by the Stiftung Volkswagenwerk.

References

13.1 P.J. Coane, D.P. Kern, A.J. Speth, T.H.P. Chang: Proc. Microcircuit Engineering Grenoble 373 (1982)
13.2 A.N. Broers: Proc. Microcircuit Engineering Amsterdam 9 (1980)
13.3 M. Hatzakis, J. Paraszezak, J. Shaw: Proc. Microcircuit Engineering Lausanne 386 (1981)
13.4 H. Koops: Electron beam projection techniques in Materials Processing Theory and Practices, Vol. 1: "Fine Line Lithography", (R. Newman ed. North Holland Publ. Comp., Amsterdam 1980) 233 - 336
13.5 B. Lischke, J. Frosien, K. Anger, W. Münchmeyer: Optik 54, 325 (1979)
13.6 M.B. Heritage: J. Vac. Sci. Technol., 12, 1135 (1975)
13.7 T. Asai, S. Ito, T. Eto, M. Migitacka: 11th Int. Conf. on Solid State Devices, Tokyo 15, (1979)
13.8 R. Speidel, M. Mayr: Optik 48, 247 (1977)
13.9 F.H. Plomp, J.B. Le Poole: 3rd European Reg. Conf. on Electron Microscopy 447 (1964)
13.10 K.H. Löffler: Dissertation, Technische Universität Berlin (1963)
13.11 K.H. v. Grote, G. Möllenstedt, R. Speidel: Optik 22, 252 (1965)
13.12 H. Koops: Optik 36, 93 (1972)
13.13 M. v. Ardenne: Nachrichtentechnik 10, 427 (1960)

13.14 H. Boersch, H. Hamisch, K.H. Löffler: Naturwiss. <u>46</u>, 596 (1959)

13.15 W. Glaser: <u>Grundlagen der Elektronenoptik</u> (Springer Verlag Wien, 1952)

13.16 M. Mayr: Dissertation Tübingen (1980)

13.17 S. Maruse, Y. Sakaki: Optik <u>15</u>, 485 (1958)

13.18 H. Bohlen, J. Greschner, J. Keyser, W. Kulke, P. Nehmitz: IBM J. Res. Develop., Vol. <u>26</u>, 568 (1982)

13.19 L. Reimer: <u>Elektronenmikroskopische Untersuchungs- und Präparationsmethoden</u> (Springer Verlag Berlin, 1967)

14. Grazing Incidence Optics for X-Ray Microscopy

A. Franks and B. Gale

Division of Mechanical and Optical Metrology, National Physical Laboratory
Teddington, Middlesex, TW11 0LW, UK

14.1 Introduction

Considerable enthusiasm for x-ray microscopy was generated in the 1950s after some successful pioneering work in the subject [14.1 - 3]. This initial enthusiasm slowly evaporated after it became evident that the achievement of high resolution was beset with experimental difficulties related to the problems of producing very precisely figured and extremely smooth mirrors. However, since 1975 [14.4] there has been a renewal of interest in x-ray microscopy as a means of studying emissions from laser-generated plasmas and also as a technique for examining the microstructure of materials, to complement optical and electron microscopy [14.5].

The intervening years have seen advances in surface finishing technology coupled with enhanced measuring capabilities at the nanometre and subnanometre levels, as well as advances in our theoretical understanding of the factors which limit resolution. Nevertheless, the practical limit of resolution is still set by the inadequacies of the manufacturing processes, although with our present knowledge, we can anticipate that the achievement of near theoretical resolution is no longer beyond our grasp.

14.2 The Basis of the Grazing Incidence Microscope

High-resolution x-ray microscopy is based on the well-established geometry of Wolter I optics, shown in Fig.14.1. The microscope consists of two confocal conicoidal surfaces, usually an ellipsoid and an hyperboloid, although, in principle, a pair of ellipsoids could also be used, to give a less compact configuration.

The principal features of the Wolter configuration are that the resolution of an image on the optical axis is not degraded by any geometrical aberrations, the resolution is thus diffraction limited only, and that the

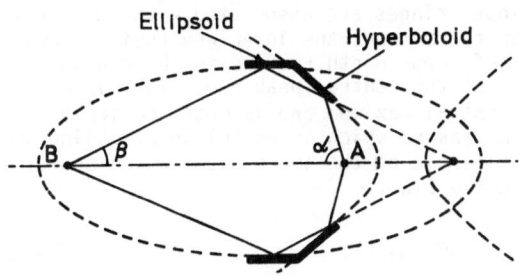

Fig.14.1 Geometry of the Wolter I microscope

off-axis aberrations are sufficiently small to allow the microscope to have an adequate field of view [14.5].

If in Fig.14.1, the object is located at the focal point A, then an enlarged image is formed at the conjugate focus B, the magnification being equal to α/β. This configuration, in which the image is recorded photographically, is the basis of the simplest Wolter microscope system. Despite its simplicity, this is, however, not the preferred configuration.

In the scanning microscope, the object and image planes are interchanged. The mirror system now forms a demagnified image of the object, which may be either an x-ray source or an illuminated pinhole. The specimen to be examined is placed in the image plane and the x-ray spot is scanned in a raster pattern over the surface. The radiation transmitted through the specimen is collected by a detector, such as a proportional counter, the output of which may be fed directly to a video display or to a computer, for image processing purposes.

The two principal advantages of the scanning microscope over the simple, static magnifying configuration are that the output is digital, which facilitates analysis (both of the image and the specimen composition), and that optimum optical performance can be achieved if the beam is kept stationary on axis and scanning is effected by motion of the specimen. If some compromise in optical performance is permissible, then it may be simpler to scan the x-ray beam, keeping the specimen stationary, but at the expense of a degradation of resolution resulting from the off-axis aberrations, as discussed in Section 14.3.

14.3 Parameters Affecting Resolution

The parameters affecting resolution fall in the following categories: (i) the aberrations of a perfect microscope, (ii) the aberrations resulting from imperfect manufacture and (iii) the aberrations arising from the microscope system as a whole. The optimum design of the microscope is based on an evaluation both of these factors and of the constraints (particularly on size), imposed by the manufacturing conditions.

14.3.1 The Resolution Limit of a Perfect Wolter I Microscope

In the geometrical optical approximation, the conjugate image of a point at the first focus is also a point; there is thus no aberration. The ultimate resolution is set by the size of the diffraction image which can readily be calculated using the Debye asymptotic plane wave approximation [14.6]. A typical image is shown in Fig.14.2, and is essentially the fringe structure produced by an annular converging spherical wave front. The fine scale fringes are associated with the minimum value of β (Fig.14.1), i. e. the angle between the optical axis and the rays from B to the intersection of the ellipsoid and hyperboloid. The envelope fringes are associated with the finite length of the ellipsoid: the longer the mirror, the lower are their intensities. Two quite distinct measures of image width require to be considered. The first is simply the radius r_1 of the central peak where the intensity falls to some fraction α_1 of the central maximum and is most commonly used in defining resolution. The second measure which is useful in assessing image quality is the radius r_2 of a circle on the image plane which contains some fraction α_2 of the total reflected flux.

$$r_1 = \lambda(1 + 2M\cos i + M^2)^{\frac{1}{2}}(1 - \alpha_1)^{\frac{1}{2}}/(\sqrt{2}\pi\sin 4i) \qquad (14.1)$$

$$r_2 = 4M\lambda L(1 + 2M\cos 4i + M^2)^{\frac{1}{2}}/[(1 - \alpha_2)(1 + M)^2\pi^2 L_2 \sin 4i] \qquad (14.2)$$

where λ is the wavelength, L is the distance between the conjugate planes, L_1 and L_2 are the lengths of the first and second mirrors, M is the magnification and i is the grazing incidence angle at the common plane of the mirrors.

For the microscopes· discussed in Section 14.4, where M=0.3 and for λ=2.5 nm, α_1=0.5 and α_2=0.80, values of $2r_1$ (FWHM) and $2r_2$, the full width at 80 % flux (FW80%F), are given in Table 14.1 for various angles of incidence.

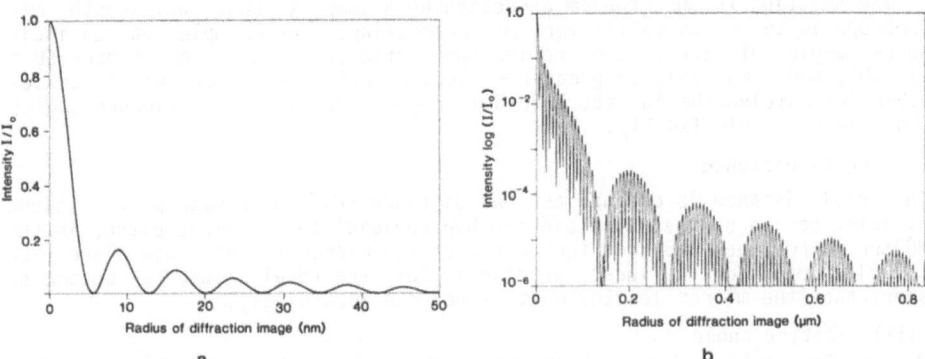

Fig.14.2 Diffraction pattern of Wolter microscope (a) near centre, (b) extended image; λ=2.5 nm, i=3^0, magnification 0.3x, ellipsoid length = 30 mm, bore diameter = 30 mm

Table 14.1 Parameters of 30 mm diameter microscopes, having magnifications of 0.3x and resolutions of 0.1 μm

Grazing incidence angle	[^{0}]	0.17	2	3	5	10	20	25
Object-image distance	[mm]	7260	606	403	242	121	61	50
Focal distance	[mm]	1677	139	92	54	25	7.21	1.9
Scatter angle	["]	0.006	0.07	0.1	0.2	0.4	1.4	5.4
Ellipsoid length	[mm]	30	30	30	30	30	20	4
Hyperboloid length	[mm]	29.7	26	24	21	16	7.05	1.7
Optical aperture	[sr]	2.65 $\times 10^{-7}$	4.8 $\times 10^{-4}$	1.7 $\times 10^{-3}$	8.3 $\times 10^{-3}$	7.5 $\times 10^{-2}$	3.2 $\times 10^{-1}$	6.8 $\times 10^{-2}$
FWHM ($2r_1$)	[nm]	89	7	5	3	1.5	0.9	0.8
FW80%F ($2r_2$)	[μm]	47	0.39	0.19	0.08	0.03	0.002	0.005
Off-axis tolerance	[μm]	8789	63	28	10	2	0.28	0.05
Axial profile error	["]	0.002	0.03	0.04	0.06	0.11	0.09	0.03
Axial profile error and Surface roughness	[nm]	34	2.8	1.9	1.1	0.6	0.3	0.2

Notes on Table 14.1

(i) Grazing incidence angle and object image distance

With a gold-coated mirror, i = 10 arcmin (0.17°) for a wavelength of 0.1nm. From the Table it can be seen that in this case, the object to image distance L would be over 7m which is impracticably large. For i greater than 2°, the values of L are substantially reduced and a microscope designed for these grazing incidence angles would be comparable in length to a conventional optical microscope. At these angles, all wavelengths greater than 2nm will be reflected with high efficiency from a gold surface and the microscope would thus be very suitable for examining biological materials.

The solution to the problem of designing a compact, short wavelength microscope is to employ multilayer mirror coatings. For example, at an incidence angle of 1.6°, the reflection efficiency at 0.15nm is over 50 % [14.10], and comparable or greater reflection efficiencies are easily achievable for wavelengths in the 1 to 5nm region for all the incidence angles shown in the table [14.11].

(ii) Focal distance

The focal distance is defined as the distance from the common plane (plane of intersection of the ellipsoid and hyperboloid) to the image plane, in the Wolter microscope, and from the centre of the mirror to the image plane for the ellipsoid. For the large grazing angles, the focal distances become so short that the mirror lengths must be reduced accordingly.

(iii) Scatter angle

The scatter angle is the maximum angle at which radiation may be scattered without impairing the 0.1 μm resolution, measured from the common or central plane. It is clear from the table that the scatter tolerances are considerably relaxed at the larger incidence angles.

(iv) Ellipsoid and hyperboloid lengths

The final microscope parameter which must be specified is the length of the mirror. The greater the mirror length, the greater is the optical aperture and the smaller the diffraction blur. On the other hand, the off-axis aberrations also increase with mirror length, but relatively slowly only. Present microscope manufacturing technology can reasonably be extended to enable a length to diameter ratio of 2:1 to be achieved, and thus the total length of the microscope can be up to 60mm. At larger incidence angles, the mirror lengths must be reduced to keep the mirror in front of the specimen.

(v) Optical aperture

This is a solid angle measured in steradians and increases steadily with increasing angle, but reaches a maximum at grazing incidence angles of about 20°, because of the reductions in mirror lengths at the higher angles.

(vi) Off-axis tolerance

This equals the maximum permissible radial displacement of a point source from the optical axis, for a resolution of 0.1 μm.

(vii) Axial profile error [arc sec]

The axial profile slope error which was discussed in Section 3, approximates fairly closely to the more readily calculable scatter angle, except at large grazing angles where the microscope is close to the image plane.

(viii) Axial profile error and surface roughness

It is shown in this section that, on the basis of the diffraction theory employed, the same manufacturing tolerances apply to the amplitude deviations of the axial profile and of the surface roughness.

14.3.2 Effect of Finite Source Size

Calculation of the effect of displacing a point source from the focus in a plane normal to the symmetry axis will yield both the alignment tolerance for the source and microscope and also information about the effect of using a finite source.

The first-order optical behaviour is determined calculating the first and second moments of the intensity distribution in the image plane, resulting from the displacements in the object plane. The first moment is a measure of position, namely the centre of gravity, and the second moment (the radius of gyration) is a simple measure of image size, and hence of resolution.

If x'_ρ are the coordinates ($\rho=1,2$) of an object point in a plane normal to the symmetry axis at the first focus, then it can be shown that the centroid and radius of gyration R of the image on the conjugate plane produced by a Wolter microscope of vanishingly small width is given, to the first order, by

$$\langle x_\rho \rangle = - Mx'_\rho \tag{14.3}$$

$$R = r'M|M - 1|\sin^2 2i/(M + \cos 4i) \tag{14.4}$$

where r' is the distance of the object point from the axis of symmetry.

It can be seen, from (14.3) and (14.4), that the displacement of the image centre is equal to the magnification, as would be expected. In Table 14.1 values of r' have been calculated for a range of values i, for a resolution of 0.1 μm, corresponding to a value of R of 0.05 μm. The table shows that for the smaller grazing angles, a limited departure of the source from the optical axis would not affect the resolution unduly.

14.3.3 Calculation of Some Manufacturing Tolerances

The image of a point source is blurred by departures of the microscope from its theoretical form. These perturbations include the relative position of the two mirrors, displacement of the source and image plane and errors in the mirror profiles. A full analysis of the effects of these perturbations has been made and it is clear that the profile errors dominate, under careful manufacturing conditions. Hence a simplified discussion of the mirror profile effects only is given here.

Let z denote the normal displacement of a surface element from the ideal shape and let z' denote its axial slope deviation. The effects of azimuthal slope errors are usually small and will be ignored here. The image blur can be calculated using simple scalar diffraction theory which tends asymptotically to geometrical optical theory as the wavelength becomes small compared with the perturbations.

For a statistically rough surface, the scattering has two components [14.7]: a sharp peak which diminishes exponentially as the rms roughness z_R increases, and a diffuse peak with a width inversely proportional to a correlation length l, and which has an integrated intensity which increases exponentially with z_R. The image radius defined by the diffuse peak is given approximately by

$$R_D \simeq (\sqrt{2} \; \lambda \; A \; \cosec \; i)/\pi l \qquad (14.5)$$

where A = 2LM/(M + cos 4i), and is a factor associated with Wolter I geometry. If the scattering from the slowly varying component of the surface perturbation can be described by geometrical optics, then the image radius is given by

$$R_G \simeq [\tfrac{1}{2} (5M^2 - 6M + 5)z^2 + 2Az'^2]^{\frac{1}{2}} \; . \qquad (14.6)$$

This expression describes the case where the perturbations of the two mirrors are identical. It is convenient to put z=a cos (2πx/d), where x is the distance along the axial arc of the mirror. If each mirror contains an integral number of periods of surface wavelength d, then

$$R_G \simeq \tfrac{1}{2} \; a \; [5m^2 - 6M + 5 + 16A^2/d^2]^{\frac{1}{2}}$$

$$\simeq 2Aa/d \quad \text{(for the microscope discussed in Section 14.4)} . \qquad (14.7)$$

For a resolution of 0.1 μm, R_G = 0.05 μm, and then the maximum allowed slope error, 2πa/d, is given by $\pi R_G/A$. Values of theses axial profile error slopes are given in Table 14.1, for various grazing angles. For all values of the surface wavelength d shorter than the mirror lengths, the corresponding values of the amplitudes are smaller than the critical height z_c which marks the boundary region between the geometrical and diffraction optical regimes, given by the Rayleigh criterion : $z_c=(\lambda \cosec \; i)/4\pi$. It therefore follows that the geometrical estimate of image width given in (14.7) is valid only for values of d much larger than the microscope. Slow variations in the surface which appreciably exceed the length of the microscope are equivalent to a dilation of the mirror. If the dilation is b, then

$$R_G = (5M^2 - 6M + 5)b/2 \qquad (14.8)$$

so that for all glancing angles and for M=0.3 and for R_G less than 0.05 μm, b must be less than 27 nm.

For all values of d less than the microscope length (i. e. for the roughness as well as the longer surface wavelength perturbations), the scattering must be calculated using diffraction optics. The resolution criterion to be satisfied is that the diffuse peak shall not be dominant and this requires that $z_R \leq z_c/2$, in which case the sharp peak is diminished by no more than about 20 %. The maximum axial profile error and surface roughness amplitudes are thus equivalent and their values are given in Table 14.1.

15.3.4 The Microscope System

The factors which have a significant effect on resolution are the relative mechanical stabilities of the source, mirror and specimen, the source size and focal length.

(a) Mechanical Stability

Displacements of the specimen relative to the microscope mirror in a direction normal to the axis must not exceed the required resolution. The stability of the source is less demanding by a factor equal to the demagnification. For a given grazing incidence angle, the object to image distance is proportional to the microscope diameter, and since it is easier to ensure stability in a smaller rather than a larger system, it is advantageous to make the mirror diameter as small as possible. However, the manufacturing and measuring procedures become increasingly difficult with decreasing bore size. Enough experience has been gained in working with a 40mm diameter bore to feel confident that a reduction to 30mm would not present undue difficulties. This diameter has therefore been chosen for the current programme on microscope development. Reductions in size by a factor of 2 to 3 might be expected in the future. Much smaller diameter microscopes of less than 10mm in diameter might be made by replication of a male master, but it does not seem likely, at present, that sufficiently stable replication materials are available to enable this to be done for the highest quality microscopes.

(b) Source Size

For any given resolution, the demagnification required decreases with decreasing source size, and this is advantageous, because the lower the demagnification the shorter will be the system length and the greater will be the optical aperture. The source employed hitherto [14.5] has a diameter of about a micrometre, while the more refined version now under development has been designed to give a spot size about ten times smaller than this.

(c) Focal Distance

The degradation of image contrast and resolution due to scatter from the mirror surface will decrease in direct proportion to the focal distance (at least for the second mirror). There is a conflict between the requirements of short focal length and low demagnification; a good compromise is to make the focal distance sufficiently short so that the demands on the quality of the surface finish, which determines the scattered intensity distribution, are kept within practicable limits.

14.4 Specifications of an X-Ray Microscope

Current Wolter x-ray microscopes have resolutions which are about a micrometre ([14.8,9] and section 14.5). The immediate objective of the present work is to improve the resolution by a factor of 10. The specifications of the new microscope are derived from the considerations presented in Sections 14.3 and 14.4.

The three parameters which determine the main geometrical features of the microscope are its bore diameter, demagnification and grazing incidence angle. For the reasons previously discussed, the bore diameter at the plane of intersection of the mirror components has been made equal to 30mm. Because the x-ray source size will be close to the required resolution, the demagnifying factor can be kept fairly small and a reasonable value for the magnification is thus 0.3x. The grazing incidence angle determines the minimum wavelength which can be employed, and for microscopy to be undertaken on a wide range of materials from metals to biological samples, the wavelength range should span the spectrum from 0.1nm upwards. Another factor which should be taken into account is the object to image distance L, which for the selected parameters of the microscope varies with the grazing incidence angle i, as shown in Table 14.1, which also lists other important optical parameters and manufacturing tolerances.

14.5 Some Experimental Developments

The work done so far on the development of a Wolter I microscope is for a microscope which has entirely different optical parameters from those previously discussed. This microscope is designed as a conventional enlarging microscope for examining the x rays generated by the interaction of a laser beam with matter, and is shown in Fig.14.3. The work was undertaken as part of a collaborative project with the Lawrence Livermore National Laboratory and the UK Atomic Weapons Research Establishment.

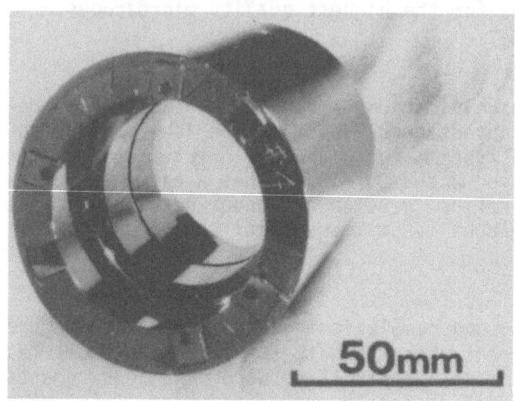

Fig.14.3 Wolter I microscope

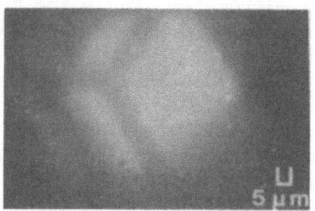

Fig.14.4 X-ray micrograph of an hexagonal gold grid taken with Al K radiation (λ = 0.83 nm)

14.5.1 Production Technique and Measurement Instrumentation

The tolerances on axial profile and surface roughness for this microscope were about 1 nm, for a resolution of 1 µm. Taking into account the cumulative effects of various errors - the error budget - it was therefore necessary to develop polishing and lapping techniques to achieve surface roughness of a few tenths of a nanometre. Peak-to-valley roughnesses of this order of magnitude have now been achieved on various relevant materials [14.12], using techniques which are immediately applicable to the x-ray microscope configuration.

Three machines have been developed at NPL to enable surface roughness, figure and diameter to be measured to the required accuracies. Surface roughness measurements at the subnanometre level are made directly on the polished surface of the microscope with a diamond stylus in the Nanosurf machine [14.13]. The axial profile is measured to nanometre accuracy by a laser autocollimation method [14.14], and diameter with a 2-axis measuring machine [14.15].

14.5.2 Experimental Results

The microscope was designed for laser fusion research. The main specification has previously been published [14.16], but minor modifications had been made to increase the magnification to 22x [14.9]. The grazing incidence angle is 1⁰, the bore diameter is 40mm, the object-image distance is 6.9m and the focal distance is 6.6m.

This configuration is dictated by needs of the experiment. It is, however, unfavourable in the extreme for microscopy of the type envisaged in this paper. The large focal distance demands that to achieve a resolution of 0.1 µm, the intensity of scattered radiation should be small at angles no greater than 0.0015 arc sec! This figure should be compared with the scatter angles quoted in Table 14.1, for the microscope now being developed where the tolerances may be relaxed by a factor of 100.

Nevertheless, valuable experience has been gained in developing the surface finishing techniques and measurement instrumentation. Surface profile was controllable to within 3 to 4nm of the theoretical curve along the major part of the axial profile. A surface roughness of 0.5nm peak-to-valley was achieved. A major problem which was not overcome was that the out-of-roundness of the microscope, as received prior to lapping, exceeded 0.5 µm. Under these conditions it was not possible to correct the profile of more than a narrow azimuthal strip. In the light of this experience, machining processes are now being developed which will obviate this problem.

Figure 14.4 shows the resolution which has been achieved at present. The object was an hexagonal gold grid illuminated with Al K radiation (wavelength = 0.83nm) from a microfocus x-ray source and was enlarged 22x directly by the microscope; it is not a scanning x-ray micrograph. It is clear, even under the very unfavourable conditions referred to, that the resolution is somewhat better than 1 µm, but with a poor peak-to-background ratio. This result shows significant promise for the future.

14.6 Conclusions

Grazing incidence x-ray microscopy can be used over a spectral range from 0.1nm upwards and can thus be employed for examining a wide range of materials from metals to biological samples.

On the basis both of the calculated manufacturing tolerances and of the experimental results achieved so far, it is concluded that resolutions of better than 0.1 µm should be achievable in the next phase of the project without requiring developments beyond our present state of the art.

Acknowledgements

Important contributions to the x-ray microscope development and evaluation were made by Mr. T.H. English, Mr. A.E. Ennos, Mr. K. Lindsey, Mr. C.J. Robbie, Dr. M. Stedman and Mr. M.S. Virdee.

References

14.1 P. Kirkpatrick, A.V. Baez: J. Opt. Soc. Amer. 38, 766 (1948)
14.2 H. Wolter: Ann. Phys. 10, 94 (1952)
14.3 H. Wolter: Ann. Phys. 10, 286 (1952)
14.4 F.D. Seward and T.M. Palmieri: Rev. Sci. Inst. 46, 204 (1975)
14.5 A. Franks, B. Gale, K. Lindsey, D.J. Pugh, C.J. Robbie and
      M. Stedmann: Ann. N. Y. Acad. Sci. 342, 167 (1980)

14.6 R.K. Luneburg: <u>Mathematical Theory of Optics</u> (University of California Press, 1964)

14.7 W.T. Welford: Opt. Quant. Electron. <u>9</u>, 269 (1977)

14.8 J.K. Silk: Ann. N. Y. Acad. Sci. <u>342</u>, 116 (1980)

14.9 R.H. Price: AIP Conf. Proc. <u>75</u>, 189 (1981)

14.10 E. Spiller: AIP Conf. Proc. <u>75</u>, 124 (1981)

14.11 T.W. Barbee: AIP Conf. Proc. <u>75</u>, 131 (1981)

14.12 K. Lindsey: Nucl. Inst. Meth. 1983 (in the press)

14.13 A. Franks: AIP Conf. Proc. <u>75</u>, 179 (1981)

14.14 A.E. Ennos, M.S. Virdee: Proc. SPIE. <u>398</u>, 1983 (in the press)

14.15 M. Stedman: Proc. SPIE. <u>316</u>, 2 (1981)

14.16 M.J. Boyle, H.G. Ahlstrom: Rev. Sci. Inst. <u>49</u>, 746 (1978)

15. Use of Multilayers for X-UV Optics: Their Fabrication and Tests in France

P. Dhez

ERA 719 et L.U.R.E., Bat. 209 C, Université Paris-Sud
F-91405 Orsay, France

15.1 Introduction

New artificial layered media on a quasi atomic scale can now be obtained by sequential evaporation of two materials. Several evaporation techniques have been used successfully and each one has its own advantage. The qualities required in the layering are dependent on the requirement of the basic or applied needs. A good control of the layer periodicity is important when the periodicity of the material must match with some wavelengths. For interference mirrors adapted to x-uv range, or slow neutron, amorphous or epitaxied materials can equally be used. On the contrary, semiconductor or semimetal supergratings with a low number of defects are necessary for the microelectronic applications. Layered media are also developed to study superconductivity and Young's modulus. In all cases a full description of the material is needed to understand the observed phenomena, but also for the progress in the evaporation methods.

The LURE laboratory is undertaking research in collaboration with several laboratories to produce and to test layered materials, taking into account the numerous potential applications of these new materials. The first goal in the near future is to improve the x-uv optics for synchrotron radiation, plasma diagnostics or x-uv astronomy. A synchrotron radiation center is an appropriate place to organize such research as it receives most of the present or potential users of x-uv optics and also scientists engaged in material research. It can also provide a continuous range of wavelengths to test layered materials. The diffraction techniques are powerful tools to study such materials and are continuously developing around synchrotron radiation sources.

15.2 Reflectivity Measurements and other Tests

The first step of our collective work has been to adapt diffraction systems, to learn more about the structure of layered materials, and simultaneously to develop the layering evaporators. Our main instrument for layering material studies is an existing high-precision x-ray goniometer ($\lambda = 1,54$ A) at the Institut d'Optique in Orsay [15.1]. An example of such test at the beginning of this study is shown below in Fig.15.1 using a multilayer kindly provided by T.W. BARBEE. This measurement revealed the qualities of sample layering by the high-precision angular and flux measurements achieved over a large range of incidence angles, including the total reflection range.

Figure 15.2 compares the rectangular profile, assessed by supposing a perfect W/C interface and bulk electronic density, with the quasi-sinusoidal

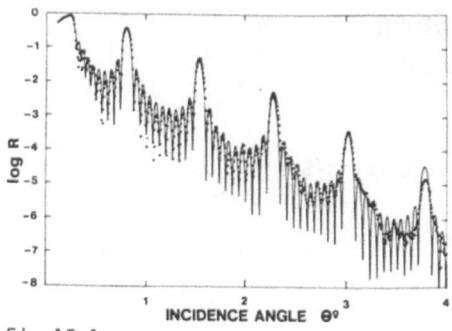

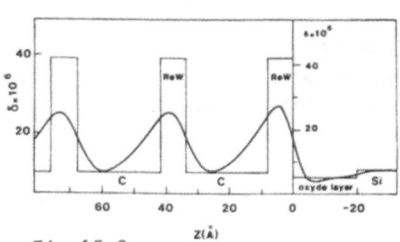

Fig.15.1   Fig.15.2

density fitting the experimental points. This does not permit a distinction between interdiffusion and nucleation contributions. A second similar, but less precise, x-ray goniometer has been built recently and is being used for preliminary tests of samples provided by the different participants in the programme.

Some commercial Θ-2Θ goniometers, using soft x-ray fluorescence lines, placed close to the evaporators are used. A dedicated windowless x-ray tube goniometer is also being constructed at the Laboratoire de Chimie Physique in Paris.

Reflectivity tests with the ACO synchrotron light were performed between 8 to 300 A in an existing x-uv reflectometer [15.2,3] which can be set behind a double crystal or grating incidence monochromators. Figure 15.3 is an example of such tests, obtained from the same multilayers as used for Fig. 15.1. In this case we have taken into account the polarized nature of the light from a toroidal grating incidence monochromator. So for two perpendicular directions of the incidence plane, we have measured the band pass of the multilayer for a set of incidence angles, as indicated on the figures. The largest difference in peak reflectivity, between the two sets of measurements, appears at 45° as predicted by the diffraction theory, taking into account the effect of polarization. This directly demonstrates the possibility of using multilayers as a linear polarizer throughout the x-uv range, if we are able to fabricate the right 2d needed. This new possibility was recently successfully used to test the circularity of x-uv emitted at 130 A by a helical undulator [15.4]. It is necessary to note that we must be careful with the comparison of theoretical and measured absolute reflectivity, specially around 45° when partially polarized flux is used.

We must also mention the new tests in progress at Marseille University. One of the groups is using a scanning transmission electron microscope on small wedges of W/C multilayers prepared by several sputtering methods. Good images of layers, within the limit of the resolution of the electron microscope used, were obtained [15.5]. In a next step, the periodicity of the layers was determined from electron diffraction patterns of the edges [15.6].

The 2d values deduced from these studies agree well with those deduced from a 1.54 A reflectivity test and confirm that the diffraction is from the artificial layering period. Further investigations on the merit of this new method are in progress.

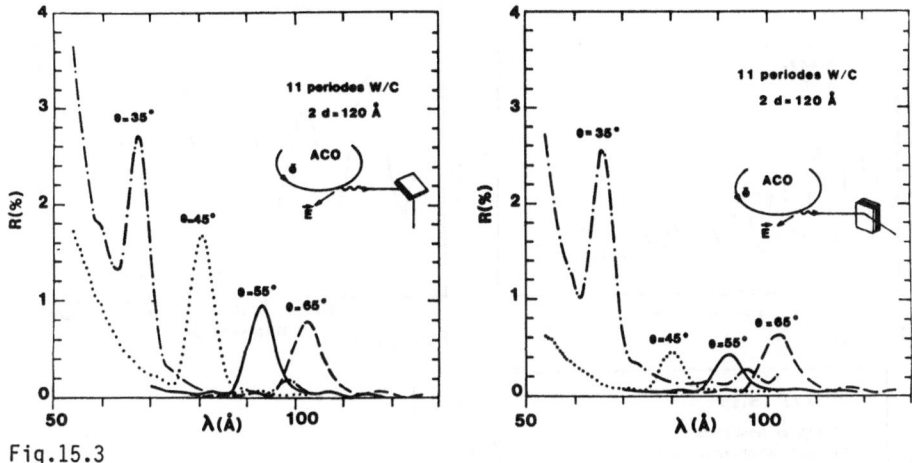

Fig.15.3

In many applications of layered materials the importance of the role of the interface between the layers is obvious and has attracted attention of several research workers in Orsay. Multilayered structures of the alternate C and W films, a few nanometers in thickness, have been analysed in situ by Auger electron spectroscopy. A special device performing the analysis during continuous layer deposition by electron guns is developed to obviate the lack of precision of conventional depth profiling by sputtering removal of very thin films. Comparison between computer simulation and experimental curves of Auger amplitude variation give information about the growth process [15.7].

15.3 Evaporators

As for the other aspects of this work, different evaporators are being developed in the different laboratories and to date layered structures have been obtained by electron guns, diode, triode or magnetron sputtering and tested by 1.54 A reflectivity.

An electron bombardment evaporation unit containing an in situ x-ray reflectometer, like the first one developed by E. SPILLER, has been designed at the Institut d'Optique in Orsay. Some improvements, such as the free choice of the incidence angle from 20⁰ to 60⁰, have been brought to the original design. By employing this set-up and different x-ray lines we were able to prepare a large range of 2d multilayers. The effect of temperature on the roughness and diffusion of the multilayers can be controlled by varying the temperature of the substrate holder. Figure 15.4 shows an example of reflectivity variation obtained during the continuous evaporation of a Si layer and in Fig.15.5 we have given a comparison of measured and calculated reflectivity variation. We have used for these calculations Henke's optical index values for bulk Si and suppose that the growth of the layer was perfect.

Figure 15.6 is a comparison of reflectivity variation calculated at the end of each optimised layer, shown by triangles, and the measured gain observed, indicated by circles. Multilayers with desired fractional values of β have been prepared and the ratio obtained has been checked by the observed extinction of the predicted Bragg peaks in the 1.54 A reflectivity tests.

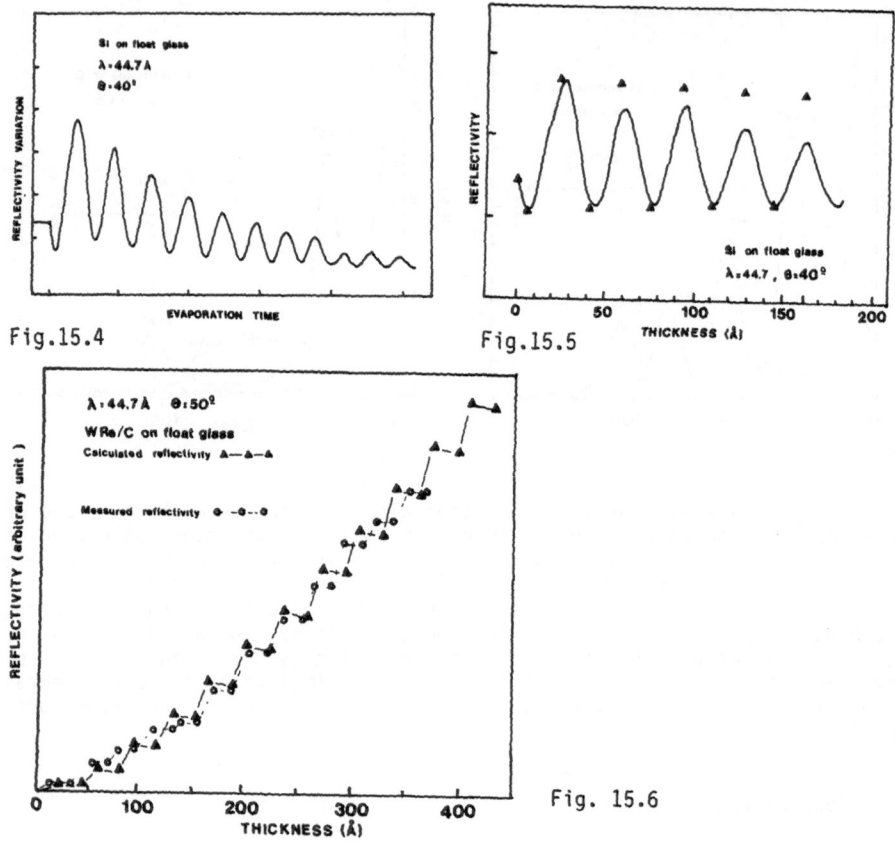

Fig.15.4

Fig.15.5

Fig. 15.6

The interface qualities obtained by different ion bombardment evaporators are being tested specially for the short periods and large number of layers. RF diode sputtering is used in Marseille [15.8] and for the first tests a magnetron sputtering at the LAAS in Toulouse has been kindly made available to us for starting experiments. These two evaporators use a rotating sample holder and fixed cathodes, similar to T.W. BARBEE's evaporator.

A triode sputtering is also adapted to layering at the CNRS laboratory in Bellevue. It has flip-flop cathodes and motionless substrate holder. With this motionless substrate it is planed to develope in situ thickness and test measurements in view of fabricating multilayers with graded index period. Such multilayers seem interesting, especially in the synchrotron field where the suppression of annoying higher orders is needed.

At the LEP (Philips France) other diode and triode sputtering methods are in progress where elipsometry control will be used.

15.4 Theoretical Reflectivity Predictions

Several participant groups are developing different formalisms stemming from their own specialities. Certain formalisms arise from dielectric multilayer calculation and in situ band pass measurements in the visible [15.9] or from reflectivity measurements on single layers or surface oxidation [15.10].

Other calculations have been done specially to study reflectivity in the anomalous region on each side of the atomic threshold of one of the layered elements [15.11]. Formalisms closer to the x-ray diffraction are also developed.

Comparison between results obtained by these different ways and the experimental results help us to test the precision and the validity of each formalism and to solve specific questions. The frequently used Debye-Waller factor and its limit of validity to describe the observed reflectivity is being tested. The use of the transition layer theory is also in progress to model different roughness and diffusion cases. Attention is also given to slight irregularity in the periodicity and roughness.

One of the French computer centers has been chosen to stock and to centralize programs useful to all groups. The same data bank has an x-uv optical index table and test results obtained from different equipment . This computer center is also used by the different groups when calculations are too long for local desk computers.

15.5 Conclusions

The joint program organized in France to fabricate multilayers is a means to have permanent collaboration with two kinds of workers. The first kind are the specialists having developed techniques useful for layering and who are ready to give only partial time. The others are the main potential users. This facilitates all participants to be in direct contact with the mutual problems in x-uv optics.

Informal meetings, at two or three months' interval, help us to exchange information and to coordinate the activities; each participant being free to develop his own application.

The use of a single computer center helps us to a quicker comparison between predictions and tests achieved in the different participant laboratories. The principle of free access to all information and useful programs also leads us to develop a common language.

References

15.1 P. Croce and L. Nevot: Revue Phys. Appl. 11, 113 (1976)
15.2 M. Berland et al.: SPIE Vol. 315 (1981) Reflecting Optics for Synchrotron Radiation pp. 155-159
15.3 M. Berland et al.: SPIE Vol. 316 (1981) High Resolution Soft X-Ray Optics pp. 169-172
15.4 E.S. Gluskin et al.: Abstract Nb 60 pp 29 - 3rd Nat. Conf. on Synch. Rad. Instrumentation - Brookhaven (USA) Sept. 1983
15.5 Y. Lepetre and A. Charai: Thin Solid Films 105, 71 (1983)
15.6 Y. Lepetre and A. Charai: To be published in Thin Solid Films
15.7 J.P. Chauvineau: To be published in Thin Solid Films
15.8 Y. Lepetre, L. Nevot and B. Vidal: Proc. 4th Col. on Plasmas and Cathodic Pulverisation, Nice, Sept. 13 (1982) in Vide, Couches Minces, Suppl. 212, 303 (1982)
15.9 B. Vidal, A. Fornier and E. Pelletier: Appl. Optics 17, 1038 (1978)
15.10 L. Nevot and P. Croce: Revue Phys. Appli. 15, 761 (1980)
15.11 R. Marmoret and J.M. Andre: Appl. Optics 22, 17 (1983)

16. Multilayers for X-Ray Optical Applications

T. W. Barbee, Jr.

Materials Science and Engineering, Stanford University
Stanford, CA 94305 USA

16.1 Introduction

Optical elements encompass a wide range of devices and structures all directed to the efficient manipulation of light so that specific physical ends may be achieved. Advances in several areas over the past decade have presented both the scientific and technological communities with the prospect of a new generation of optics for the x-ray and VUV. These advances have required significant technological effort resulting in a degree of microstructural control not heretofore attainable.

In this paper one particular class of microstructures is considered: Layered Synthetic Microstructures (LSM). LSM based optic elements are direct analogues to standard or classical optic elements with the restriction that absorption must be considered. A short review of methods for LSM synthesis is presented first in this paper. The general nature of the behavior of LSM's as x-ray dispersion elements is then considered and a summary of maximum predicted reflectivities over the spectral range 100 to 10,000 eV is reviewed. Results for specific optic devices are then presented and simple extrapolations to new and unique optic elements made.

Layered synthetic microstructures are manmade [16.1] periodic layered structures of high enough quality to be considered synthetic crystals. Layers of materials A and B having a significant difference in their scattering powers for x rays and uniform thicknesses t_a and t_b are combined to form a sample of uniform period d = t_a + t_b in the Z direction. The layers are numbered in the order of deposition starting at the substrate with the A layer closest to the substrate, i.e., substrate, A1, B1, A2, B2, . . . An, Bn, where n is the number of layer pairs deposited. In such a structure the ratio Γ is defined as the thickness of high scattering power layer divided by the period of the LSM.

Layer thicknesses may be varied in several ways during synthesis. The thicknesses of A layers can be varied linearly with the depth in the LSM but held uniform in the X-Y plane. Layer B is uniform in the depth and in the X-Y plane. Note that the thickness of layer A may be largest at the substrate or at the LSM surface (i.e., $t_{A1} > t_{An}$ or $t_{A1} < t_{An}$). Generally, the layer thicknesses can be independently varied in depth in an LSM. Component layers in an LSM may also be uniform with depth Z but vary in thickness in the X-Y producing a wedge crystal.

As discussed further in the following pages, many factors determine the character of the LSM response to an incident spectrum. The important parameters are the substrate quality relative to atomic roughness and figure, the thicknesses of the component layers, the x-ray optical constants of the component elements, the number of layers in the structure, the interface width or abruptness in atomic position and composition, and the interface roughness.

16.2 Synthesis Processes

Synthesis processes applied to LSM x-ray optic multilayers are based on physical vapor deposition using sputter [16.2,3] or thermal [16.4,5] sources and moderate vacuums. This is an atom by atom synthesis process approach that allows, in a limited sense, atomic engineering. The objective is to control the basic growth process so that uniform layers varying in thickness from one monolayer to hundreds of atomic layers may be reproducibly deposited. In the case of thermal sources the substrates are generally stationary and individual layer thickness is determined by shuttering, with shutter control activated by thickness monitors. The most effective monitoring system appears to be an in situ reflectometer [16.4] using a characteristic emission line at fixed angle of incidence to the substrate. The interference maxima and minima observed as layer thickness increases are used to determine thickness and in fact, optical thickness at the wavelength used. Additionally, given an independent determination of thickness, it is possible to characterize the deposit as to optical constants and the surface of the growing film as to roughness. Accurate determination of film thickness is the most effective control mechanism since the exponential temperature dependence of the vapor pressure of the component materials limits the control of evaporation rate from thermal sources.

In sputter deposition synthesis moving substrates [16.2,3] have been used with both traditional glow discharge and enhanced sources. The most effective work appears to have been performed using magnetron sources and moving substrates. Substrates are typically mounted with the deposition surface parallel to the sputter source plane, the center of the substrate passing under the source center. Many substrates are mounted in a synthesis experiment and approximately 90 cm^2 of uniform multilayer is deposited at one time.

Individual layer thickness is determined by the excitation of the sputter source, the source/substrate geometry (examples are given later), the velocity of the substrate past the source and the sputter gas pressure in the chamber. With laboratory research systems typical deposition rates are greater than 1 A/sec onto moving substrates so that a 20 A period 100 layer pair sample requires 2000 sec ($\sim$34 minutes) to deposit. The thinnest layers that can be effectively deposited are approximately 5 A or 2 atomic diameters thick.

I also note that laser evaporation [16.6] has been developed for the synthesis of multilayer structures for x-ray optic applications. Preliminary results indicate this may be an effective approach allowing work with unique materials which would not be usable with both sputter and thermal sources.

16.3 Component Materials and Reflectivity Dependence on Optical Properties

In quite general terms there is now cause to project that LSM's may be synthesized using a large fraction of the ninety-two naturally occurring elements for technical purposes. In many cases the layers would be in the form of compounds and alloys. A recent compilation [16.7] of the optical constants of the elements over the spectral range 100 to 2000 eV has made pos-

sible modeling calculations which allow optimization of LSM structural parameters for a given set of component elements and spectral ranges.

The optical constants are defined in terms of a complex refractive index
given by:

$$n_A = 1 - [\delta_A + i\beta_A] = 1 - \frac{r_o \lambda^2}{2\pi} N_A (f_{A1} + if_{A2}) \tag{16.1}$$

where r_o is the classical radius of an electron, N_A is the volume concentration of A atoms, λ the wavelength, and f_{A1} and f_{A2} the atomic scattering
factors. Tabulated values [16.7] of f_{A1} and f_{A2} are used to calculate the
reflectivity. Note that δ_A is directly related to the scattering power of a
given element, and β_A to its linear or mass absorption coefficient.

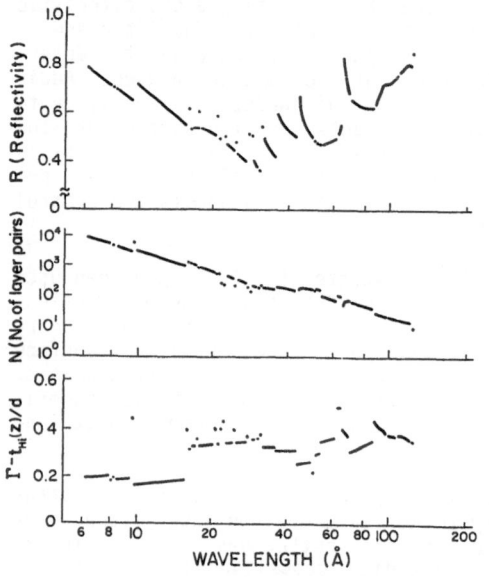

Fig. 16.1 Calculated optimum normal incidence peak reflectivity R,
the number of layer pairs N necessary to achieve R, and Γ, the ratio of the thickness of the Hi(Z)
scattering layer to the LSM period
are plotted as a function of diffracted light wavelength (6.2
A $\leq \lambda \leq$ 124 A). These calculations
were performed by A. ROSENBLUTH
[16.8] who used the optical constants reported by HENKE et al.
(16.7) for photon energies of
100 eV $\leq$ E $\leq$ 2000 eV

A summary of optimized normal incidence reflectivities calculated by A.E.
ROSENBLUTH [16.8] using tabulated optical constants for elements Z > 5, reported by HENKE et al. [16.7] is shown in Fig.16.1 for the spectral range
6.2 to 124 A. In Rosenbluth's calculations the materials were selected specifically on the basis of peak reflectivity so that the discontinuities
shown as a function of wavelength represent absorption edges at which discontinuities in the optical constants are observed. The predicted reflectivities are surprisingly large over this full spectral range - between 40 and
80%. Also shown in Fig.16.1 are the number of layer pairs N needed to attain
the reflectivity (R) of figure A at a given wavelength. It is also necessary
to optimize the relative thicknesses of the component materials Γ, which is
plotted in Fig.16.1C. Note that Γ has questionable physical meaning for
wavelengths less than approximately 20 A. From Bragg's equation we know
d $\simeq \lambda/2$ for normal incidence reflection. Therefore, for optimum reflectivity
$t_{Hi}(Z) < 2$ A for $\lambda < 20$ A, a value smaller than typical interplanar spacings
of most solids. Irrespective of this simple criticism, the important result
in this work is that surprisingly large reflectivities are attainable

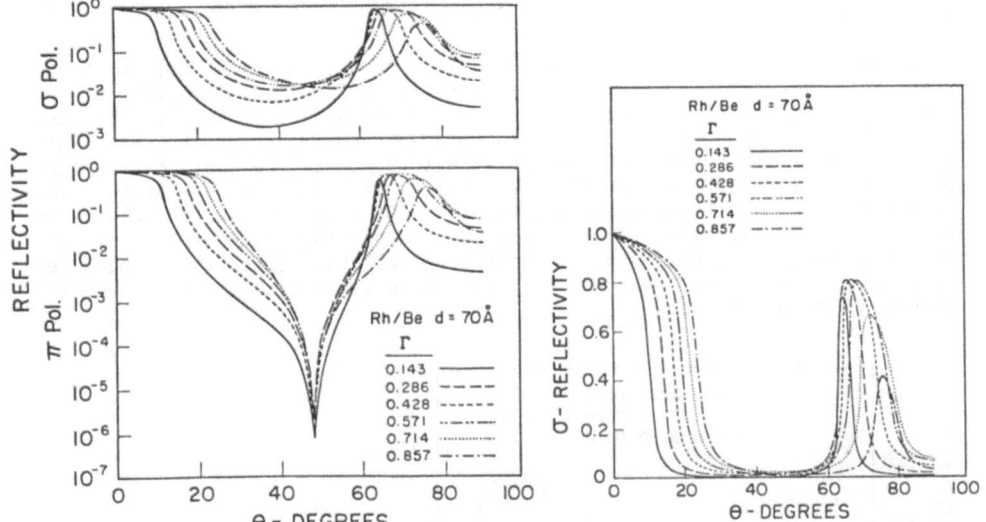

Fig.16.2 The reflectivities for 124 A light of six model (d = 70 A, N = 300) rhodium/beryllium LSM's having $0.143 \leq \Gamma \leq 0.857$ over the angular range $0° \leq \Theta \leq 90°$ are shown. π and σ polarizations are plotted separately. Optical constants are those of HENKE et al. [16.7]

Fig.16.3 The reflectivities for 124 A σ polarized light of the six Rh/Be model samples of Fig.16.4 are linearly plotted for $0° \leq \Theta \leq 90°$

throughout this spectral range, assuming the optical constants used in the calculations are correct. Also, the resolution (ΔE) ranges from 1 to 10 eV for 20 A $\leq \lambda \leq$ 124 A ensuring high throughput for LSM optics, but limited resolution.

The reflectivities of a series of six hypothetical Rh/Be LSM's having a period of 70 A containing 300 layer pairs with $0.143 \leq \Gamma \leq 0.857$ over the angular range 0 to 90° are shown in Fig.16.2. The sigma and pi polarization reflectivities are plotted separately. Note, that ignoring dispersion shifts, the Bragg angle should be 62.3°. As is obvious from these curves the position of the Bragg peaks were shifted to angles of 65 to 75 degrees as Γ varies from 0.143 to 0.857 due to refraction effects. Resolution also varies in a systematic manner, being largest for the lowest Rh layer thickness. Reflectivity is affected by Γ, being maximum (> 80%) for $\Gamma \simeq 0.4$ in accord with optimization analysis [16.9,10] reported by other authors. As is known, LSM's act as polarimeters, the reflectivity of pi polarized light falling to very low levels at the Brewster angle, typically taken to be 45°. In this case refraction shifts the Brewster angle to approximately 48°. It is also important to note that the critical angle Θ_c for specular reflection increases from approximately 10° for $\Gamma = 0.143$ to 24° for $\Gamma = 0.857$.

Though much of the physics of LSM's is shown in Fig.16.2, the point of this figure is the magnitude of reflectivities predicted (> 80%) and the range of Γ for which this can be approached. This is better demonstrated in Fig.16.3 where the reflectivity of sigma polarized light is plotted as a function of Θ for the Γ values 0.143 to 0.857. All the same effects described in the previous paragraph are seen but the range of Γ for which $I/I_0 >$ 80%

is more clearly demonstrated, $0.29 < \Gamma < 0.55$, showing that a substantial range in rhodium layer thickness is accessible.

Normal incidence reflectivities for 124⁰ radiation by Rh/Be, Rh/B and Rh/C LSM's containing 300 layer pairs is shown in Fig.16.4. The peak reflectivities vary from 40 to 74% in these calculations, being largest for the sample having the lowest absorption coefficient spacer component, Be. This is also the lowest energy resolution structure, since the scattering power per layer pair is largest for this combination of materials. Therefore, low resolution or a limited number of LSM periods is necessary in the soft x-ray regime so that absorption effects are minimized. Other means for achieving high resolution must be found. Irrespective of this limitation, these reflectivities are large, exceeding by far those typical of optics currently used in this energy range.

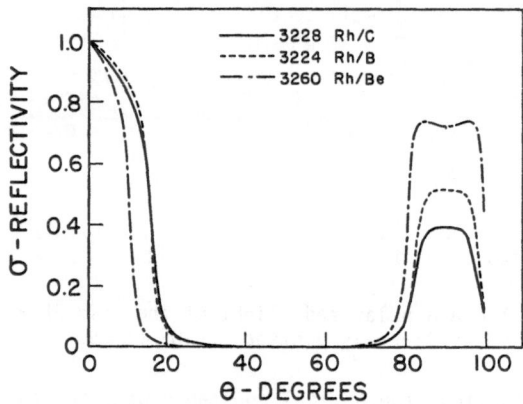

Fig.16.4 Normal incidence reflectivities for 124 A σ polarized light of Rh/Be, Rh/B, and Rh/C LSM's ($\Gamma = 0.84$, $d \simeq 62$ A, $n = 300$) are plotted as a function of Θ ($0^0 \leq \Theta < 100^0$)

16.4 Substrate Roughness and Figure

Use of LSM's as dispersion elements does not necessarily require high optic quality. Applications in which focussing is envisioned, including imaging, are limited by optic quality in the same manner that normal optics are affected. Such applications are x-ray microprobes, broad band pass high throughput optics, and high flux insertion device optics. It is also possible to envision graded period LSM's which will allow focussing of monochromatic light through large angles, as well as unique dispersion element structures.

In the case of LSM optics primarily two sources of error must be considered. The first is the quality of the substrate upon which the multilayers are deposited. Here, two effects are of importance: the figure and the substrate roughness. The second type of error has to do with the nature of the interfaces beween LSM constituent layers as well as the uniformity in thickness of the layers.

The optic quality of multilayer elements has been investigated in three reported studies, which have demonstrated that the primary error is in sub-

strate quality. J.P. HENRY et al. [16.11] have tested an LSM normal incidence astronomical telescope mirror reporting a resolution of 1 arc sec.
They indicate this resolution was limited by the spatial resolution of the
detector and by geometrical constraints in the test equipment. More recently
D.H. BILDERBACK et al. [16.12] tested the optic quality of LSM's deposited
onto commercial single-crystal silicon wafer surfaces and as received substrates, concluding that substrate defects resulting in slope errors > 10
arc sec's were the limiting optic defect. There was no evidence in this work
that the LSM's degraded the optic quality of the light beam. This indicates
that LSM coatings mimic the undulations of the substrate so that substrate
quality must be substantially improved, for applications requiring high optic quality.

A similar conclusion is reached by WARBURTON et al. [16.13] in a paper
presented as part of this conference (no.17). In these experiments diffracted beam monochrometers Si (220) having a measured 4.05 sec F.W.H.M.
($E/\Delta E \simeq 2\times10^4$) at 8 keV were used. It was possible, with this experimental
arrangement, coupled to either a scatter slit with white light incident or
to an incident light monochromator, to demonstrate that substrate defects
are limiting LSM performance and that nonspecular scattering appears to be
small. These results are discussed further in the referenced papers in this
volume.

More recently preliminary results for a Kirkpatrick-Beaz x-ray microscope
[16.14] designed to operate at 4.75 keV have been reported. The microscope
uses a 2-deg grazing angle configuration, the mirrors (6A rms roughness;
16.45 m radius cylindrical quartz substrates) are coated with tungsten/car-
bon LSM's having periods of 38.4 and 40.8 A each containing 50 layer pairs.
LSM reflectivities at 8keV are 65 to 73% absolute for LSM on the quartz substrates and single-crystal silicon substrates coated at the same time. The
microscope band pass is experimentally observed to be 250 eV (5.32%) centered at 4.75 keV in agreement with design. Preliminary imaging experiments
have demonstrated a resolution better than 5 µm which is to be compared to
the design value of 3 µm. Further characterization is underway.

The primary conclusion of this work is that the optic quality of substrates is singularly important in determining the optic quality of LSM based
reflecting optics. Evidence at this point indicates that for LSM's having
individual layer thicknesses greater than 7 A, this substrate characteristic
will control the LSM behavior for applications requiring high optic quality.
It is also possible that substrate quality dominates the response of smaller
period LSM's (d < 15 A) since the LSM acceptance angles are approximately 60
arc sec compared to substrate slope errors of greater than 10 arc secs for
small surface areas, and considerably more over large surface areas.

The preceding discussion centered on substrate figure effects. Substrate
roughness effects are also important. If the substrate roughness is a large
fraction of the period to be synthesized [16.13] it is clear that significant errors in the definition of the angle of incidence of the light, as
well as of the depositing species will be made. This will cause a mosaic structure type response of the scale of the facetting due to roughness which is
several times the wavelength of the incident light and the period of the
LSM. When roughness is of atomic dimension and substantially less than the
LSM period it results in nonspecular scattering and a reduction in the
diffracted intensity. This has been of significant interest since it is
likely to limit the resolution once substrate figure limitations are removed.

16.5 Interfacial Roughness

The prior analysis provides a general framework for assessing the effectiveness of a given set of materials as LSM components for a given photon energy. If an appropriate combination of elements or compounds is chosen, structural perfection is the next important consideration. It has been shown [16.3,4] that interfacial roughness between constituent layers appears to be the primary structural error. Roughness is typically of atomic scale and assumed to be noninteracting, interface to interface. Under such conditions the reduction in reflectivity for a single interface [16.15,16] caused by an rms Gaussian roughness ΔZ is a good approximation and is given by:

$$\frac{I}{I_0} = \frac{I}{I_0}\Big|_T \cdot \exp\left[-\left(4\pi \frac{\sin\Theta}{\lambda} \Delta Z\right)^2\right] \tag{16.2}$$

where $I/I_0|_T$ is the reflectivity calculated assuming ideal interfaces. Rearrangement and use of Bragg's equation yields:

$$\frac{I}{I_0} = \frac{I}{I_0}\Big|_T \cdot \exp\left[-\left(2\pi n \frac{\Delta Z}{d}\right)^2\right] \tag{16.3}$$

where n is the order of the Bragg reflection from an LSM of period d. This explanation has been investigated by means of curve fitting [16.3] and by study of single surface roughness [16.4]. In Fig.16.5 I/I_0 (n = 1) for a series of tungsten/carbon LSM's designed to have the same $I/I_0|_T$ is compared to calculations made assuming ΔZ = 3.15 A. The agreement is semiquantitative, but sufficiently good to support the hypothesis that interfacial roughness is a critical structural parameter. These data [16.3] were taken in 1981. A more recent datapoint [16.12] (Sample 82-010; d = 21.6 A) exhibited a reflectivity for 8 keV photons of 68% compared to a calculated value of 84%. This can be rationalized by an interfacial roughness $\Delta Z \simeq 1.5$ A, a

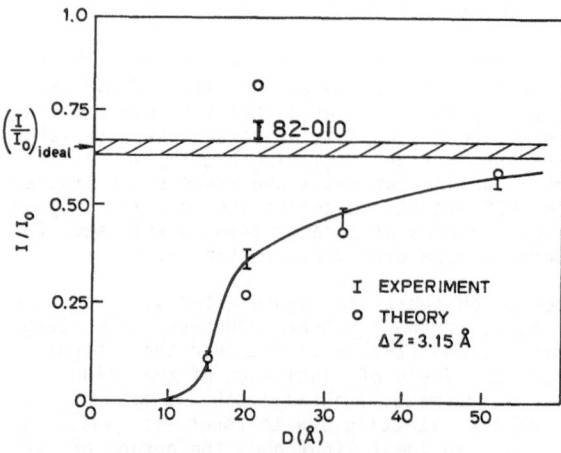

Fig.16.5 The observed peak reflectivity in first order for a series of W/C LSM's, for which the product $N^2 d^4 (f_A - f_B)^2$ has been maintained constant, is compared to model predictions assuming an interfacial roughness ΔZ = 3.15 A. Also shown is data for a W/C LSM (82-010) which has substantially improved performance, indicative of improved synthesis procedures

substantial improvement over earlier efforts. Therefore for 20 A period or larger it is routinely possible to synthesize W/C LSM's which, for hard radiation, perform at 80% of calculated reflectivities.

16.6 Stability

Most data relating to the long-term stability of LSM x-ray optic elements is the result of annealing studies of tungsten/carbon [16.17] or tungsten-rhenium/carbon LSM's. My experience with W/C LSM's is that once synthesized they are stable under ambient conditions for periods in excess of two years. The accuracy of measurement is 0.1A for LSM's of period 20 A < d < 50 A.

Isothermal annealing at 500°C in vacuum results in changes in reflectivity over a period of three hours with no further discernable changes for a period of up to 15 hours. In Fig.16.6 reflectivities of two samples of LSM 81-118 and 81-118 A (10 hours at 500°C) are plotted for CuK$_\alpha$ radiation (λ = 1.5418 A). Four important changes upon annealing are observed in this figure. First, the reflectivity of the annealed sample in first order increases from 70% to 83%. Second, the reflectivity in second order is greatly increased for the annealed sample (from 8% to 28%). Third, the period of the annealed sample is approximately 10% larger than that of the unannealed sample. Fourth, the fine structure of the secondary Fresnel oscillations are much better defined and of higher amplitude for the annealed sample relative to the unannealed sample. The four effects of annealing can be qualitatively explained by smoothing of the interfaces between the tungsten and carbon (assuming compositional abruptness) or by increased compositional abruptness (assuming smoothness) at the W/C interfaces, and/or by increased uniformity of the layers so that variations in layer thickness are decreased. Although it is not clear which of these mechanisms is controlling, it seems likely that

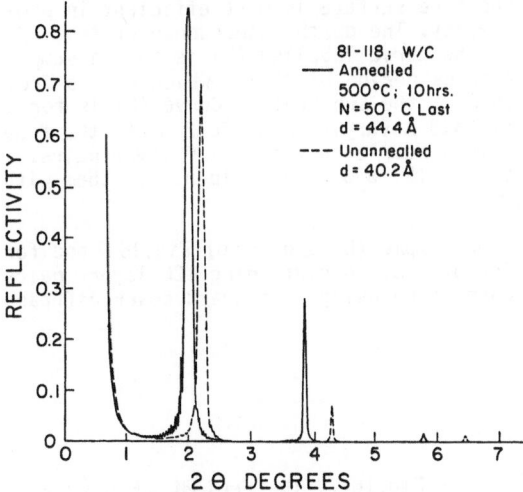

Fig.16.6   Measured reflectivities for CuK$_\alpha$ radiation (λ = 1.5418 A) for LSM 81-118 (W/C, N = 50) in the as synthesized (d = 40.2 A) and annealed (P < 2 x 10^{-6} Torr, T = 500 °C, t = 10 hrs, d = 44.4 A) are plotted as a function of 2 Θ through the third-order Bragg peaks. Note the dramatic increase in reflectivity in second order (6% to 28%) upon annealing

interfacial roughness must be strongly decreased. A simple argument based on
(16.3) indicates that the interfacial roughness should decrease from 2.8 A
for the unannealed sample to less than 1.1 A upon annealing. Detailed examination of the (110) tungsten reflection from the tungsten layers showed that
order in the individual tungsten layers was substantially increased by annealing. Also, ReflEXAFS measurement indicates that the order in the tungsten
layers is greatly increased by annealing. Irrespective of the specific mechanism W/C LSM's improve in reflectivity upon annealing for up to 3 hours at
500°C and are stable for subsequent annealing of up to 12 hours more at 500°C.
I note that similar studies are underway with other combinations of component
layers.

16.7 Graded Period LSM's

It has been experimentally demonstrated [16.18] that both depth graded period and laterally graded period LSM's may be synthesized using sputter deposition techniques. Depth grading has been proposed as a means for increasing the integrated reflectivity [16.4,18,19] of a particular LSM and lateral
grading as a means for focussing [16.20] and fixed sample spectrometers
[16.18]. In this section both depth graded and laterally graded multilayers
will be described and optic applications outlined. Particular emphasis will
be given to combining these technologies with other microstructure technologies to form a new and interesting class of optics.

Depth graded LSM's may be synthesized by varying the thickness of a given
component layer with depth or by varying the thickness of both component
layers with depth. This may be done either continuously or discontinuously
with the period having its largest value at either the substrate surface or
the free surface of the LSM. FUKUHARA and TAKANO [16.21] demonstrated with
boron gradient doped silicon that a negative gradient (substrate to free
surface) with a smaller period at the free surface is most efficient in producing increased integrated reflectivity. The depth dependence of two W/C
LSM's periods is shown in Fig.16.7. The curve labelled (B) is for a sample
containing 40 layer pairs with an average period of 25.5 A which varies from
24.1 A at the free surface to 26.9 A at the substrate. Curve (C) is for a
W/C LSM having an average period of 24.3 A varying from 26.2 A at the free
surface to 22.4 A at the substrate. This sample contained 120 layer pairs. I
note here that the curves are labelled (B) and (C) to identify them with
data presented in Fig.16.8 and 16.9.

Experimental diffraction curves for sample (B) and (C) of Fig.16.7 and for
a uniform period LSM having a period of 28.5 A containing 120 layer pairs
are shown in Fig.16.8. These data were taken using a standard power diffrac-

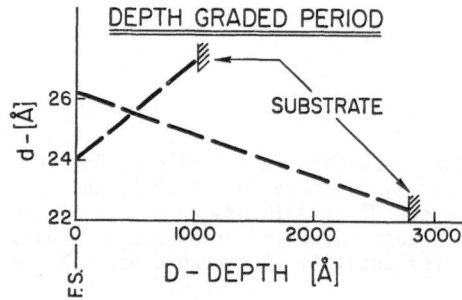

Fig.16.7 The periods of two Depth
Graded Period LSM's are plotted as
a function of depth from the LSM
free surface into the structure.
These two samples represent the
specific cases: $d_{F.S.} > d_{sub}$;
$d_{F.S.} < d_{sub}$

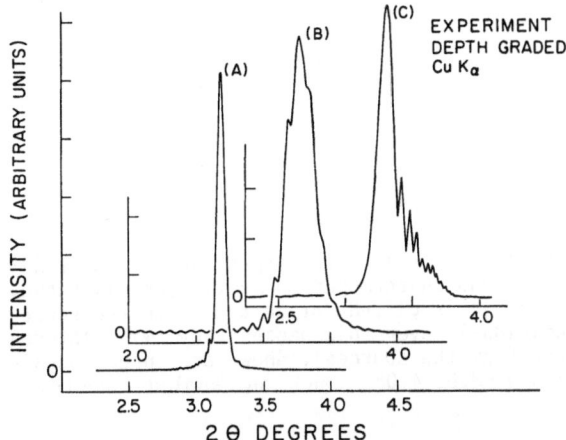

Fig.16.8 Experimental Θ/2Θ scans taken using CuK$_\alpha$ radiation (λ = 1.5418 A) for a uniform period (Curve A) W/C (81-205; d = 28.54 A, N = 120), for a depth-graded (Curve B) W/C LSM (81-207; d = 25.51 A, Δd/d $\simeq$ 0.1, N = 40) and for another depth-graded (Curve C) W/C LSM (82-145; d = 24.4 A, Δd/d $\simeq$ 0.1, N = 120) are shown. The curves are normalized to have the same intensity. The observed peak reflectivities are 81-205 (Curve A) I/Io = 70%; 81-207 (Curve B) I/Io = 32%; and 82-145 (Curve C) I/Io = 52%

tometer equipped with a diffracted beam monochromator and Cu - K$_\alpha$ radiation. Observed peak reflectivities are (A) 70%, (B) 32%, (C) 52%. Reflectivity curves calculated using a standard Fresnel code are shown in Fig.16.9. The general behavior of the reflectivity is certainly well reproduced by the calculations though the calculated reflectivities are all somewhat larger: (A) 85%, (B) 37% and (C) 58%. Curves 2B and 3B show results for a sample which is largely kinematic while 2C and 3C show results for a sample which is in the dynamic range with both extinction and absorption limiting the reflectivity at higher Θ from the smaller period regions deeper in the sam-

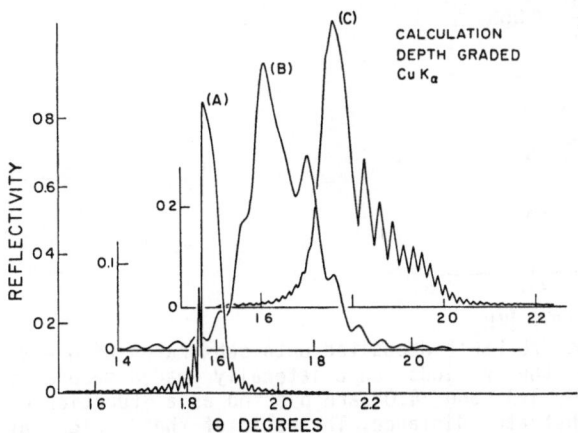

Fig.16.9 Calculated Θ/2Θ scans for CuK$_\alpha$ radiation for structures corresponding to those for which experimental curves are shown in Fig.16.8

LATERALLY GRADED PERIOD

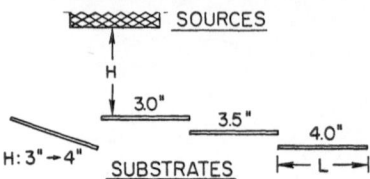

Fig.16.10 The experimental synthesis arrangement for deposition of laterally graded period multilayers and uniform period multilayers is shown. The laterally graded sample is placed at an angle to the source plane so that the individual layer thicknesses vary with lateral position (i.e.,distance from the sources). Shown are three parallel substrates at H = 3.0, 3.5, 4.0" and an angled substrate $3" \leq H \leq 4"$

ple. It is concluded from this work that depth graded LSM's can be synthesized and exhibit diffraction behavior in accord with calculation.

Laterally graded period LSM's may also be synthesized using sputter deposition techniques. The deposition rate varies inversely as the square of the source to substrate spacing (i.e.,as $1/H^2$). Therefore, if a flat substrate is placed at an angle to the plane of the deposition source the deposition rate and film thickness vary with position along the substrate. Such an arrangement is seen in Fig.16.10 where an array of four silicon single-crystal substrates is schematically shown. Three were parallel to the sputter sources (and laterally uniform), situated at H=3, 3.5, and 4". One substrate was at an angle to the source, 3" < H < 4", over the three inch diameter of the wafer. These LSM's contained 120 tungsten/carbon layer pairs and were characterized using CuK_α radiation in a standard powder diffractometer equipped with a diffracted beam monochromiter.

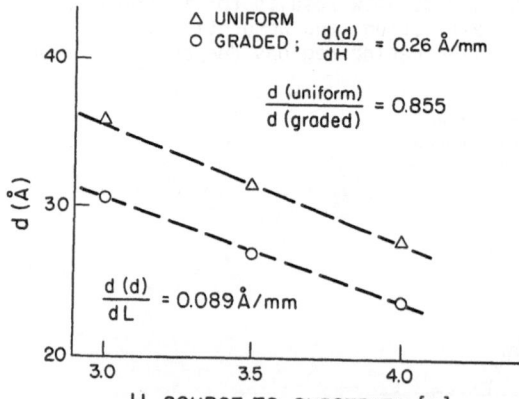

Fig.16.11 The periods of three W/C LSM's deposited onto substrates at H = 3, 3.5 and 4.0" and the periods on a laterally graded sample at points where H = 3.0, 3.5, and 4.0" are plotted as a function of H, the source to substrate distance. The ratio of the period, at fixed H, of the laterally graded to the uniform samples is 0.855. The lateral gradient in period is 0.089 A/mm

In Fig.16.11 the measured periods of the uniform samples (3, 3.5, and 4" from the sputter sources) and the periods on the tilted sample for areas at the same distances from the sputter sources are plotted versus H, the source to substrate distance. The ratio of the period on the uniform samples to those of the graded sample was 0.855, primarily as a result of the reduced solid angle subtended by a unit area on the graded substrate relative to the uniform substrate. The gradient in period along the graded sample was measured to be 0.089 A/mm, the period varying from 23.8 to 30.3 A. Reflectivities were measured for all four samples and were greater than 70% for CuK_α radiation.

As has been pointed out by several authors, such laterally graded period samples represent a unique optic element. Such gradient LSM's coupled with slight substrate curvature allows focussing and light gathering capabilities better than those currently available. Further, a fixed detector coupled to linear motion of a laterally graded period LSM will allow rapid spectrographic analysis of fluorescent sources. There are also specific applications in x-ray monochromators for soft x rays.

In Fig.16.12, a schematic of a fixed extrance/fixed exit position soft x-ray monochromator using a pair of laterally graded period LSM's $(d_1 < d_2)$ is shown. The two LSM's are mounted parallel with the gradient in period, d_1 to d_2 parallel to the propagation direction of the incident light. The angle chosen for this example is 45° since this allows the monochromator to operate as a polarimeter passing sigma polarized light, and specularly scattered light contamination is minimized. In operation two motions are envisioned to provide wavelength and position control.

Wavelength or energy control is attained by coupled vertical motion of the two laterally graded period LSM's as indicated by $\circledA$. The illuminated region moves along the LSM's in a coupled fashion so that regions having the same average period d are simultaneously used, though d varies with the il-

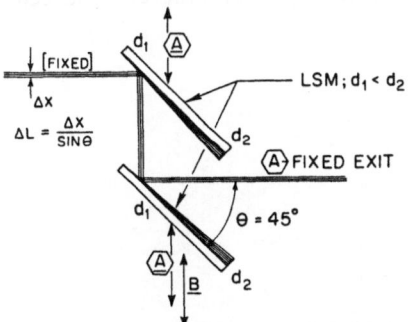

Fig.16.12 A schematic of a monochromator constructed using two laterally graded period LSM structures is shown. The substrates of these two LGP's are parallel, facing one another with the period gradients having the same sense relative to the incident light direction. $\circledA$, a vertical coupled motion, causes the illuminated zone on the LGP to move so that transmitted wavelength is varied. Motion $\circledA$ results in a fixed exit beam configuration while motion $\underline{B}$ of the lower LGP causes a vertical motion of the exit beam. The illuminated length in the LGP's is $\Delta L = \Delta X/\sin\Theta$, where ΔX is the height of the incident beam

luminated position. If the spacing between the two LSM's is held constant then the relationship between the entrance and exit beams is held constant (i.e. a fixed exit beam). If the second graded period LSM is moved while the first is held fixed, the vertical position of the exit beam is varied, allowing controlled scanning in one direction.

Attainable characteristics for such laterally graded period LSM's are $d_1/d_2 > 0.5$ which would allow an energy range of approximately a factor of two to be scanned. Energy resolutions attainable will all probably be in the range $50 \leq E/_{\Delta E} \leq 100$. This will be limited by the range of period Δd illuminated which is related to the vertical height of the incident light beam ΔX and the angle of incidence Θ

$$\Delta d = \Delta L \cdot \left(\frac{\Delta d}{\Delta L}\right) = \frac{\Delta X}{\sin \Theta} \cdot \left(\frac{\Delta d}{\Delta L}\right) \ . \tag{16.4}$$

For a sample with $d_1/d_2 \simeq 0.5$ the expected value of $\Delta d/\Delta L \simeq 0.02$ A/$_{mm}$. Assuming $\Delta X = 0.5$ mm and $\Theta = 45^0$ the illuminated range of period will be approximately $\Delta d \simeq 0.014$ A, which, assuming the minimum value d is 15 A, is $10^{-3}d$. Therefore, resolution will be primarily defined by the resolution of the LSM's acting as if they were laterally uniform.

16.8 Solid Fabry-Perot Etalons for X Rays

A more complex LSM structure [16.22] in which two identical LSM Bragg diffractors are separated by a low absorption layer much thicker than a single period of the LSM's is exactly analogous to optical Fabry-Perot etalons. In Fig.16.13 a schematic diagram of an x-ray etalon is shown and compared to a typical solid etalon for the visible spectral region. The direct correspondence between the two LSM mirrors (LSM$_1$ and LSM$_2$) and the mirrors (M$_1$ and M$_2$) on the surface of the optical flat spacer for the long wavelength etalon is indicated. A low absorption carbon film was used as the resonance cavity in the x-ray etalon and is directly analogous to the optical flat of the conventional solid etalon.

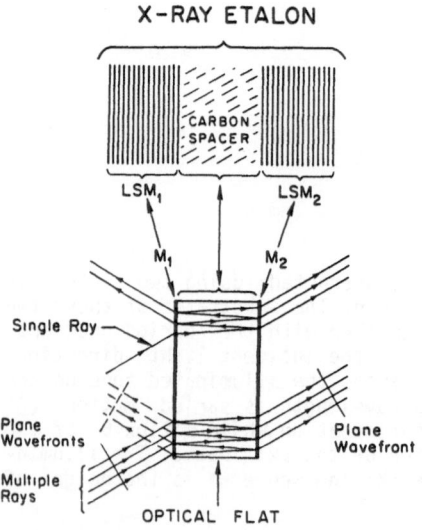

Fig. 16.13 Schematic diagrams of a solid Fabry-Perot x-ray etalon and a solid etalon for the visible spectral region are compared. LSM$_1$ and LSM$_2$, the x-ray Bragg mirrors, are directly analogous to the optical mirrors M$_1$ and M$_2$ which can be either single metal films or dielectric multilayers. The carbon spacer acts as a resonance cavity in the x-ray etalon and is directly analogous to the optical flat

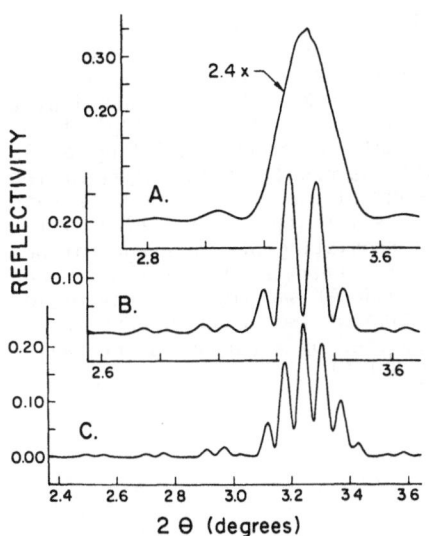

Fig. 16.14 The experimentally observed reflectivity of CuK$_\alpha$ (λ = 1.5418 A) in the region of the first-order Bragg peak of the LSM mirrors (d = 27.6 A) are plotted as a function of 2 Θ for (A) 81-205, a single Bragg mirror LSM structure consisting of 15 tungsten layers (t$_w$ = 8.5 A) separated by carbon layers (19.1 A) with carbon as the top layer; (B) 81-203, an x-ray etalon consisting of two Bragg mirrors (as in A) separated by a carbon spacer t$_{Sp}$ = 496 A thick; and (C) 81-201, an x-ray etalon consisting of two Bragg mirrors (as in A) separated by a carbon spacer t$_{Sp}$ = 981 A thick

The spectral resolution of a Fabry-Perot etalon is conventionally expressed [16.23] in terms of the fringe sharpness or finesse, F, defined as the separation between two adjacent fringes (measured in phase difference Δ) divided by their half intensity width (in Δ) . F is given by:

$$F = \frac{\pi\sqrt{R}}{1 - R} \qquad (16.5)$$

where R is the reflectivity of a single LSM mirror. Equation (16.5) gives the correct value for F if Θ and λ are held constant and t$_{Sp}$ varied. However, when scanning in angle Θ, holding both t$_{Sp}$ and λ constant (16.5) gives only a local approximation for F. Therefore, this definition for F is useful here as an approximation only and provides estimates of the upper limits of the finesse expected in our Θ/2Θ experiments. Detailed numerical computations are necessary for accurate estimates of the finesse to be made.

Experimental diffraction spectra from three samples in the region of the first-order Bragg peak of the component LSM mirrors is shown in Fig.16.14. Curve 16.14-A shows the observed reflectivity (multiplied by 2.4) from a single Bragg mirror LSM (82-005, N$_w$ = 15, t$_w$ = 8.5 A, t$_c$ = 19.1 A) identical to the mirrors in the etalons studied (Curves 16.14-B and 16.14-C). This data is plotted in the expanded form as the reflectivity of the single LSM Bragg mirror forms an envelope for the diffraction spectra of the etalons whose absolute reflectivities are larger since two LSM mirrors are active in the interference/reflection process. The value of the multiplier (2.4) was arbitrarily chosen so that Fig.16.14 (and Fig.16.15) would clearly demonstrate that the calculated diffraction reflection response of a single LSM mirror forms an envelope for the diffraction reflection behavior of the two Fabry-Perot etalon structures examined.

Curve 16.14-B shows the reflectivity of an etalon (82-003) structure consisting of two Bragg mirrors as in (A) separated by a carbon spacer cavity thickness t$_{Sp}$ = 496.6 A. Curve 16.14-C shows the reflectivity of an etalon (82-001) structure consisting of two Bragg mirrors as in (A) separated by a carbon spacer of thickness t$_{Sp}$ = 981 A. Experimentally observed finesses were approximately 2.3 for both etalons studied.

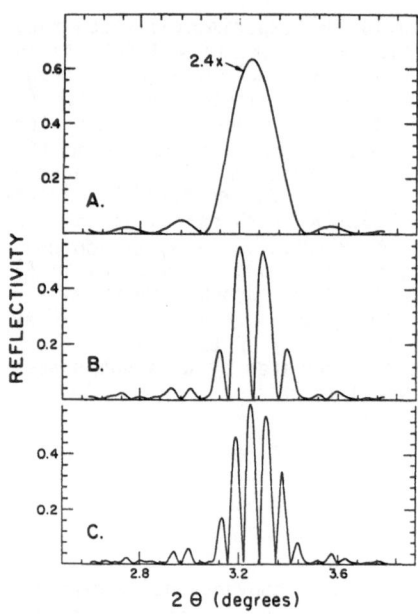

Fig. 16.15 Calculated reflectivities of CuK$_\alpha$ (λ = 1.5418 A) in the region of the first-order Bragg peaks of the LSM mirrors (d = 27.6 A) are plotted as a function of 2 Θ for (a) a single Bragg mirror LSM structure consisting of 15 tungsten layers (t_W = 8.5 A) separated by carbon layers (t_C = 19.1 A) with carbon as the top layers; (b) an x-ray etalon consisting of two Bragg mirrors (as in A) separated by a carbon spacer (t_{S_p} = 496 A) and (c) an x-ray etalon consisting of two Bragg mirrors (as in A) separated by a carbon spacer cavity (t_{S_p} = 981 A)

It is clear in this figure that the observed reflectivity of a single component LSM forms an envelope for the reflectivity of the etalons when appropriately scaled. It is also clear that the periods of LSM mirrors in these three samples are not identical, since the angular positions of the first-order Bragg peaks vary by approximately 0.088 deg in 2 Θ. The periods of the LSM mirrors in the three samples vary by approximately 1.0 A about an average value of 27.6 A. This variation is the result of synthesis process variation since these samples were deposited in three separate synthesis runs.

Calculated reflectivity curves for three LSM structures whose structural parameters match those of 82-005, 82-003 and 82-001 as described earlier are shown in Fig.16.15. The period of the LSM mirrors was held constant at 27.6A in these calculations with t_W = 8.5 A and t_C = 19.1 A, each LSM mirror containing 15 tungsten layers. Curve 16.15-A is again the reflectivity of a single LSM Bragg mirror (multiplied by 2.4). Curve 16.15-B is for an etalon structure having a carbon spacer 496.6 A thick. Curve 16.15-C is for an etalon structure having a carbon spacer 981 A thick. Resonance cavity thicknesses were chosen to give best agreement between the experimental and calculated angular dependences of the reflectivities. Calculated finesses for both etalons were approximately 2.4.

Comparison of the experimental results presented in Fig.16.14 and the calculated reflectivities presented in Fig.16.15 demonstrate that multiple beam interference of x rays in a solid Fabry-Perot etalon has been experimentally observed. The primary areas of disagreement between experiment and modelling calculations are (1) the absolute reflectivities observed are approximately 50% of the predicted values and (2) the contrast of the fringes of equal inclination observed are not as large as predicted. At the present time both these areas of disagreement can be explained in terms of interfacial roughness and the divergence of the incident light.

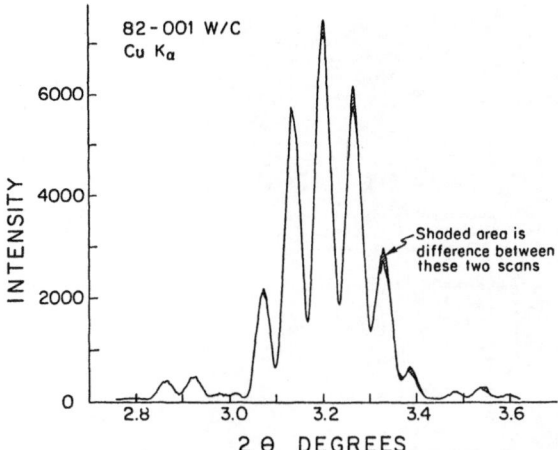

Fig.16.16 An overlay of two experimental θ/2θ scans taken from regions separated by 1.0 cm on sample 82-001, the Fabry-Perot etalon shown
in Fig.16.1 C. Shaded areas show the difference between the two
scans. These differences, if they arise from changes in the carbon
spacer thickness, may be explained by a variation of $\leq$ 0.4 A in
the spacer thickness of $\sim$ 981 A

In Fig.16.16 an overlay of diffraction curves taken from two areas (1mm x
5mm) separated by 10 cm on sample 82-001 (t_{Sp} = 981 A) are plotted. The differences observed are shown as shaded areas. If the differences are assumed
to be the result of nonuniformity in the carbon spacer, modeling calculations indicate these cannot be larger than 0.4 A. Therefore, the spacer is
t_{Sp} = 981 ± 0.2 A or uniform to 0.04%. This is a unique result in that the
interfacial roughness appears to be substantially larger than the apparent
nonuniformity in average thickness of the spacer layer, indicating that the
number of atoms per unit area is conserved.

16.9 Combined Manufactured Microstructures

Combination of differing methods of producing microstructures will result in
the ability to synthesize new types of optic elements, allowing for very
high spatial and wavelength (energy) resolution. For example, let us assume
that an LSM uniform both in depth and laterally has been synthesized. Performing lithography [16.24] typical of integrated circuit processing allows
the formation of gratings having bars consisting of the remaining unetched
LSM. Etched resolution test patterns formed in a tungsten/carbon LSM containing 40 layer pairs and having a period of 28.5 A are shown in
Fig.16.17-A and 16.17-B. In both figures the remaining unetched LSM appears
light. In Fig.16.17 the finest linewidths are 1.25 μm and are seen at the top
of the figure. At the upper left of the figure 1.25 μm x 1.25 μm etch patterns remain but show evidence of over-etching. In Fig.16.17-B a different
test pattern is shown. The upper area shows square light regions of unetched
LSM on a dark background, the largest of which are 10 μm on a side. The
smallest are 1.25 μm on a side. The bottom pattern consists of holes etched
into the LSM film, the largest of which are 10 μm on a side, the smallest of
which is intended to be 1.25 μm on a side. These test pattern results indi-

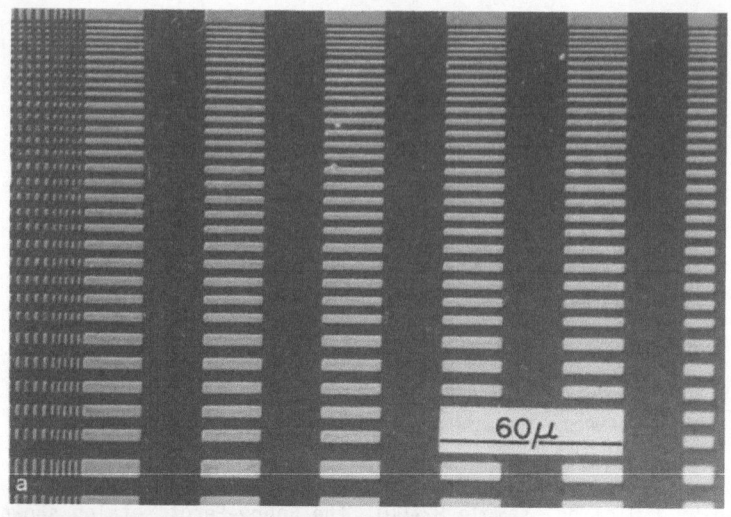

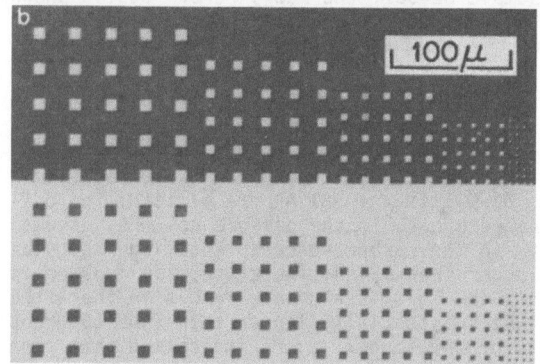

Fig.16.17-A An optical photomicrograph of a test pattern etched structure processed into a W/C LSM (d = 28.5 A, N = 40) by plasma etching is
shown. This pattern was produced using standard integrated circuit
optical lithography and reactive plasma etching. The finest lines
shown at the top of the figure are 1.25 μm wide

Fig.16.17-B An optical micrograph of a test pattern produced using standard
integrated circuit lithography and reactive plasma etching is
shown. This sample was a W/C LSM (d = 28.5 A, N = 40). The light
areas are the remaining LSM material which is columnar in the upper area and is a film with square holes etched in the lower region

cate that good resolution may be achieved using optical techniques and plasma etching processing. The optic implications of these results are discussed
in the following.

As a simple example let us consider a beam of parallel light ABCD of
wavelength incident on LSM grating structures of period d, at an angle θ
satisfying Bragg's equation, so that the diffracted beam A'B'C'D' is reflected at an angle θ by the remaining LSM bars as shown schematically in
Fig.16.18. These bars will act as virtual plane wave sources much like slits

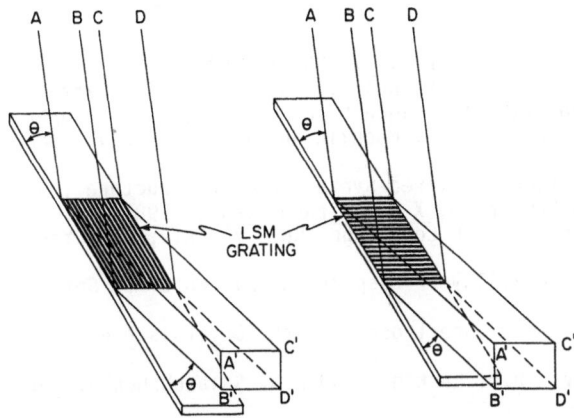

Fig.16.18 Light (ABCD) incident onto LSM gratings at Θ, the Bragg angle of the LSM's, is schematically shown to be scattered (A'B'C'D') by two orientations of etched grating structure. These gratings cause dispersion normal to the direction of the light and, depending on Θ, in a direction nearly parallel to the direction of the Bragg diffracted light. The light is scattered by the LSM grating bars with an intensity I_0R, where I_0 is the intensity of the incident light and R the LSM reflectivity

in a transmission structure. The intensity of these sources will be RI_0 where R is the LSM reflectivity and I_0 the incident intensity. The multiple reflected beams will interfere to form a diffraction pattern in essentially the same manner that traditional gratings perform. In the soft x-ray regime such structures will operate at reasonable angles of incidence varying from 10 deg. for Al-K to normal incidence at longer wavelengths.

The resolution will, in the simplest sense, be equal to the number of bars illuminated, and for sufficiently perfect structures will reach the diffraction limit. If the period of the bars is 2 μm and the illumination width 1mm, a resolution of 500 is achievable in the first diffracted order. If an order separation of 2mm is desired for AlK$_\alpha$ radiation then the detector must be set off approximately 2.2 meters. With 100A radiation this is relaxed to 200mm with working Bragg angles of up to 90°.

Let us now assume that an LSM grating such as shown in Fig.16.18 is not uniform but is laterally graded. This will, when coupled with an unetched matched laterally graded LSM, provide a simple high-resolution monochromator which will operate over a spectral range primarily determined by magnitude of the lateral gradient of the LSM period. If the bar spacing is varied with position along the grating a fixed exit beam monochromator can be constructed. I note here that if the +1, -1 orders of this grating structure are used with specular mirrors, off-axis holography can be experimentally attempted. This will give a simple straightforward method for initiating holography studies using x rays.

I also note that zone plates, phase gratings, echelles, echettes, and other more demanding grating structures can be constructed. Such attempts will meet with increasing success as microstructure technology progresses and resolutions of 10^4 to 10^5 will be attainable in the 50-150 A spectral range. Such efforts will be very dependent on substrate preparation. Additionally, transmission structures will become available, making likely the generation of whole new experimental techniques in the field of x-ray physics.

References

16.1 J. Dumond and J.P. Youtz: J. Appl. Phy. $\underline{11}$, 357 (1940)

16.2 Troy W. Barbee, Jr., and Douglas L. Keith: "Synthesis of Metastable Materials by Sputter Deposition Techniques" in Synthesis and Properties of Metastable Phase (ed. by E.S. Machlin and T.J. Rowland) AIME, New York, N.Y. 93 (1980)

16.3 Troy W. Barbee, Jr.: "Sputtered Layered Synthetic Microstructure (LSM) Dispersion Elements", in Low Energy X-Ray Diagnostics in 1981 (ed. by D.T. Attwood and B.L. Henke), AIP Conf.Proc. No. 75, AIP, New York, 131 (1981)

16.4 E. Spiller: "Evaporated Multilayer Dispersion Elements for Soft X-Rays", Ibid., 124

16.5 B.L. Henke: "Low Energy X-Ray Spectroscopy with Crystals and Multilayers", Ibid., 85

16.6 S.V. Gaponov, S.A. Gusev, B.M. Luskin, and N.N. Salaschchenko: Opt. Comm. $\underline{38}$, 7 (1981)

16.7 B.L. Henke, P. Lee, T.J. Tanaka, R.L. Shimabukuro, and B.K. Fujikawa, Atomic and Nuclear Data Table 27 (1982)

16.8 A.E. Rosenbluth, Thesis, Univ. of Rochester (1982)

16.9 A.V. Vinogradov and B.Ya. Zeldovich: Appl. Opt. $\underline{16}$, 89 (1977)

16.10 A.E. Rosenbluth and J.M. Forsyth, in Proceedings of the Topical Conference on Low Energy X-Ray Diagnostics (ed. by B.L. Henke and D.T. Attwood) AIP, New York (1981)

16.11 J.P. Henry, E. Spiller and M. Weisskopf: Appl. Phys. Lett. $\underline{40}$, 25 (1982)

16.12 D.H. Bilderback, B.M. Lairson, T.W. Barbee, Jr., G.E. Ice, and C.J. Sparks, Jr.: Nucl. Instrum. Methods, $\underline{208}$, 251-261 (1983)

16.13 W.K. Warburton, Z.U. Rek, and T.W. Barbee, Jr.: "Performance Tests on Layered Synthetic Microstructures for X-Ray Optical Elements", this volume no.17

16.14 R.H. Price, T.W. Barbee, Jr., M. Gerassimenko, W.M. Cook, W.M. Ploeger, M.W. Kobierecki, W.B. Laird: "A Kirkpatrick-Baez X-Ray Microscope (KBXRMS) with W/C Multilayer X-Ray Reflecting Coatings" to be presented at the 25th Annual Meeting, Division of Plasma Physics, 11/7 to 11/11 1983, Los Angeles, CA

16.15 H.E. Bennett and J.M. Bennett: in "Physics of Thin Films", Vol. 4, ed. by G. Hass and R.E. Thun, Academic Press, 1969, 1

16.16 P. Beckmann: "The Scattering of Electromagnetic Waves by Rough Surfaces" Pergamon Press, New York, 1963, Part 1

16.17 T.W. Barbee, Jr.: Unpublished results

16.18 D.J. Nagel, T.W. Barbee, Jr., and J.V. Gilfrich: Nucl. Instrum. Methods, $\underline{195}$, 63 (1982)

16.19 Ping Lee: Applied Optics $\underline{22}$, 1241 (1983)

16.20 R.K. Smither: Rev. Sci. Instrum. $\underline{53}$, 131 (1982)

16.21 A. Fukuhara and Y. Takano: J. Appl. Cryst. $\underline{13}$, 391 (1980)

16.22 T.W. Barbee, Jr. and J.H. Underwood: "Solid Fabry-Perot Etalons for X-Rays" to be published in Optics Communications

16.23 M. Born and E. Wolf: Principles of Optics 1st Edition, Pergamon Press (1959) Chapter VII

16.24 R.E. Howard, P.F. Liao, W.J. Skocpol, L.D. Jackel, and H.G. Craighead: Science $\underline{221}$, 117 (1983)

17. Performance Tests on Layered Synthetic Microstructures (LSM's) for X-Ray Optical Elements

W. K. Warburton and Z. U. Rek

Stanford Synchrotron Radiation Laboratory, SSRL/SLAC Bin 69, P.O. Box 4349 Stanford, CA 94305, USA

T. W. Barbee, Jr.

Department of Materials Science and Engineering, Stanford University Stanford, CA 94305, USA

17.1 Introduction

The advent of Synchrotron Radiation (SR) sources, which are capable of producing intense photon beams with energies ranging from ultraviolet to hard x ray, has created a need for new optical elements capable at once of focussing, collimating, or passing a restricted energy band of these photons and yet being durable enough to withstand extremely high power loads. Layered Synthetic Microstructures (LSM's) are currently being intensively evaluated in this context [17.1-4]. The range of applications considered has included normal incidence imaging systems [17.5-7] and monochromators [17.8] in the soft x-ray regime, and wideband focussing monochromators or premonochromators in the hard x-ray regime [17.9,10]. In all of these applications, whether acting either as focussing or condensing elements, the large separations between source and image plane create rigorous demands for optical quality. Usually surface slope accuracies in the sub-arcsec region are required.

Recognizing these issues, we have attempted to develop techniques for evaluating LSM's for these uses, and are reporting our initial results in this paper. Such tests will clearly be necessary if LSM's are to find widespread uses, yet to date little work of this sort has been done. Mancini and Bilderback reported results on a single LSM studied at CHESS by reflection of a slit diffraction pattern [17.11], and Henry et al. reported results from imaging a point source using an LSM coated mirror for satellite x-ray astronomy [17.12]. The philosophy which we followed in developing these tests was to employ methods which would simulate as closely as possible the intended LSM end uses, so that the test results would be directly applicable in choosing structures to meet the desired needs. In this context, we singled out two principal uses: focussing elements and monochromators, and devised the tests described in the following section to evaluate the LSM's in these capacities.

Considered generally, the optical performance of LSM's will depend on three things: the quality of the layers in the LSM, the quality of the substrate surface on which it is deposited, and the degree to which it conforms to the substrate surface. Of these, only the topic of layer quality has received much attention, either experimentally [17.1,3,13] or theoretically [17.14]. In our initial experiments, we therefore examined LSM's deposited on three different surfaces in order to obtain initial experience concerning the importance of substrate surface quality. These surfaces were: Si(111), Si(100), and Hoya glass. The Si surfaces were wafers from the electronics industry which had been polished to the standards required for circuit work. The Hoya glass is used in the electronics industry for mask deposition and is flat to 2 microns over any local 12 mm circle. Surface roughness is only known to be less than 500 A. Some of the samples were codeposited on both Si(100) and Hoya glass. The properties of all samples are given in Table 1.

17.2   Diffractometer Analysis

17.2.1 Experimental Technique

Tests of LSM optical quality were done on the Materials Diffractometer at
SSRL using an exit beam monochromator (EBM). The SR incoming beam was
monochromatized by Si(220) crystals to 8 keV and collimated by a 0.05 mm
slit 20 meters from the 0.24 mm FWHM SR source. The beam divergence is thus
about 3.05 arc sec. Light scattered from the LSM was analysed by a Si(220)
EBM in antiparallel geometry. The convolution of source divergence and
Si(220) Darwin width gives an angular resolution of about 5-6 arc sec for
these measurements.

Three sorts of scans were employed in these studies and they are shown
schematically in Fig.17.1. Calling the angle of incidence on the LSM ω.
and the angle to the EBM 2Θ, the various scans were: the coupled $2\Theta-\omega$
scan, the 2Θ scan at fixed ω, and the ω scan at fixed 2Θ. Each scan produces some distinct information about scattering from the LSM. If we con-
sider the LSM surface to consist of local facets of perfect material oriented parallel to the local substrate surface, then we would expect the following scattering behavior. For the 2Θ scan at fixed ω, the incident angle
is fixed so, since the LSM Darwin width exceeds in incident divergence, all
scattering outside this range will be nonspecular, and is the best measure
of this property. The ω scan would bring successive facets into the Bragg
condition with the incident beam and so this rocking curve produces a description of the range of facet angles on the surface. Outside this range,
it also samples nonspecular reflection, similarly to the 2Θ scan. The
coupled $2\Theta-\omega$ scan, on the other hand, maintains angle of incidence equal to
angle of reflection for the same set of facets throughout the scan. The results obtained are those of the LSM itself, once the angle of incidence is
high enough so the bulk LSM is sampled. In order to simplify discrimination
between the various sorts of scans, a small version of Fig.17.1 will be in-
set into each data figure.

The major experimental difficulty to be overcome was to obtain the initial alignment of the sample. Since the experimental resolution is so good,
lack of precise correlation between Θ and ω can cause the entire scattering
behavior to be missed, as is shown symbolically by scan #4 in Fig.17.1.
This problem was solved by using the total external reflection behavior of
the samples at low angle (the long ridge of contours in Fig.17.1). If a scan
in ω was made at fixed Θ, as is shown in Fig.17.2 (for sample Pt 83-031),
reflection would be seen when Θ and ω are in exact correspondence. Typical
offsets are a few hundredths of a degree, as seen in Fig.17.2, and result
either from dust particles on the sample surfaces or long-range undulations
in the substrates. After these misalignments were corrected, a simple $2\Theta-\omega$
scan would suffice to locate the Bragg Peak (the concentric contours in
Fig.17.1).

17.2.2 Results

The results of omega scans through the Bragg peaks for samples on Si(111),
Si(100), and Hoya glass are shown in Fig.17.3, 17.4 and 17.5, and have been
chosen to be representative for the general results. The horizontal scales
of the three figures are identical for ease of comparison. Qualitatively,
the Si(111) sample shows a narrow, semi-Gaussian reflectivity, which is less
than one-third the width of the other two cases. The Si(100) reflection is
much wider and shows structure, which is evidence of a nonrandom distribu-

tion of surface slopes. Outside the region of specular reflectivity the intensity drops rapidly to zero, similarly to the Si(111) case. The Hoya glass sample shows a broad, fairly Gaussian peak, overlaid on what appears to be a uniform background whose intensity is about 25% of the peak maximum. Nonspecular reflection appears to be a serious matter in this sample. Qualitatively, the respective peaks show FWHMs of 29 arcsec, 92 arcsec, and 126 arcsec. Even the best of these samples, therefore, is not good enough to be an optical element in a SR beam line, where values of order 1-2 arcsec are required.

Figures 17.6, 17.7 and 17.8 show the results of coupled 2 Θ-ω scans on the same three samples. The general features of all three are similar. The spike at 0 degrees is the direct beam, showing the system resolution. The peak at low angles is total external reflection, which drops to zero at zero degrees because the samples are of finite length. The details of these peaks vary extremely from sample to sample and appear to depend on the exact details of the surface topography. When similar scans are done with an open detector, instead of an EBM, these peaks are always smooth and featureless below the critical angle. The Bragg peaks, on the other hand, are well behaved and in all cases show FWHM values in agreement with modeling calculations based on their materials and layer thicknesses. The FWHM in Fig.17.6 (Sample W/C 82-153) is 0.098 degrees in 2Θ, corresponding to an energy resolution of $\Delta E/E$ = 2.2% (or 1/46). Since the sample had 120 layers, extinction is occurring as predicted by the modelling. Figures 17.7 and 17.8 are from codeposited samples (W/C 83-022S and W/C 83-022H), but show different resolutions of 4.5% and 3.1%. The causes of this behavior are not currently understood.

17.3   Monochromator Performance

17.3.1 Experimental Procedure

These measurements were designed to test the performance of the LSM's as premonochromators. White SR in the SSRL Topography Line BL II-4 was directed onto the LSM's and then analysed with a standard SSRL Si(111) two-crystal monochromator. A slit between LSM and the monochromator selected the portion of reflected beam to analyse. A 5 micron Ni foil was inserted in the line to produce Ni EXAFS spectra for energy calibration and resolution measurements. This procedure therefore directly models the way a LSM would be used as a premonochromator in an SR beam line.

17.3.2 Experimental Results

Results are shown in Figs.17.9, 17.10 and 17.11 for the same samples measured in section 17.2.2. The multiple curves in a given figure each represent the analysis of intensity versus energy for a given fraction of the reflected beam, as determined by the slit position. The width of an individual curve is the energy resolution of the LSM, corresponding in energy space to the Bragg peak widths seen in the diffractometer 2 Θ-ω scans (Figs.17.6, 17.7 and 17.8). The correspondence between the two is within the accuracy of the experiments in all cases. The progression from essentially specular reflection (Fig.17.9 Si(111) substrate) to largely nonspecular reflection (Fig.17.11, Hoya glass) is both obvious and dramatic. The envelope of curves in Fig.17.11 shows the same wings as the omega scan in Fig.17.5.

Figures 17.12, 17.13 and 17.14 show EXAFS spectra taken using LSM premonochromators under various conditions. The smooth background curves, excluding glitches, are the LSM transmittance without the Ni foil inserted. The first, Fig.17.12, is from sample W/C 81-204A, which is on Si(111) and annealed for 10 hours at 500 °C. This is the highest resolution result we obtained, with smallest feature size corresponding to about 12 eV (which is still about 3-4 times the finest features on a high resolution Ni spectrum). Figure 17.13 is data from our earlier sample W/C 83-022S on Si(100). The resolution here is about 25 eV. Figure 17.14 repeats Fig.17.13 with a wider slit. The energy resolution is now worse than 14 eV and displays the degradation of resolution from nonspecular reflection. No attempt was made to obtain EXAFS from a Hoya glass specimen.

17.4 Discussion of Results

Our results clearly indicate the influence of substrate quality on LSM optical performance. By all indicators, Si(111) was superior to Si(100), which was superior to Hoya glass. The radically different behavior of codeposited samples demonstrates how profound this influence can be. We note, however, that none of these substrates was specifically prepared to be an x-ray optical surface. Both Si(111) and Si(100) are typically cut slightly off axis and may be expected to facet at the submicron level under etch-polishing. Hoya glass is manufactured to have small slope errors but roughness well below the wavelength of visible light is unimportant.

These results suggest a logical next step our research should take: to prepare LSM samples on ultrasmooth substrates for further tests. We are currently investigating this possibility. Such results will also be of importance in making ultimate determinations of the LSMs' internal quality. To date, all indications are that the substrates are still the limiting factor, not the quality of the layered structures themselves. The other major area of planned research is to devise means to test LSM robustness in the face of the extreme conditions which they will encounter in actual SR beamline use, which will include high power loads in ultrahigh vacuum and the presence of extremely energetic photons.

Acknowledgments

We would particularly like to express our appreciation to Dr. Alain Fontaine, who assisted in the collection of the premonochromator data. One of us (TWB) was supported by the Department of Energy through Los Alamos National Laboratory. The work reported herein was performed at SSRL, which is supported by the Department of Energy, Office of Basic Energy Sciences; the National Science Foundation, Division of Materials Research; and the National Institutes of Health, Biotechnology Resource Program, Division of Research Resources.

Table 17.1 Properties of the LSM samples

SAMPLE	METHOD		MATERIAL		d SPACING		N_{pairs}	SUBSTRATE
	EBM	PM-M	A	B	d_T[Å]	d_A[Å]		
W/C 82-153	X	X	W	C	21.4	8.3	120	Si(111)
Pt 83-031	X		Pt	-	268.8	---	1	Si(111)
W/C 83-022S	X	X	W	C	35.6	13.2	100	Si(100)
W/C 83-022H	X	X	W	C	35.6	13.2	100	Hoya Glass
W/C 81-204A		X	W	C	30.2	11.2	103	Si(111)

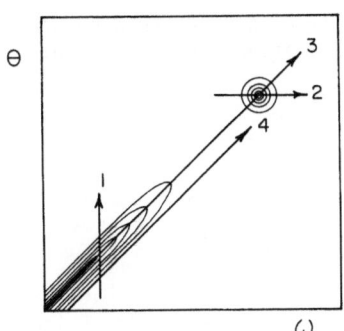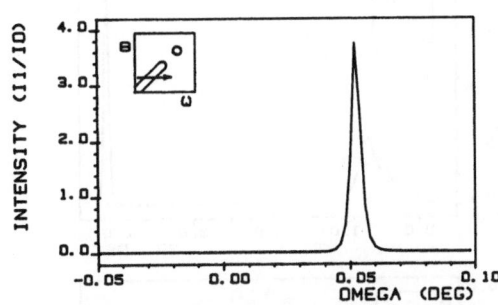

Fig.17.1 A contour map of scattered intensity in the 2θ-ω plane showing different possible scans. These are: 1: total reflection vs 2θ at fixed ω; 2: ω rocking curve through the Bragg peak; 3: complete 2θ-ω scan; 4: similarly with an offset error

Fig.17.2 An ω scan through the total external refflection region on sample Pt 83-031

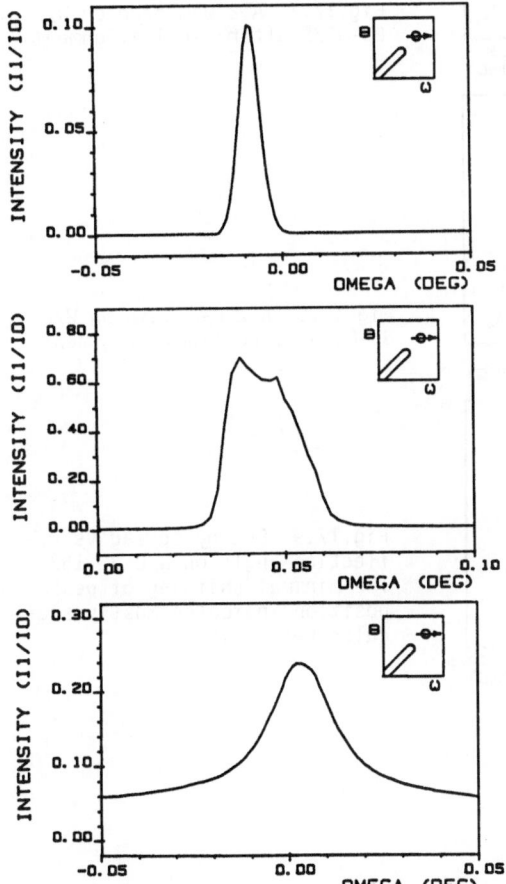

Fig.17.3 An ω rocking curve through the Bragg peak on W/C 82-153

Fig.17.4 An ω rocking curve through the Bragg peak on W/C 83-022 S

Fig.17.5 An ω rocking curve through the Bragg peak on W/C 83-022H

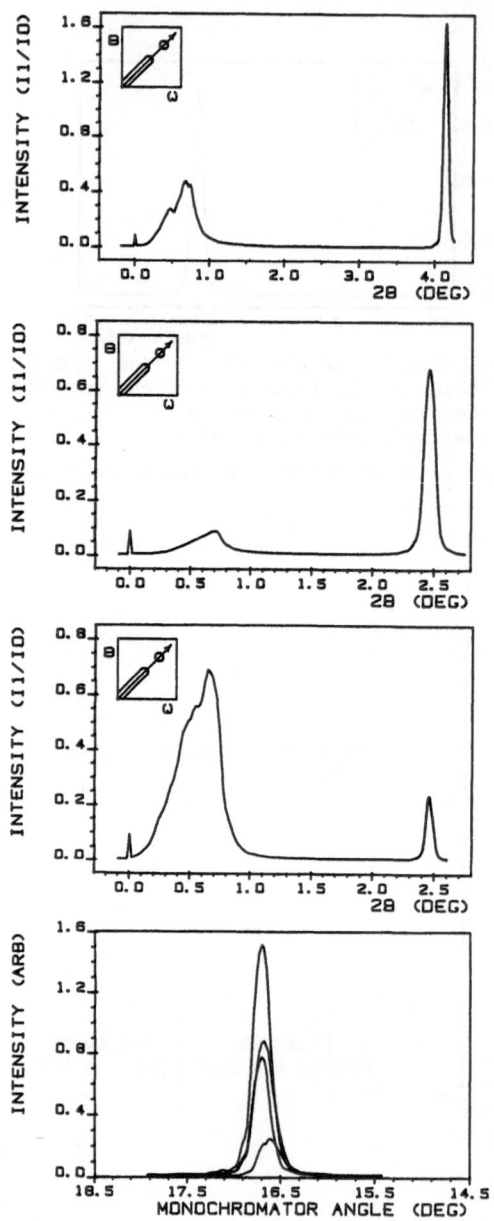

Fig.17.6 A 2 θ-ω scan on W/C
82-153 after final alignment

Fig.17.7 A 2 θ-ω scan on W/C
83-022S after final alignment

Fig.17.8 A 2 θ-ω scan on W/C
83-022H after final alignment

Fig.17.9 Energy spread vs reflection angle on W/C 82-153.
The minimal shifting of peak
position indicates mostly specular reflection

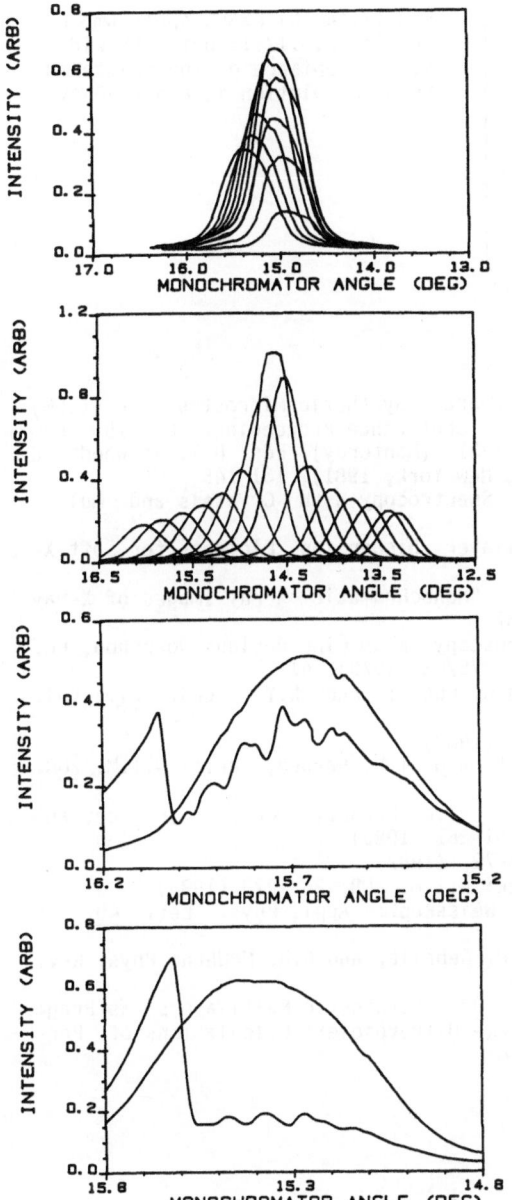

Fig.17.10 Energy spread vs reflection angle on W/C 83-022S. The curve maxima shift to smaller angles (higher energy) as the slit is raised

Fig.17.11 Energy spread vs reflection angle on W/C 83-022H. The broad wings correspond to intense nonspecular scattering

Fig.17.12 Ni EXAFS spectrum using sample W/C 81-204A, compared to the direct reflection spectrum. Resolution is about 12 eV

Fig.17.13 Ni EXAFS spectrum using sample W/C 83-022S, compared to the direct reflection spectrum. Resolution is about 25 eV

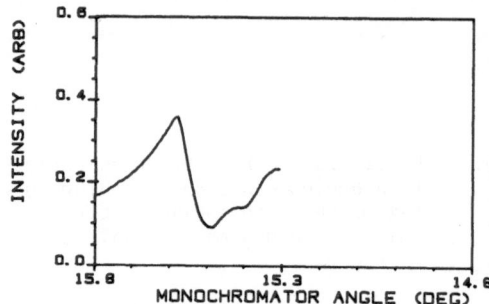

Fig.17.14 Ni EXAFS spectrum, repeating Fig.17.13 but with wide slit acceptance of the scattered beam. Resolution is about 40 eV

References

17.1 T.W. Barbee, Jr.: "Sputtered Layered Synthetic Microstructure (LSM) Dispersion Elements", in A.I.P. Conference Proceedings No. 75, Low Energy X-ray Diagnostics - 1981 (Monterey). Eds. D.T. Attwood and B.L. Henke, (Am. Inst. Physics, New York, 1981), 131-145

17.2 B.L. Henke: "Low Energy X-ray Spectrocopy With Crystals and Multi-layers", Ibid. 85-96

17.3 E. Spiller: "Evaporated Multilayer Dispersion Elements for Soft X-rays", Ibid. 124-130

17.4 R.R. Whitlock and D.J. Nagel: "Monochromatic X-ray Images of X-ray Emitting Sources", Ibid. 334-337

17.5 E. Spiller: in Lithography/Microscopy Beam Line Design Workshop, Ed. C.R. Dannemiller(SSRL Report No. 79/02, 1979), 41

17.6 R.-P. Haelbich, W. Staehr, and C. Kunz.: Ann. N.Y. Acad. Sci. Vol. 342, 148-157 (1980)

17.7 G. Schmahl: N.I.M. 208, 361-365 (1983)

17.8 R. Day, J. Grosso, R. Bartlett and T.W. Barbee, Jr.: N.I.M. 208, 245-289 (1983)

17.9 D.H. Bilderback, B.M. Lairson, T.W. Barbee, Jr., G.E. Ice, and C.J. Sparks, Jr.: N.I.M. 208, 251-261 (1983)

17.10 D.H. Bilderback: N.I.M. 195, 67-72 (1982)

17.11 D.C. Mancini and D.H. Bilderback: N.I.M. 208, 263-273 (1983)

17.12 J.P. Henry, E. Spiller, and M. Weisskopf: Appl. Phys. Lett. 40, 25 (1982)

17.13 W.P. Lowe, T.W. Barbee, Jr., T.H. Geballe, and D.B. McWhan: Phys. Rev. B24, 6193 (1981)

17.14 J.H. Underwood and T.W. Barbee, Jr., "Synthetic Multilayers as Bragg Diffractors for X-rays and Extreme Ultravoiolet: Calculations of Performance", in [17.1], pp.170-178

Part III

X-Ray Detectors

18. Charge and Scintillation Gaseous Detectors for Low Energy X-Rays

A. J. P. L. Policarpo

Departamento de Fisica, Universidade de Coimbra
P-3000 Coimbra, Portugal

A brief review is made of low energy gaseous x-ray detectors. Special relevance is given to devices in which the photon flux plays an important role.

18.1 Introduction

The interaction of x-rays with gaseous detecting media gives rise to excited
and ionized atoms or molecules and free electrons. Most of the detection
methods referred to in this work rely essentially on information associated
with these primary electrons.

For lower energies, shall we say below 1 keV, the electronic noise is
dominant, even using cooled electronics with small detection areas, and some
kind of internal amplification is needed. This amplification is achieved
using an external source of energy, an applied electric field.

Two extreme modes can be considered, depending on the intensity of the
electric field and on the nature of the gaseous medium: either the drifting
of primary electrons leads only to excitations and the detection is achieved
using the photon flux associated with the deexcitation of the medium or the
drifting of the primary electrons gives rise to excitations and ionizations,
leading to the production of secondary electrons and avalanche development.
With the first mode is associated the gas scintillation proportional counter
(GSPC); with the second, proportional counters, multiwire chambers, geiger
tubes, etc.

Intermediate situations, in which there are either rather low charge
gains or very large avalanches but the message is, in both cases, essentially carried by the associated photon flux, have characteristics that
favour special instrumental parameters and lead to interesting detector developments. This is the case, for example, for the electron counting
technique (ECT) to determine the x-ray energy.

An essential feature of detectors that use the photon message is that
this is associated with electron migration, while charge devices rely, in
most cases, on ionic motion.

In this work some relevant detection techniques based on the principles
outlined above will be considered. Performances concerning energy, position,
time definition, counting rate capabilities, etc., will be referred to. Some
of the physical processes and mechanisms involved will be briefly indicated,
as they provide intrinsic limits to future developments. Some data reported
concern energies of a few keV, rather than below 1 keV, as in many cases
results are yet of a preliminary nature, the techniques being under development.

18.2   Gas Scintillation Proportional Counters

18.2.1 General

Following the interaction of x-rays in a noble gas medium its deexcitation gives rise to the primary scintillation, a fraction of 20% of the energy deposited being dissipated radiatively. For energies lower than about 10 keV this information cannot be used with efficiency.

If an electric field just below the threshold for charge multiplication is applied to the noble gas, the drifting of the primary electrons originates inelastic collisions and the field energy is very efficiently converted into these processes ($\sim$80%). Very large photon fluxes arising from deexcitation of the medium can then be obtained without charge multiplication and space-charge effects, the total photon flux is proportional to the number of primary electrons, and the energy information is preserved as the statistical fluctuations in the number of photons associated with the drift of the primary electron cloud are small.

Although other characteristics of the GSPC are of interest and will be referred to later, its energy resolution, independent of electronic noise and detecting area, was the main feature that triggered its use. As a general reference see [18.1].

Noble gases do not feature low Fano factors. They are $\sim$0.15 and for many gaseous mixtures reach values as low as 0.05. The good performance of the GSPC arises rather from large photon yields with low statistical fluctuations. Both these features are associated with the absence of collective excitation modes and the Ramsauer dip, such that under adequate electric fields, only a small fraction of the supplied energy is used in a very large number of elastic collisions, most of the field energy originating useful excitations.

In a uniform electric field for an electron drifting across V volts let H excitations be produced at an average energy expenditure U. $g = \overline{H}\,\overline{U}/V$, the fraction of the energy supplied by the field that is used in excitations, is a function only of E/p, the reduced electric field. Let us define the fluctuations in H through $(\sigma_H/\overline{H})^2 = J/\overline{H}$. It can then very simply be shown [18.2] that $J = J_1 + J_2$, where $J_1 = (1-g)^2$ and $J_2 = g^2\,(\sigma_u/\overline{U})^2$, where σ_u corresponds to fluctuations in U. Both $\overline{U}$ and σ_u can be estimated in such a way that the corresponding uncertainties in J are small, $\overline{U}$ being the mean value of the excitation energies of the lower set of states $^3P_2$, $^3P_1$, $^3P_0$ and $^1P_1$ and σ_u as arising from its uniform feeding. Using then only experimental data concerning $\overline{H}$, the above formulae allow the calculation of the statistical features of the process.

For xenon, a well-studied gas and the most used filling medium in GSPC, within a reasonable approximation, Fig.18.1 shows some calculated relevant information. Several hundred photons, shall we say $\overline{H}\sim$500 can be obtained per migrating electron, with $J \sim 0.02$, i.e. $\sigma_H/\overline{H}\sim$0.6%, below the threshold for charge multiplication. For detailed information concerning excitations of noble gases by drifting electrons see [18.3].

These intrinsic statistical fluctuations are indeed very small and larger effects can easily arise from mechanical accuracies or cell dimensions of the grids usually used to define the scintillation region of the detector. Essentially then the energy resolution is determined by the Fano factor and the statistics associated with the detector of the photons, this last parameter being favoured by the large yield. The energy resolutions obtained are

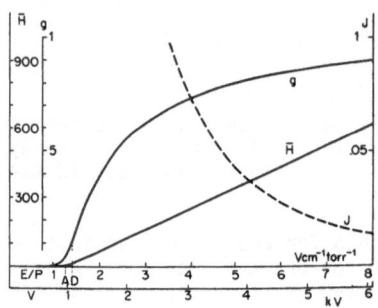

Fig.18.1 For xenon, g and J are represented as a function of E/p. For a gap of 1 cm and 760 torr, $\bar{H}$ is also represented as a function of the voltage V across the gap (data from [18.2])

typically a factor of two better than from proportional counters, with similar contributions arising from the Fano factor and from the photon detecting devices, in general photomultipliers or photoionization detectors.

18.2.2 Scintillation Mechanisms and Timing

A brief summary of the main processes associated with the secondary scintillation of pure noble gases around atmospheric pressure will be given here (for more detailed information see [18.1,4]).

For fields just below charge multiplication, only the first set of excited states of the noble gas are essentially excited by the drifting electrons. Through three-body collisions the molecular states $^1\Sigma_u^+$ and $^3\Sigma_u^+$ are formed in high vibrational levels, that are rapidly relaxed by collisions, and their decay to the repulsive ground state $^1\Sigma_g^+$ originates the observed secondary scintillation emission, the second continuum. The emission bands are narrow ($\sim$1 eV) and are centred at 7.3 eV, 8.3 eV and 9.8 eV, for Xe, Kr and Ar respectively.

The direct percursors of those molecular states are the atomic levels $^3P_2$ and $^3P_1$, with rate constants K_1 and K_2 respectively, and through rather fast and complex collisional mechanisms the states $^3P_0$ and $^1P_1$ also contribute to the formation of those dimers. The rate constants K_1 and K_2 [torr^{-2}s^{-1}] are for argon $K_1 \simeq K_2 \simeq 11$ and for xenon $K_1 \simeq K_2 \simeq 44$, and at atmospheric pressure the formation time of the molecular states is for xenon $\simeq$40 ns; argon is four times slower. As examples, the radiative decay times associated with the second continuum are, respectively for $^1\Sigma_u^+$ and $^3\Sigma_u^+$ [ns] : for Ar$_2^*$ 4.2 and 3200; for Xe$_2^*$ 6 and 110.

Although higher pressures very quickly decrease the formation time, as this varies with $1/p^2$, the radiative decay times of the molecular states are a fundamental limitation to the time response of devices based on these processes.

Another limitation to the time response of GSPC is the low electron drift velocities associated with noble gases, as shown in Fig.18.2, and usually the drift time in the scintillation gap essentially determines the light pulse duration, in general >1 μs.

For timing purposes, using the first scintillation photon after x-ray absorption in a drift region where some secondary scintillation is produced, a previous treatment [18.6,7] gives good agreement with the experimental data for 5.5, 8.2 and 1.5 keV. Following that work (formation time and mean

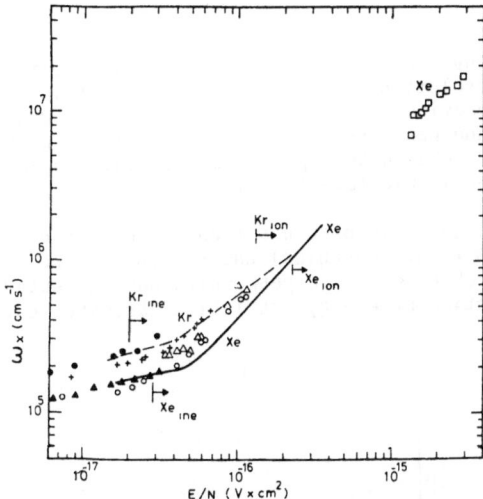

Fig.18.2 Compilation of experimental data for drift velocities in Xe and Kr. The solid and dashed lines are computed (data from [18.5])

life time of dimers 40 and 100 ns respectively) the FWHM timing resolution is determined by the rate of detection of the VUV photons produced by the cloud of primary electrons, a function of the rate of excitations and the efficiency of VUV photon detection Ω. In Fig.18.3, where FWHM time resolutions are displayed as a function of energy, the curve labeled 3 $Vcm^{-1}torr^{-1}$ corresponds to experimental data from [18.6], with $\Omega = 1.9\%$. If the detection region is also the scintillating gap a clear improvement of the timing performances is obtained, arising both from higher rates of excitation and higher drift velocities (see Fig.18.1 and Fig.18.2): calculated data for 4.4 and 6.5 $Vcm^{-1}torr^{-1}$, and for the same Ω, are also shown in Fig.18.3. It should be noted that, in general, detection in the scintillation region would cause only a small deterioration of the energy resolution if the total duration of the light pulse is also measured [18.6].

For energies lower than about 1 keV there would be practically no energy resolution deterioration even without the determination of this quantity, because the mean free paths for the x-rays are small (a few tenths of a mm for Xe at 1 atm. [18.8]).

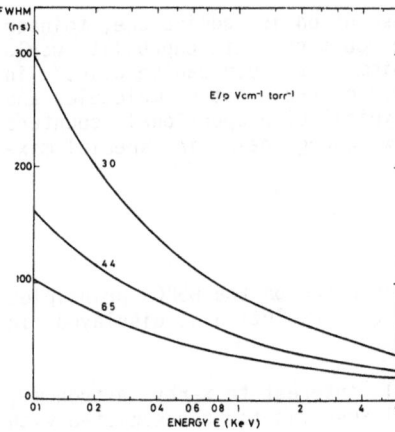

Fig.18.3 Expected time resolution as a function of energy for Xe at 1 atm., for three values of E/p

18.2.3 Counting Rate

The spherical anode counter [18.9], that due to good light collection efficiency features the typical energy resolutions of GSPC with a thin intrinsically built scintillation region, provides light signals of about 300 ns
total duration (Xe, 1 atm., 5 kV) and was used to test rate effects with
5.9 keV. The effect of the counting rate on the peak position and on the
energy resolution [18.10] is displayed in Fig.18.4 and Fig.18.5.

At about 10^5 c/s, the experimental shift in peak position is less than 3%
(to be compared with much larger FWHM at low energies) and the energy deterioration is less than 14% (no detectable intrinsic peak shift was detected
within 0.4%). At 10^4 c/s no deterioration of energy resolution was observed.

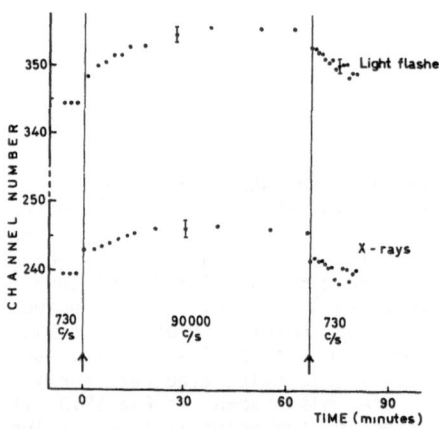

Fig.18.4 Experimentally determined
peak position as a function of counting rate. The light flashes from a
gallium phosphide diode were used to
measure the photomultiplier contribution (data from [18.19])

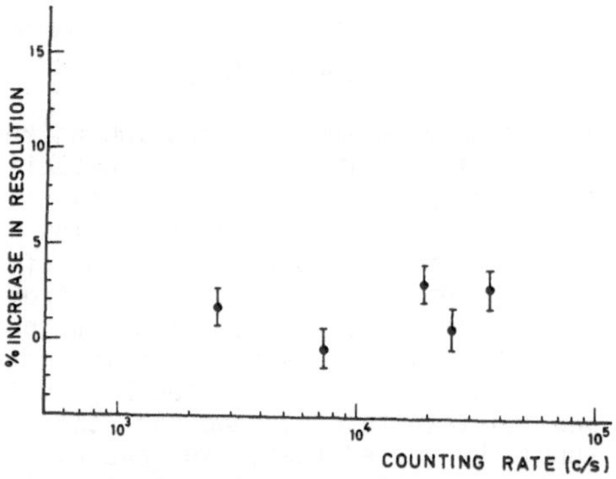

Fig.18.5 Increase of energy resolution (percent) as a function of
counting rate. The open dot is
only an upper limit (data from
[18.10])

If some deterioration of the energy resolution is admissible, thinner
scintillation regions can be used and the counting rate capability would
then be limited by the scintillation mechanism. Not much can be gained in
this direction due to the long mean life time of the excited molecules and
probably a better solution, if resolutions typical of proportional counters
can be tolerated, would be to allow for low charge gains in special mixtures, as will be referred to later.

18.2.4 Instruments and Position Resolution

Two rather simple and two elaborated detectors based on the GSPC principles
are shown in Fig.18.6, and data on its energy resolution is displayed in
Fig.18.7.

For other interesting set-ups, of special interest to x-ray astronomy,
see, for example [18.13]. Indeed special specifications associated with

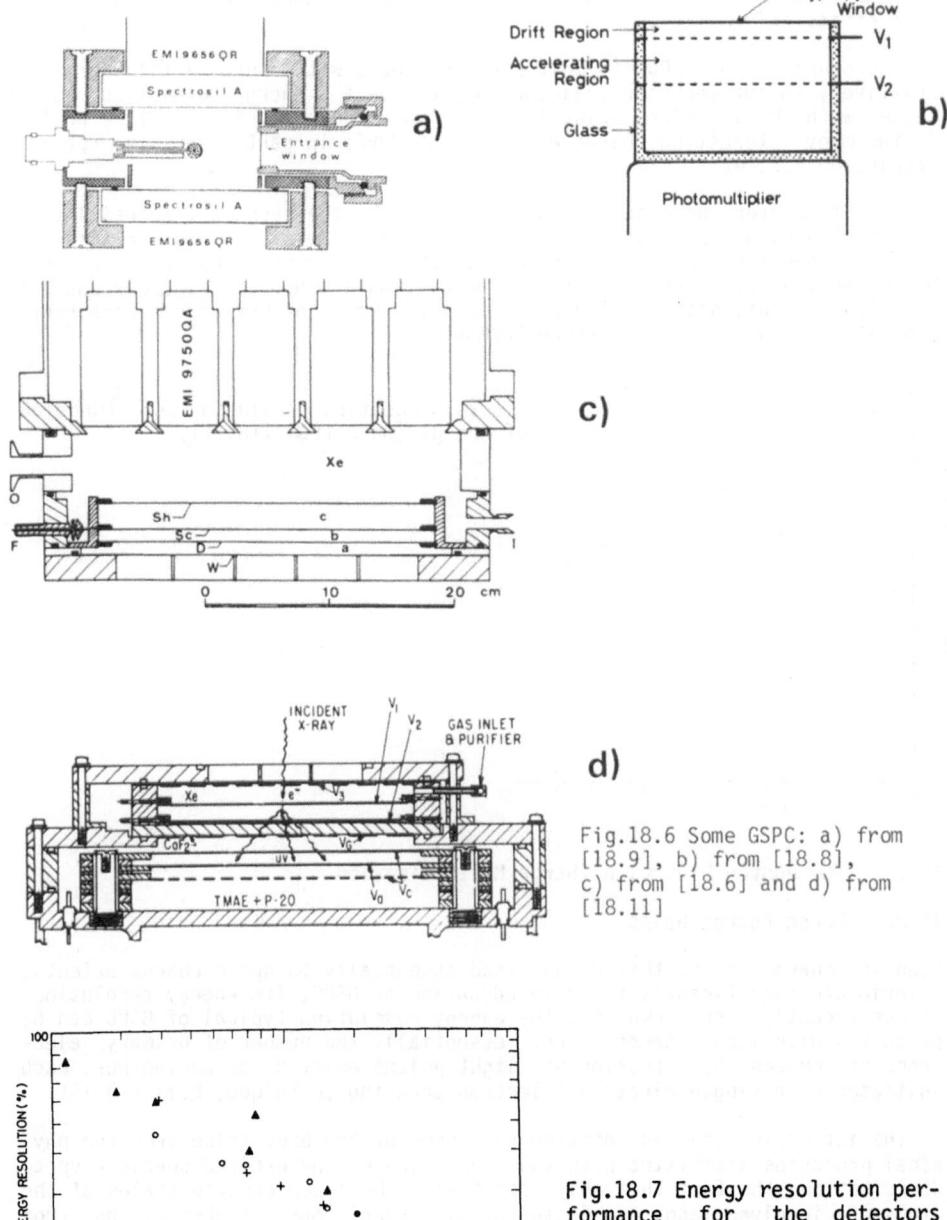

Fig.18.6 Some GSPC: a) from [18.9], b) from [18.8], c) from [18.6] and d) from [18.11]

Fig.18.7 Energy resolution performance for the detectors shown in Fig.18.6. Compilation from: +[18.11], ▲[18.8], •[18.6], o[18.9], *[18.7] and ▽[18.12]

astrophysics lead to important developments of the GSPC and to better handling of its information.

The coupling of GSPC to multiwire chambers with special fillings is a relatively recent technique, aiming to preserve the energy resolution associated with the secondary scintillation without charge gain, and take profit of the many interesting characteristics of MWPC, in particular its spatial resolution [18.14].

In the counter shown in Fig.18.6 d), [18.11], the secondary scintillation is detected with a photoionization chamber in place of photomultiplier tubes. Designed for a broad energy range, between 0.1 and 10 keV with reasonable detection efficiency, its energy resolution is shown in Fig.18.7 and the position resolution is displayed in Fig.18.8. Rates of $\sim 4 \times 10^4$ c/s were handled without degradation of its performance.

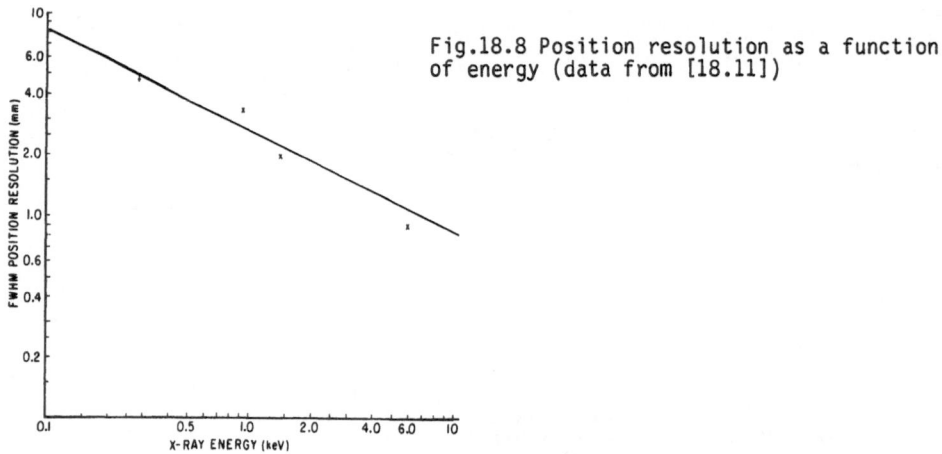

Fig.18.8 Position resolution as a function of energy (data from [18.11])

18.3 The Photon Flux with Charge Multiplication

18.3.1 Large Charge Gains

Even low charge gains, that do not lead essentially to space charge effects, deteriorate significantly the main advantage of GSPC, its energy resolution. It has recently been shown that the energy resolution typical of GSPC can be preserved with large charge gains: essentially the number of primary electrons is counted by detection of light pulses emitted by avalanches, each initiated by a single electron (electron counting technique, ECT) [18.15].

The intrinsic time and position responses of the GSPC arise from the physical processes associated with the evolution of the excited species up to deexcitation, the fact that only essentially the lower excited states of the atoms are involved, and then faster atomic transitions are absent, and from low drift velocities and large diffusion coefficients. Charge multiplication leads to excitations of the medium extending up to the continuum and to much larger photon yields per primary electron. It is then possible to suppress all the slow emission mechanisms, using additives that lead to fast collisional processes, and still have photon yields that can be used profitably. At the same time the cooling effect of these additives originates drift velocities and diffusion coefficients that are much lower than in noble gases.

A basic mixture Ar-CH$_4$ with additions of CO$_2$ and N$_2$ [18.16] that reduce diffusion and increase the light output have such properties. The main emission lines arise from atomic states of Ar, around 7000 A and from the second positive band of N$_2$, around 3500 A, and the dominant mechanisms are collisional quenching of the Ar and N$_2$ excited states by CH$_4$ and CO$_2$. It had been shown [18.17] that the mean life time involved is of the order of 1 ns.

Using large voltages on the thin avalanche region of a parallel plate proportional counter, a similar time is associated with the development of the avalanche.

By choosing low fields (E=30 V/cm) in the drift region, see Fig.18.9, the size of the cloud of primary electrons is such that they arrive separated by about 50 ns at the avalanche region and the light pulse associated with each individual avalanche (a few ns) can be counted electronically, using fast photomultipliers and electronics.

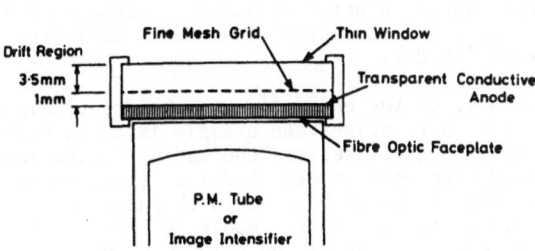

Fig.18.9 Schematic of the imaging proportional counter from [18.16]. Copyright © 1981 IEEE

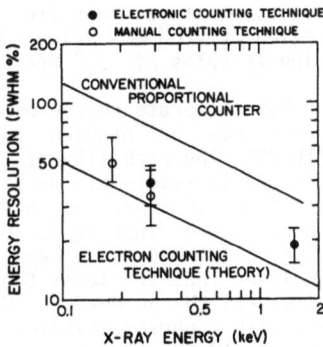

Fig.18.10 Energy resolution versus energy for conventional proportional counters and using the electron counting technique from [18.18]. Copyright © 1983 IEEE

The results obtained concerning the energy resolution are shown in Fig. 18.10, essentially the same as those of GSPC for lower energies. The performance of the multiplier and the electronic counting system limits the ECT to energies below ∿2 keV: for Al-K the energy resolution is 19% with ECT to be compared with 15% for GSPC [18.18,19].

The nature of the ECT leads to a relatively long overall light emission duration per x ray, that can be made smaller for lower energies (∿0.5 µs for 280 eV) and this is a limitation to its performance concerning time definition and counting rate.

The intrinsic position resolutions of this device, as with current parallel plate proportional counters, is essentially limited by transverse diffusion effects, essentially independent of the x-ray energy for energies lower than ∿1.5 keV (photoelectron range < 70 µm) and then determined by the absorption depth. With large fields in the drift region such that diffusion is minimized (∿300 V/cm across 3.5 mm), and using the light information, resolutions of 350 µm FWHM have been obtained for C-K and Al-K x rays [18.18] to be compared with a best value of 250 µm [18.20] and a current value of

~400 μm at C-K with parallel plate proportional counters. Further improvements in position resolution are expected with ECT, provided that a suitable centroid technique is developed [18.18].

As was mentioned, the ECT requirements for good energy resolution lead to a deterioration of its timing capabilities. Again they contribute also to a deterioration of the position resolution due to increased diffusion (a deterioration by a factor of ~3 was indicated [18.15]). But being a recent method no doubt further improvements of its intrinsic capabilities are to be expected [18.21].

18.3.2 Low Charge Gains

For the reasons mentioned previously the photon flux can be large and fast if there is charge multiplication in suitable fillings, but high rate capabilities are naturally associated with low charge gains due to space-charge effects: positive ions from a proportional counter avalanche create a dead region of ~200 μm during ~20 μs, leading to substantial loss of amplification at rates of 10^4 per mm of wire [18.22].

With moderate charge amplifications, of the order of a few hundreds, high rates can be handled using the light pulses. An example is taken from [18.23]. The stability of operation at high fluxes was checked by measuring the single-rate plateaux at a fixed threshold of detection: a space-charge effect would appear as a shift in the plateaux position with rate. At the maximum rate that could be achieved, instant rates above 10^6 c/s (8 keV well-collimated 1 mm^2 beam from an x-ray generator), no shift was observed, implying that at least two orders of magnitude have been gained over the classical charge devices. More recently [18.24], looking into a MWPC (Ar+10% N_2) with a photomultiplier, a shift of about 10% in peak position was obtained at ~10 MHz, for a charge gain of 10. The energy resolution at this high rate was ~56% for 6.9 keV x rays, but of course this parameter is heavily dependent on the light collection efficiency.

The field of very high counting rates has not yet been fully exploited. See also [18.21].

18.4 Multiwire Chambers

It was shown that position accuracies in the direction orthogonal to the anode wires of MWPC could be better than wire spacing, by determination of the azimuth of the avalanche associated with a punctual cloud of primary electrons arising from a low-energy x-ray interaction [18.25]. By computing the charge centroid of induced pulses through direct determination of pulse amplitude from each relevant electrode, an essentially local measurement and then independent of chamber size, good accuracies have been obtained. With 2 mm wire spacing, for 1.5 keV x rays, along the anode wire the measured accuracy was ~85 μm FWHM. In the orthogonal direction, if information from neighbouring anode wires is used, the position resolution is ~360 μm FWHM; reading only cathode strips, it deteriorates by a factor of two. An illustration of these capabilities is shown in Fig.18.11 using avalanches only on one anode wire.

Although the very large charge gains associated with the transition to charge saturation cannot be used because the azimuthal information is lost, this system can be used with much lower energies, as the electronic contribution to the position accuracy was small, ~15 μm FWHM.

Fig.18.11 Bidimensional radiograph with Al-K x rays. Letters are 1.5 mm in height, 50 μm in width, cut out in a copper mask (data from [18.25])

Parallel-grid structures allowing for single-step and multistep amplifying gaps have remarkable properties. Among others and only with a mild degradation of the energy resolution they feature specific localization characteristics arising essentially from the fact that the pulses induced in successive gaps are due to electron motion only. For a survey see [18.26] and references therein.

18.5 Spectrum Reconstruction

The method of soft x-ray spectrum reconstruction that will be presented relies on the determination of the spectral distribution by measuring the absorption of radiation [18.27]. The detector is a multiwire chamber with 10 anodes (distance between anode wires and between cathode plates was 1 cm) filled with hydrogen at a pressure variable between 10 torr and 20 atmospheres. The x-ray beam enters the counter orthogonally to the anode wires and in its plane. In this way such a detector may be considered as a set of consecutive single wire counters each one detecting the photons that are absorbed in its volume. Although the use of the pulse-height distribution could also improve the energy resolution, and work in this direction goes on [18.27] (see also [18.28]), the data to be shown correspond simply to the counting of the number of pulses in each wire. The mathematical technique used to reconstruct the spectrum is essentially a maximum likelihood method, able to handle continuous, discrete and superimposed spectra; moreover it allows to estimate the restoration accuracy.

A striking example is shown in Fig.18.12 where the insert displays the data of three independent experiments. The x-ray beam to be analyzed was produced using an x-ray tube and a Be filter and, in the insert, the data corresponds to three different tube voltages 410, 448 and 470 V for 1, 2 and 3, respectively. The number of photons is of the order of 10^6 in each experiment. In the same figure the reconstructed energy spectra show the narrow band radiation due to x-ray absorption in the filter, extending up to an energy that is in very good correspondence with the x-ray tube voltages, and the superimposed carbon emission lines, that for the three independent measurements, coincide.

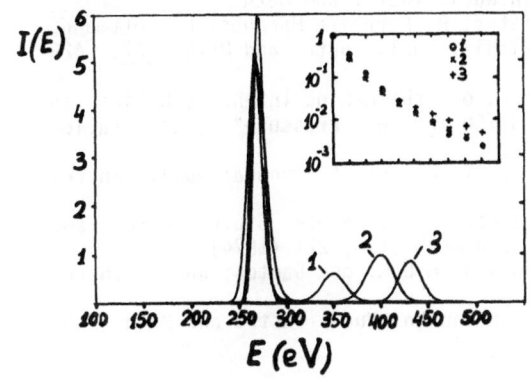

Fig.18.12 Spectrum reconstruction. The energy resolution is 5% for the C-K line (15 eV) (data from [18.27])

The accuracy of determination of the photo-absorption function and the number of detected photons limit the energy resolution that can be obtained. The useful energy range of operation is given by $0.5\ \varepsilon < E < 3\varepsilon$, ε being the optimum energy for resolution restoration; for the counter described and taking profit of the pressure variation, ε can be varied between 50 and 300 eV. As a further example and with $\sim 10^6$ photons in each line, for one line at 100 eV the energy resolution is $\sim 1\%$; for three lines at 70, 110 and 160 eV the energy resolution is $\sim 20\%$. More elaborated mathematical methods can improve these values.

18.6 Conclusion

A change of approach to the field of detector research, less empirical and relying more and more on understanding the fundamental physical processes, has made possible some significant advances, in particular through appropriate control of the role associated with the photon emission and propagation and its relationship to amplifying mechanisms in gaseous detectors.

From the data that were presented it seems that the use of the photon flux, associated in several ways with the interaction of x rays, its coupling to charge processes, the evolution of structure of wire and grid chambers, the evolution of the associated electronics and the application of modern mathematical procedures to data analysis, assures an important role for gaseous counters in the field of low energy x-ray detection.

Acknowledgements

Thanks are due to E.P. de Lima, M. Alegria Feio, M.A.F. Alves and R. Ferreira Marques. Financial support from Instituto Nacional de Investigação Cientifica-Centro de Fisica da Radiação e dos Materiais da Universidade de Coimbra is acknowledged.

References

18.1 A.J.P.L. Policarpo: Phys. Scripta 23, 539 (1981)
18.2 A.J.P.L. Policarpo: Nucl. Instr. and Meth. 196, 53 (1982)
18.3 Teresa H.V.T. Dias, A.D. Stauffer and C.A.N. Conde: IEEE Trans. Nucl. Sci. NS-30, 389 (1983)
18.4 M. Salete S.C.P. Leite: Portgal. Phys. 11, 53 (1980)
18.5 M. Alegria Feio and A.J.P.L. Policarpo: "Secondary Light Emission by Noble Gases", to be published in Nucl. Instr. and Meth.
18.6 H.P. von Arb, J. Böcklin, F. Dittus, R. Ferreira Marques, F. Kottmann, R. Schaeren, D. Taqqu and M. Wälchli: Nucl. Instr. and Meth. 207, 429 (1983)
18.7 R. Ferreira Marques: "Measurements of the K-Line Intensity Ratios in Muonic Hydrogen Between 0.25 and 150 Torr Gas Pressure" (Dissertation ETH n° 7111, Zürich 1982)
18.8 H. Inoue, K. Koyama, M. Matsuoka, T. Ohashi, Y. Tanaka: Nucl. Instr. and Meth. 157, 295 (1978)
18.9 A.J.P.L. Policarpo, M.A.F. Alves, M. Salete S.C.P. Leite and M.C.M. dos Santos: Nucl. Instr. and Meth. 118, 221 (1974)
18.10 M.A.F. Alves, A.J.P.L. Policarpo and M.C.M. dos Santos: Nucl. Instr. and Meth. 111, 413 (1973)
18.11 C.J. Hailey, W.H.M. Ku and M.H. Vartanian: Nucl. Instr. and Meth. 213, 397 (1983)

18.12 A.J.P.L. Policarpo, M.A.F. Alves, M.C.M. dos Santos and M.J.T. Carvalho: Nucl. Instr. and Meth. 102, 337 (1972)

18.13 M.R. Sims, G. Manzo, A. Peacock and B.G. Taylor: Nucl. Instr. and Meth. 211, 449 (1983)

18.14 A.J.P.L. Policarpo: Nucl. Instr. and Meth. 153, 389 (1978); G. Charpak, A. Policarpo and F. Sauli: IEEE Trans. Nucl. Sci. NS-27, 212 (1980); D.F. Anderson: Nucl. Instr. and Meth. 178, 125 (1981)

18.15 O.H.W. Siegmund, J.L. Culhane, I.M. Mason and P.W. Sanford: Nature 295, 678 (1982)

18.16 O. Siegmund, P. Sanford, I. Mason, L. Culhane, S. Kellock and R. Cockshott: IEEE Trans. Nucl. Sci. NS-28, 478 (1981)

18.17 O.H.W. Siegmund: "Study of an Imaging Gas Proportional Counter Used for the Detection of X-Rays" (Ph.D. Thesis, University College, London 1981)

18.18 O.H.W. Siegmund, S. Clothier, J.L. Culhane and I.M. Mason: IEEE Trans. Nucl. Sci. NS-30, 350 (1983)

18.19 M.A.F. Alves, A.J.P.L. Policarpo and M. Salete S.C.P.Leite: IEEE Trans. Nucl. Sci. NS-22, 107 (1975)

18.20 J.A.M. Bleeker, H. Huizenga, A.J.F. Den Boggende and A.C. Brinkmann: IEEE Trans. Nucl. Sci. NS-27, 176 (1980)

18.21 D. Anderson and G. Charpak: Nucl. Instr. and Meth. 201, 527 (1982)

18.22 G. Charpak, F. Sauli and R. Kahn: Nucl. Instr. and Meth. 152, 185 (1978)

18.23 G. Charpak, S. Magewski and F. Sauli: Nucl. Instr. and Meth. 126, 381 (1975)

18.24 S.S. Al-Dargazelli, T.R. Ariyaratne, J.M. Breare and B.C. Nandi: Nucl. Instr. and Meth. 200, 341 (1982)

18.25 G. Charpak, G. Peterson, A. Policarpo and F. Sauli: Nucl. Instr. and Meth. 148, 471 (1978)

18.26 G. Charpak: Nucl. Instr. and Meth. 201, 181 (1982)

18.27 E.L. Korasev, V.D. Peskov and E.R. Podolyak: Nucl. Instr. and Meth. 208, 637 (1983)

18.28 S.M. Kahn and R.J. Blisset: Astrophys. J. 238, 417 (1980)

19. The Detection of Soft X-Rays with Charged Coupled Detectors

P. Burstein and J. M. Davis

American Science and Engineering, Inc., Fort Washington
Cambridge, MA 02139, USA

The characteristics of an ideal soft x-ray imaging detector are enumerated. Of recent technical developments the CCD or charge coupled device goes furthest to meeting these requirements. Several properties of CCDs are described with reference to experimental work and their application to practical instruments is reviewed.

19.1 Introduction

The development of soft x-ray sensitive, electronic imaging detectors for scientific applications is a major concern of laboratories worldwide. In contrast to purely imaging applications scientific observations require quantitative, intensity and position information from the image. Therefore the characteristics of an ideal detector must include:

(1) A spatial resolution comparable to photographic film which implies a format with a large number of picture elements. In this case a large number is of the order 10^6 to 10^7.

(2) A high sensitivity, to achieve the efficient detection of single, incident photons, to minimize the degradation caused by system noise and to provide high temporal resolution.

(3) A stable transfer function between input and output in order to achieve a photometric accuracy of 1%.

(4) Broad spectral response, or quantum efficiency, covering the energy range from 0.2 to 20 keV coupled with energy resolution for single photons over the same range.

The search for a single detector which completely satisfies all these conditions has been largely unsuccessful. It has included imaging proportional counters [19.1], which combine very large areas, good energy resolution, but only moderate spatial resolution, microchannel plates with a variety of readout systems which have large areas, good spatial resolution but extremely limited energy resolution [19.2,3], and more recently and quite promisingly charge coupled devices or CCDs.

In the following paragraphs the properties of CCDs as they apply to soft x-ray detection and their application to scientific investigations is discussed.

19.2 Properties of Charge Coupled Devices

CCDs are closely spaced, two-dimensional arrays of MOS capacitors which are laid down on a silicon substrate, shown schematically in Fig.19.1. The capacitors are electrically isolated from each other by the p- and n-type archi-

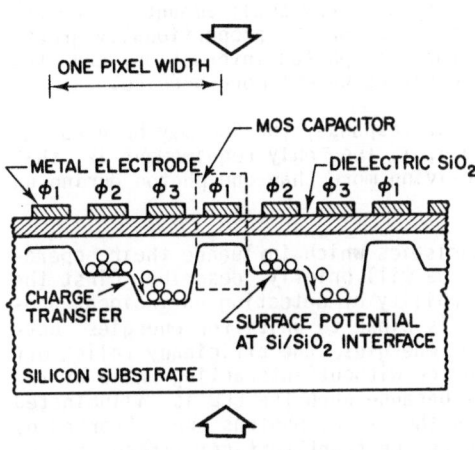

(FRONT-SIDE ILLUMINATED)

ONE PIXEL WIDTH

MOS CAPACITOR

METAL ELECTRODE
DIELECTRIC SiO$_2$
ϕ_1 ϕ_2 ϕ_3 ϕ_1 ϕ_2 ϕ_3 ϕ_1

CHARGE
TRANSFER
SURFACE POTENTIAL
AT Si/SiO$_2$ INTERFACE

SILICON SUBSTRATE

(BACK-SIDE ILLUMINATED)

Fig.19.1 Functional description of a typical three-phase CCD architecture illustrating front and back illumination, taken from MELEN and BUSS [19.7]

tecture of the device and by the applied voltages. A photosite, or picture element (pixel), consists of a set of three adjacent capacitors grouped in the columns of the array. The electrodes of the capacitors are independently controlled by "clock" or "gate" voltages. Because of this arrangement this type of CCD is known as a three-phase device.

When an incident x ray is absorbed in the silicon substrate it excites electrons into the conduction band which then diffuse into a depletion layer formed by the positive voltages applied to the electrodes. These applied voltages form a potential well which traps charge at a particular photosite. To read out the device the applied voltages are changed, or clocked, so that the charge at each photosite along a row is transferred vertically to the adjacent site in its column. The charges in the row formed by the lowest site in each column are transferred into a shift register where they are read out serially following on-chip amplification.

The key to the operation of the CCD as an x-ray detector is the use of the interaction site as the storage site. It can be thought of as an array of solid-state detectors each with its own memory for the CCD can accept photons over its entire surface simultaneously.

In "conventional x-ray detection" by the CCD [19.4,5] the detector is used exactly as for visible light detection, as a total energy detector. The output of a particular pixel is a charge which is proportional to the total amount of energy deposited in the pixel.

The spatial resolution can be determined largely on the basis of pixel-to-pixel spacing and charge localization between pixels.

When only one x-ray photon is known to have interacted in an element of the array, then the charge in that element will be a function of the photon energy, as in a solid-state detector. Thus, each pixel will have an associated energy to charge transfer curve, and hence an energy resolution curve. The energy resolution, to a first approximation, is that of a tiny solid state detector.

On the average one electron-hole pair is created in the pixel for every 3.6 eV of energy in the x-ray photon. This is a very small amount of energy when compared with other soft x-ray detectors. Thus, a proportionally greater number of electrons will be created for each photon interaction, and the associated Poisson (or Poisson-like) statistics become more precise.

Since all pixels have a similar type of response, the CCD may be used as a nondispersive x-ray spectrometer [19.5,6]. The only requirement is that the probability of any single pixel receiving more than one photon during an integration period be small.

CCDs have several important characteristics which influence their operation as x-ray imaging detectors which we will briefly describe. First the quantum efficiency, defined as the probability of detecting an incident photon, is a function of x-ray energy and is close to unity for energies between 1 and 10 kilovolts. At the higher energies the efficiency falls off because the photons pass through the device without interacting while at the lower energies the efficiency falls off because when the CCD is illuminated in the so-called front-illuminated mode the x-ray photons are absorbed by the electrode and insulating structures on the front surface. These structures form a dead layer between 0.5 to 2 microns thick, and to overcome their effect CCDs have been operated in a back-illuminated mode. In this case the silicon substrate is illuminated directly, and to maximize the efficiency its thickness is tailored to the particular application, a process known as thinning. This leads to a considerable improvement in sensitivity, and we have been able to detect carbon K_α x rays at 250 eV with RCA CCDs manufactured in this fashion.

An alternative approach has been developed by Texas Instruments, who have developed a virtual phase CCD [19.8,9]. In this device the three applied voltages of the three-phase CCD have been reduced to one. The steplike potential is created through the use of ion implants in the n-type buried channel. By reducing the number of polysilicon gates per pixel from three to one, the thickness of the dead layer can be substantially reduced. However, in our tests of such a front-illuminated device, we were unable to detect carbon K_α x rays, and so in this respect, the modification is not an adequate substitution for back illumination.

The intrinsic noise of a CCD limits both the energy resolution and the length of time a picture can be integrated. Noise levels of 30 electrons rms can be achieved corresponding to energy resolution of 250 eV. The energy resolution is essentially independent of energy [19.10], and therefore CCDs are better than proportional counters at energies above 500 eV and marginally worse at lower energies. The ultimate noise goal is of the order of 10 electrons rms, which would be set by the stray capacitance of a few hundred pFs between the on-chip preamplifier and the last transfer gate. Noise levels approaching this have been reported in the literature [19.11] which makes their energy resolution superior to proportional counters.

To achieve these noise levels, the CCD and the on-chip amplifier have been cooled. Typical operating temperatures are around -100 °C. However, the operating temperature can be made too low for there are other completing factors, of which the most important are the charge transfer efficiency and the leakage current, which are also functions of temperature. The charge transfer efficiency is the fraction of the original charge transferred from one pixel to the next during the readout process. Incomplete charge transfer results in a loss of both photometric accuracy and dynamic range and introduces smearing of the image. The leakage current is a measure of the charge of the spills from an illuminated pixel to adjacent dark pixels. In our tests with RCA CCDs we have found that an ideal operating temperature must be determined for each device, which is warm enough so that charge transfer efficiency is adequate ($\sim$ 0.99995 to 0.99999) while still cool enough to keep the leakage current acceptably low. This tradeoff has to be made on a device by device basis.

In principle the dynamic range of the CCD is limited by the readout noise and the full well capacity. The latter scales roughly as the pixel area, and for 30 μm square pixels the well capacity is $\sim$250,000 electrons. The typical dc level for the three-phase CCDs we have tested is several hundred electrons, which corresponds to a dynamic range of the order of or less than 10^3.

Although one should be able to extend the dynamic range by improving the noise characteristics of the preamplifier this is not necessarily true in the x-ray region. For a single x-ray photon produces a large number of electrons,e.g.,a one kilovolt x ray will contribute in excess of 300 electrons, and if this value is greater than the noise it will place the limit on the dynamic range which is thus energy dependent varying inversely with the incident photon energy.

Finally, CCDs have excellent linearity to increases in the incident x-ray intensity characteristics of solid-state detectors and pixel nonuniformities arising from processing variations and mask alignment errors during fabrication are generally quite small.

19.3 Applications

Although the primary incentive for our studies of CCDs has been their application to astronomical observations there are many other scientific investigations in which their sensitivity and excellent spatial resolution can be used to advantage. Examples are the in vivo examination, in real time, of biological specimens and the recording of the x-ray emission arising during the collapse of the fuel pellets used in inertial confinement fusion experiments.

In x-ray astronomy a heavy emphasis has been placed on obtaining observations with the highest spatial resolution. In this regard most electronic detectors have compared poorly to film. However, since film requires recovery which is impossible in most missions, astronomers have had to be satisfied with observations which were limited by the detector. As a numerical example, if we require one arc second resolution and we use a CCD with 15 μm pixels, a focal length in excess of 6 m is required for the optical system. Such arguments have lead to the choice of a 10 m focal length for the Advanced x-ray Astronomical Facility. The increased image size resulting from a large focal length is something of a mixed blessing. First, the field of view subtended by the detector is correspondingly reduced. For instance an 800 x 800 array of 1 arc second pixels subtends a field of just over 13 arc minutes. This can be compared with the diameter of the sun which is 32 arc minutes. Secondly, such an image contains a tremendous amount of information which has to be processed digitally. If the intensity scale is divided into 256 gray levels, the number of bits required to specify the image is in excess of half a million. Thus unless very high data rates are available, the transmission time for the image can be very much greater than the exposure time. This difficulty tends to negate the advantage provided by the high sensitivity of the CCD which is roughly three orders of magnitude better than photographic emulsions and, for solar observation, allows exposure times of a few to a few tens of milliseconds. In principle this should open up a new field of coronal observations involving the dynamics of the coronal structures, changes at boundaries due to magnetic reconnection and the mechanisms of flaring events at high spatial resolutions. However, because of the high telemetry rates that are required the full potential of these studies has yet to be realized.

An added advantage of the CCD which results directly from its ability to detect single x-ray photons is its ability to minimize the effects of blooming. Blooming is the spreading of charge which has accumulated in overexposed areas to adjacent pixels of the CCD and is common to all electronic imaging systems. In many astronomical applications, the source object has a very large dynamic range and thus the average flux from the region might dictate an optimum exposure which causes blooming somewhere else on the chip. However, if the CCD is used in the single photon or spectrometric mode, many short exposures which will not cause blooming can be summed electronically, without loosing spatial resolution because of the digital nature of the device, to provide a single image with an effective dynamic range larger than that of the CCD.

19.4 Conclusions

CCDs used for the detection of soft x rays are a relatively new technology. They hold great promise as astronomical x-ray imaging detectors combining high spatial resolution and energy sensitivity. In practical applications CCDs have both advantages and disadvantages over competing technologies. On the plus side, they are compact, low-power devices whose operation requires neither the use of high voltages or hard vacuums. On the negative side they have to be cooled to temperatures on the order of -100 $^{\circ}$C for optimum performance.

It is almost certain that in the next few years they will see wide application in a variety of space missions and their success will determine their future development.

Acknowledgements

This work has been supported by NASA through several contracts.

References

19.1 P. Gorenstein, H. Gursky, F.R. Harnden, Jr., A. DeCaprio, and P. Bjorkohlm: IEEE Trans. NS-22, 616 (1975)
19.2 R. Giacconi et al.: Astrophys. J. 230, 540 (1979)
19.3 J.M. Davis, J.W. Ting, and M. Gerassimenko: Space Sci. Instrum. 5, 17 (1979)
19.4 G. Renda and J.L. Lowrance: Proc. Symp. on CCD Technology, JPL SP 43-21, 91 (1975)
19.5 P. Burstein, A.S. Krieger, M.J. Vanderhill, and R.B. Wattson: Proc. SPIE 143, 114 (1978)
19.6 R.E. Griffiths, G. Polluci, A. Mak, S.S. Murray, D.A. Schwartz, and M.V. Zombeck: Proc. SPIE 244, 57 (1980)
19.7 R. Melen and D. Buss: Charge-Coupled Devices: Technology and Applications, IEEE Press (1977)

19.8 J. Janesick, J. Hynecek, and M.M. Blouke: Proc. SPIE 290, 165 (1981)
19.9 R.A. Stern, K. Liewer, and J.R. Janesick: Rev. Sci. Instrum. 54, 198
      (1983)
19.10 D.A. Schwartz,   R.E. Griffiths,   S.S. Murray,   M.V. Zombeck,   and
      W. Bradley: Proc. SPIE 184, 247 (1979)
19.11 S. Marcus, R. Nelson, and R. Lynds: Proc. SPIE 172, 207 (1979)

Part IV

X-Ray Microscopes

20. The Göttingen X-Ray Microscope and X-Ray Microscopy Experiments at the BESSY Storage Ring

D. Rudolph, B. Niemann, G. Schmahl, and O. Christ

Forschungsgruppe Röntgenmikroskopie, Universität Göttingen
Geismarlandstraße 11, D-3400 Göttingen, Fed. Rep. of Germany

20.1 Introduction

X-ray microscopy with zone plates as x-ray imaging elements and synchrotron radiation was started by our group in 1976 at the Deutsches Elektronensynchrotron (DESY), Hamburg [20.1]. This prototype x-ray microscope was equipped with a grazing incidence grating monochromator, a condenser zone plate, and a micro-zone-plate. The grating was a 100 mm laminar grating with 600 l/mm holographically made in our optical laboratory [20.2]. The zone plates resulted from early holographic work using commercially available optics for zone plate construction [20.1,3,4]. The obtained resolution was 200 nm [20.5].

The next step was to develop faster optics, zone plates with better f ratios. For this purpose new aplanatic optical systems were developed and used to make high-resolution zone plates holographically (micro-zone-plates and condenser zone plates) with the necessary correction of spherical aberration. A new microscope with these zone plates was established at the storage ring ACO at Paris Orsay in 1978. The obtained resolution was about 150 nm using the first and 70 nm using the second order of the micro-zone-plate [20.8].

To supply the micro-zone-plate with quasimonochromatic x radiation a slitless monochromator was used, consisting of a grazing incidence holographic laminar grating and a condenser zone plate for generating a reduced image of the synchrotron radiation source in the object field.

It is well known that a spherical wave passing a plane grating is affected with astigmatism as perpendicular to the grooves different parts of the wave meet the grating under slightly different angles. Hence the wave is dispersed differently and, therefore, the curvature of the monochromatic waves leaving the grating is not constant, the waves are no longer spherical waves but are affected with astigmatism. To eliminate this astigmatism a plane grating with variable line density was calculated and holographically built. The variation of the line density was done so that the curvature of the dispersed wave remains constant. This correction can be done exactly for one definite wavelength. After passing the grating the quasimonochromatic radiation is used to generate a demagnified image of the synchrotron source by the condenser zone plate. This image is located in the object field. Provided that the grating does not introduce aberration, the intensity mainly depends from the lateral demagnification. The aberration caused by the plane grating in our case reduced the intensity to one-third of the calculated value. This loss was avoided with the corrected grating. The exposure time with the new grating was about a factor of three shorter than before.

The next intention was to gain another factor of about 50 in the exposure time by using a zone plate linear monochromator instead of the grating monochromator. Precondition of this step was to reduce the monochromaticity necessary for operating the micro-zone-plate (compare [20.6] for details of the different micro-zone-plates and section 20.3.1 and 20.4. for details concerning the linear monochromator). To meet these requirements the aplanatic system designed for micro-zone-plate 1 (MZP1) with 625 zones was modified to generate MZP2 with only 100 zones requiring a monochromaticity of only $\lambda/\Delta\lambda = 100$, six times less than for MZP1. The gain in intensity and hence reduction of exposure time was twofold: firstly, cancelling the grating with an efficiency of about 12% yielded a factor of ~8 and second the extended $\Delta\lambda$ interval another factor of ~6. Another advantage is that the arrangement without grazing incidence optical element results in considerably simplified effort for adjustment. With the improved MZP2 a resolution of 50 nm was obtained [20.6,7].

The experiments at the ACO storage ring were suspended and the microscope was demounted in spring 1982. In April 1983 a slightly improved version of the microscope was installed at the BESSY storage ring in Berlin.

20.2 The Beam line of the Microscopic Area at BESSY

At the storage ring BESSY in Berlin at the magnet DIP 11R two beam lines, 11.11 and 11.12 are dedicated to x-ray microscopy. Figure 20.1 shows the arrangement of the beam lines of the microscopic area. Both beam lines emerge

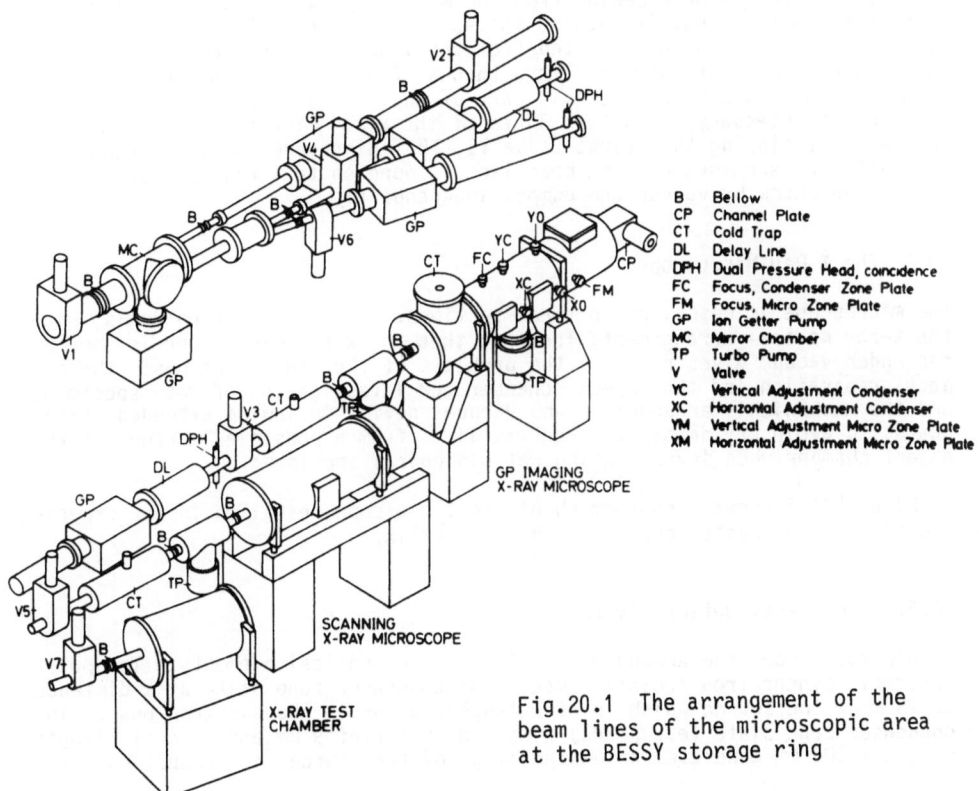

B   Bellow
CP   Channel Plate
CT   Cold Trap
DL   Delay Line
DPH   Dual Pressure Head, coincidence
FC   Focus, Condenser Zone Plate
FM   Focus, Micro Zone Plate
GP   Ion Getter Pump
MC   Mirror Chamber
TP   Turbo Pump
V   Valve
YC   Vertical Adjustment Condenser
XC   Horizontal Adjustment Condenser
YM   Vertical Adjustment Micro Zone Plate
XM   Horizontal Adjustment Micro Zone Plate

Fig.20.1 The arrangement of the beam lines of the microscopic area at the BESSY storage ring

from the tee-fitting behind the valve V1 with which the microscopic beam lines can be separated from the outlet of the storage ring. In the mirror
chamber MC the beam of the line 11.11 can be reflected into a third line
provided for an experimental chamber for testing x-ray optical elements
etc. This line as well as the line 11.11 itself which is dedicated to the
scanning x-ray microscope are not subjects of this paper. The beam line 11.12
is dedicated to the imaging x-ray microscope which has been operation since
April 1983.

20.2.1 The Beamline Safety System

The further composition of the beam line is discussed on the basis of the
safety system. An important component of the outlet system is the beam shutter and the fast-acting shutter located before the first valve V1 in
Fig.20.1. Whenever a vacuum accident in the area of the microscope occurs,
a pressure wave moves towards the ring. As soon as the wave meets the
sensor system DPH, the fast-acting shutter is activated with a reaction time
shorter than the transit time of the pressure wave. The delay line DL, a
stainless steel tube of about 60 cm length and 15 cm diameter with 9 copper
discs inside fitting exactly to the inner diameter and having a central bore
of 3 cm each, provides additional delay of the pressure wave. DPH is a dual
pressure head operating in coincidence for fail-safe reasons. Valve V2 between the two getter pumps in the line can be held open only if the getter
pump in front of the microscope is in operation and has good vacuum. The two
cooling traps CT assist in pumping the line and the microscope and prevent
especially water vapor creeping from the microscope into the beam line towards the storage ring. By closing the valve V3 between DPH and CT the whole
microscope region can be separated from the line. V3 is closed whenever the
microscope is out of duty for more than a few hours or the cooling traps
need deicing; a new vacuum in this part of the line is started by the turbo
pump TP. If necessary, all beam lines of the x-ray microscopic area can be
separated by closing the first valve V1. The distribution of the pumps to
the different sections of the beam line is done so that all sections which
can be separated by valves are pumped independently.

20.3 The X-Ray Microscope

The microscope is described in three sections. The first section deals with
the x-ray optical arrangement. The fact that the x-ray microscope is operated under vacuum gives rise to the problems of the other sections: the object preparation and the object chamber for investigation of wet specimens
and the mechanical arrangement and vacuum device including extended safe-
guards to avoid any impact on the beam line from a possible leakage of the
object chamber when investigating wet biological specimens.

Figure 20.2 shows a photograph of the microscope installed in the experimental hall for basic research at BESSY, Berlin.

20.3.1 The X-Ray Optical System

Figure 20.3 shows the arrangement of the x-ray optical elements. The polychromatic synchrotron radiation meets the condenser zone plate at a distance
of 15 m of the source which is the tangent point to the electron beam. The
condenser zone plate (e.g., KZP3) with a diameter of 9 mm and a focal length
$f_{4.5nm}$ = 304 mm generates a reduced image of the source. The reduction fac-

Fig.20.2   The x-ray microscope

tor is 48 at 4.5 nm. The diameter of the source in METRO configuration with BESSY running at 100 mA is about 1 mm. Together with the free diameter of the object chamber the condenser acts as linear monochromator. Using the micro-zone-plate MZP3 with r_1 = 1.76 µm, r_n = 27.8 µm and n = 251 [20.6], a monochromaticity of $\lambda/\Delta\lambda \simeq 250$ is necessary. The spectral resolution of a linear monochromator of this type is given by $\lambda/\Delta\lambda \simeq D/2d$ [20.4,8] with D = diameter of the condenser zone plate and d = diameter of the diaphragm which is the pinhole in the object chamber in this case. With a source diameter of 1 mm, a reduction factor of 48, and an object field of 20 µm Ø, $\lambda/\Delta\lambda \simeq 225$ approximately meets the requirements of MZP3 at 4.5 nm. The apodized central region of the condenser zone plate prevents zero-order radiation from reaching the object field and first-order radiation of the condenser arriving on the direct way, via zero-order of the micro-zone-plate, at the center of the image field. The micro-zone-plate generates a magnified image in the image field. The magnification depends on the resolution which has to be matched to the resolution of the recording medium. Using fine grain photographic layers and zone plates with a resolution up to 50 nm, x-ray magnifications between 100 and 500 are normally used.

The above-mentioned apodization of the condenser zone plate provides a magnified x-ray image free of contrast, reducing zero-order radiation of the micro-zone-plate. As a further improvement the installation of an additional stop between the MZP and the image plane is under development. This stop - as shown in Fig.20.3 - prevents radiation of all other orders than first order, which is used for image generation, from reaching the image field. This holds especially for the second order which in practice has next to the zero order the highest intensity in the image field and hence a considerable contrast-reducing influence. The enlarged image can be viewed using a channel plate (CP) for converting the x-ray image into the visible or it can be photographed directly.

195

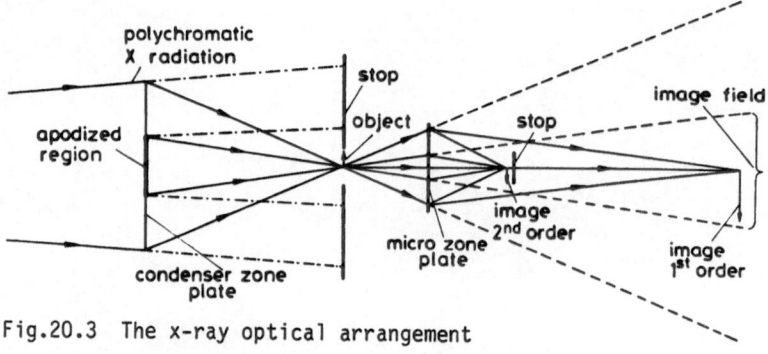

Fig.20.3 The x-ray optical arrangement

20.3.2 Object Chamber, Mechanical Arrangement, and Vacuum Device

In normal air x radiation is absorbed to I/I_0 = 1/e in a layer of 1.56 mm thickness at λ = 4.5 nm, for λ = 2.4 nm the layer is 0.57 mm thick. Therefore, the microscope has to be operated under vacuum conditions. The conditions of this vacuum, however, are very moderate (10^{-3} - 10^{-4} Torr) compared to the ultrahigh vacuum (< 10^{-8} Torr) of the storage ring in front of the microscope and to the high vacuum conditions (10^{-5} - 10^{-6} Torr) necessary for operating the x-ray image converters (CPs) at the other end. Amidst the low vacuum region intersecting the optical path between condenser and microzone-plate, the object chamber is located.

Figure 20.4 shows the specimen chamber for wet biological material in natural state. The chamber is filled and adjusted outside the microscope and then fitted into a special mount. The fitting proved to be better than a few

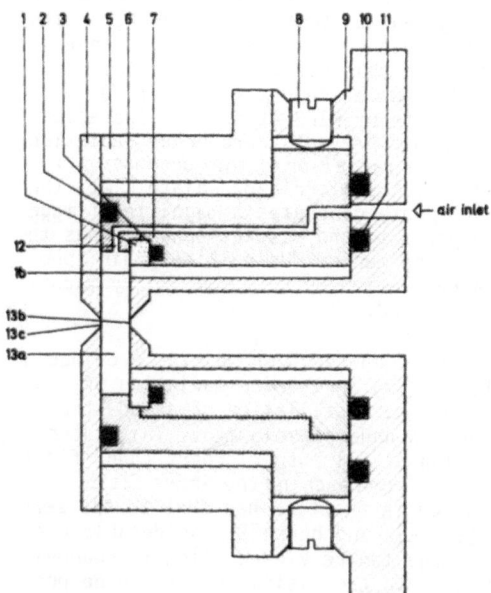

Fig.20.4 The environmental chamber for investigation of wet biological material. 1, glass ring with thin foil (1b); 4, 5, chamber closure; 2, 3, 10, 11, O-rings; 6, 7, adjustable glass ring holders; 9, base plate of the chamber; 8, adjusting screws; 12, air inlet channel; 13a, specimen chamber with radiation entrance (13b) and exit (13c) holes

microns so that only little adjustment of the focus is necessary between different fillings. The biological specimens are put onto a thin foil (1b) glued to a glass ring (1); mammalian cells are cultivated on this foil for investigation. The glass ring with the foil can be inserted into the chamber, the cells can be adjusted to the center of the pinhole (13b) with screws (8). The object chamber (13a) is separated from the outside vacuum by the object foil and a foil covering the exit hole (13c). The chamber volume is connected with a water reservoir. The volume pressure can be regulated separately up to full atmospheric pressure.

Figure 20.5 shows the schematic arrangement of the microscope. The beam enters the microscope via the entrance valve, passing a thin window consisting of a 0.5 μm polyimide foil evaporated with 40 nm Al (the Al coating towards the incident radiation). This window separates the ultrahigh vacuum of the beam line from the moderate vacuum of the microscope and the Al coating prevents visible light of the synchrotron radiation from entering the microscope. The entrance valve is closed whenever synchrotron radiation is not needed to keep radiation damage of the window and zone plates to a minimum. In the next compartment the condenser zone plate is mounted adjustable in x,y and along the optical axis to select the wavelength in combination with the free diameter of the object chamber (linear monochromator). The condenser compartment is pumped by a turbo pump to slightly better conditions as prevacuum. The next compartment separated by a valve contains the mounting for the object chamber, the micro-zone-plate, and the camera. The camera can be turned down so that the radiation enters the last compartment and meets the CP image converter after passing a 0.5 μm polyimide window and a valve. For changing the object the middle part is flooded; this can be done only when the manually operated valve to the CP compartment is closed. Flooding is done with dry nitrogen via an electromagnetic valve. Before this can be opened, the pneumatic valve to the condenser compartment is closed automatically. There are some additional fail-safe circuits: the power supply of the CP is controlled by the current of the ion getter pump, it cannot be operated - or it is switched off when in operation - under bad vacuum conditions. The whole microscope including the condenser compartment can be flooded only if the entrance valve is closed. Extremely bad vacuum in the middle compartment which may occur when the object chamber leaks closes the valve to the condenser compartment. It was already mentioned that the ion getter pump just in front of the entrance valve controls valve V2. Small amounts of water

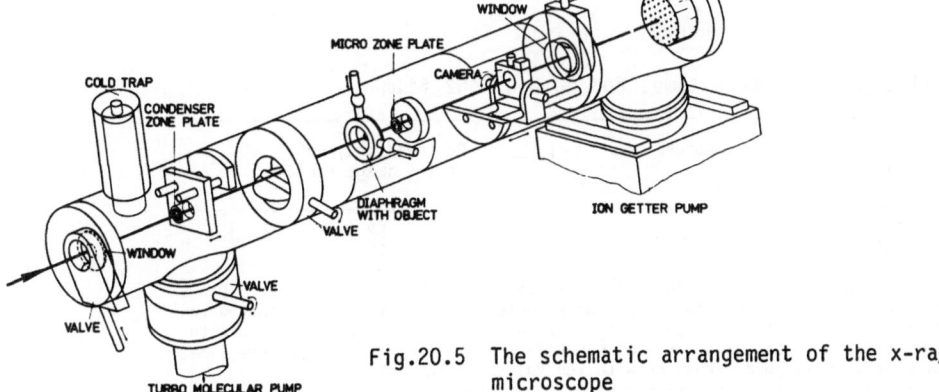

Fig.20.5 The schematic arrangement of the x-ray microscope

vapor which may escape from the object chamber are frozen in two cooling traps of the beam line, if necessary a third cooling trap in the condenser compartment can be added.

Before operating the microscope the zone plates have to be adjusted perpendicular to the optical axis, by using a preadjusted small helium-neon laser which can be flanged to the end of the microscope instead of the CP compartment. In this way the condenser is adjusted without object chamber and micro-zone-plate being in their positions. The second step is to insert and to adjust the micro-zone-plate. Thereafter the exact positions of the condenser and the micro-zone-plate are adjusted using synchrotron radiation. For x,y adjustment of the condenser the beam is at first adjusted to the fixed micro-zone-plate using an object chamber with a wide diaphragm, hereafter the object chamber is fine adjusted a small diaphragm to its exact position. All adjustments are optimized by controlling the x-ray intensity at the CP.

20.4 Experimental Results

An x-ray source must have a brilliance as high as possible to meet the requirements of x-ray microscopy. Spectral brilliance is defined as photons per (sec, sterad, bandwidth, emitting area). The BESSY storage ring is operated in different electron optical configurations. The configuration with the highest brilliance is called METRO. During our experiments up to now the source diameter was about 1 mm at an electron current of 150 mA. With these parameters we get $5 \cdot 10^{12}$ photons/ sec,$(0.64 \cdot 0.64$ mrad^2), $(\lambda/\Delta\lambda = 225)$, $(1 \cdot 1mm^2)$, (150 mA) at $\lambda = 4.5$ nm. To get an idea of the exposure times in the x-ray microscope we have to calculate the photon density in the image field. We have to assume the following parameters: the condenser generates a reduced image of the source, a reduction factor of 48 yields an image of 21 µm diameter. The condenser efficiency is 4%. The radiation has to pass three polyimide foils of 0,5 µm thickness each (1 window, 2 foils of the object chamber). This yields $2 \cdot 10^8$ phot/sec, µm^2 in the object field. Assuming a microzone plate with 3% efficiency and a magnification of 250x (500x) we get 6 (23) photons/sec, µm^2. Up to now for recording the x-ray images the Kodak film material HC and 2462 was used. To exposure HC film to a density 1 (development: Rodinal 1:40, 5^m, 18°-20°C) 58 photons/µm^2 are necessary [20.9]. Film 2462 has about the same quantum efficiency, half sized grain and needs about 120 photons/µm^2 under the same conditions. Table 20.1 shows the calculated exposure times for a magnification of 250x and 500x and for an object with 100% transparency.

Table 20.1 Exposure times for HC and 2462 film

Magnification	Exposure time [sec]		Photons/(sec, µm^2) in the image field
	HC film	2462 film	
250x	0.4	0.8	92
500x	1.6	3.2	23

The exposure time depends in addition on the absorption of the object. Diatoms with holes and highly absorbing structures are objects well suited to test the values of Table 20.1. Figure 20.6 shows diatoms photographed with an x-ray magnification of 250x on Kodak 2462 film made with an exposure time of 2 seconds in good agreement with the values of Table 20.1.

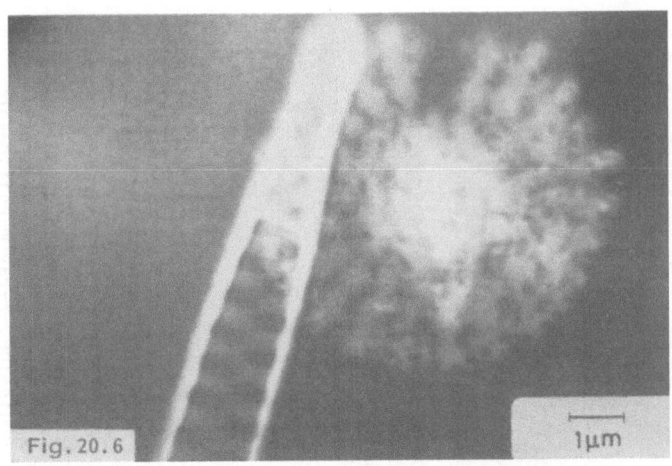

Fig.20.6 Diatoms, imaged with 4.5 nm radiation, x-ray magnification 230x

Figures 20.7-9 show different parts of giant chromosomes of cells of the salivary glands of larvae of *Chironomus thummi*.

Figure 20.10 shows a chloroplast from spinach, prepared from grinded lea- ves in a buffer solution [20.10] and glued to the object carrier foil with Poly-L-Lysine [20.11]. During the exposure the object chamber was loaded with full air pressure (760Torr) without showing any leakage. Exposure time was 7 sec with BESSY running at 145 mA. The crucial point of this experiment was to demonstrate that preparation of wet biological specimens and object chamber arrangement worked well and that exposure times of a few seconds can be obtained with such specimens. It could not be expected to achieve optimum contrast in these pictures made with 4.5 nm. For good contrast in wet biolo- gical specimens the wavelength has to be changed to 2.4 nm where the con- trast between protein structures and water is about a factor of 10 better.

Fig.20.7 Caption see next page

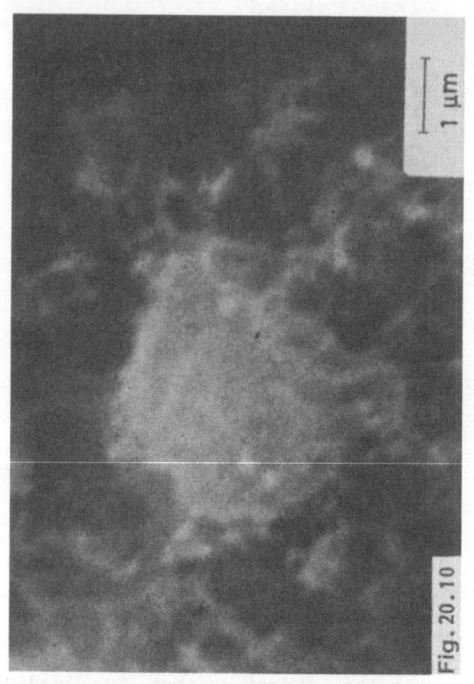

Fig.20.10 Chloroplast of spinach, imaged with 4.5 nm
radiation, x-ray magnification 200x, exposure time
7 sec, BESSY running at 152 mA

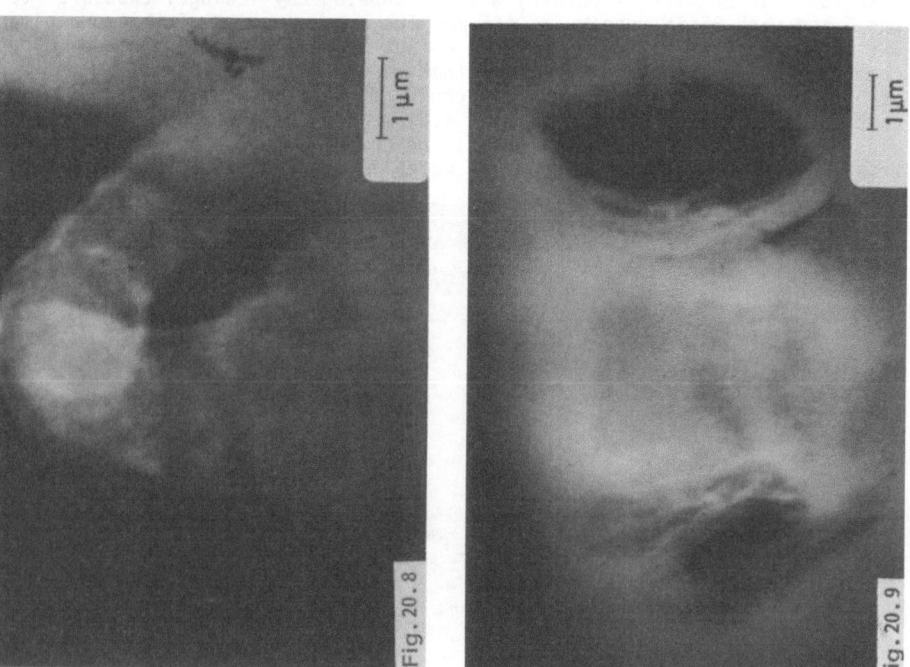

▶ Fig.20.7-9 Parts of giant chromosomes of cells of the
salivary glands of a larvae of *Chironomus thummi*, imaged with 4.5 nm x radiation, x-ray magnification 230x

It should be mentioned that during the experiments up to now damaging of zones plates occurred during direct irradiation only: in case of the condenser zone plate when the first window was broken, in case of the micro-zone-plate when used without the monochromator pinhole in the object chamber and hence irradiated with a more extended $\Delta\lambda$ interval. No zone plate has been lost during normal operation of the microscope up to now.

20.5 Further Developments and Improvements

The first experimental period in 1983 at BESSY was dedicated to the problems of installing the microscope and getting it into operation under the well known conditions with 4.5 nm radiation at which all parameters, e.g., efficiency of zone plates and the absorption of polyimide foils, were well known from laboratory measurements with C K_α radiation and from previous operation of the microscope at the ACO storage ring. The next step will be to switch over to 2.4 nm for investigation of biological specimens with optimum contrast. In addition, the experimental chamber of the second beam line will be installed to get experimental data of the x-ray optical elements and foils at different wavelengths, especially at 2.4 nm.

Improvement of zone plates, better f ratio as well as better efficiency by developing phase zone plates will be an important part of future work [20.6].

For biological specimens, or more generally for all specimens sensitive to radiation damage, the existing procedure to operate the microscope has a severe disadvantage: adjustment and prefocusing is done by CP control, fine focusing by photographing. By this procedure the specimen is loaded with a radiation dose of a factor of about 100 more than necessary for the final exposure. Preadjustment and focusing can be done with visible light as well. One possibility is to combine the micro-zone-plate with a phase zone plate for visible light on the same supporting foil. The specimen is then illuminated from behind with visible light. The wavelength must be chosen so that for this specific wavelength the focal length of the phase zone plate in the visible is the same as the x-ray focal length of the micro-zone-plate. This combination will enable preadjustment and focusing in the visible and hence reduce the radiation dose to an amount necessary to make the x-ray image. The final goal of this development will be a new arrangement specially matched to biological applications. One will be able to inspect a large area of, e.g., cultivated cells and select a small interesting object in the visible. After having adjusted this detail to a certain position in the object field it can be imaged with x rays, applying an x-ray radiation dose of not more than necessary for only this single imaging process.

Acknowledgements

The development of the x-ray microscope was supported by the Stiftung Volkswagenwerk. The experiments at the BESSY storage ring are supported by the Bundesministerium für Forschung und Technologie. We are grateful to the BESSY staff for good experimental conditions. We are indebted to the workshops of the Universitäts-Sternwarte Göttingen for good cooperation and to V. Sarafis for discussions concerning the preparation of the plant cells. We thank Dr. M. Robert-Nicoud for the preparation of the giant chromosomes.

References

20.1 B. Niemann, D. Rudolph, and G. Schmahl: Appl. Opt. 15, 1883 (1976)
20.2 G. Schmahl and D. Rudolph: "Holographic Diffraction Gratings" in Progress in Optics, ed. E. Wolf, Vol. XIV, 195-255 (1976)
20.3 G. Schmahl and D. Rudolph: Optik 29, 577-585 (1969)
20.4 B. Niemann, D. Rudolph, and G. Schmahl: Opt. Comm. 12, 160-163 (1974)
20.5 G. Schmahl, D. Rudolph, and B. Niemann: Journal de Physique C4, suppl. au no.4, Tome 39, 202-204 (1978)
20.6 G. Schmahl, D. Rudolph, P. Guttmann, and O. Christ: "Zone Plates for X-Ray Microscopy", this volume, no.8
20.7 B. Niemann, G. Schmahl, and D. Rudolph: "X-Ray Microscopy: recent developments and practical applications", SPIE Proc. Vol. 368, 2-8 (1982)
20.8 B. Niemann, D. Rudolph, and G. Schmahl: NIM 208, 367-371 (1983)
20.9 B. Niemann: "Detective Quantum Efficiency of Some Film Materials in the Soft X-Ray Region" in Ultrasoft X-Ray Microscopy: Its Application to Biology and Physical Sciences. Ann. N.Y. Acad. Sci. 342, 230-234 (1980)
20.10 D. Spencer, and S.G. Wildman: Austr. J. of Biol. Sci. 15, 599-610 (1962)
20.11 V. Sarafis: private communication

21. Recent Results from the Stony Brook Scanning Microscope

H. Rarback, J. M. Kenney, and J. Kirz Physics Department
State University of New York, Stony Brook, NY 11794, USA
M. R. Howells National Synchrotron Light Source
Brookhaven National Laboratory, Upton, NY 11973, USA
P. Chang, P. J. Coane, R. Feder, P. J. Houzego, D. P. Kern, and
D. Sayre IBM T. J. Watson Research Center, P.O. Box 218
Yorktown Heights, NY 10598, USA

21.1 Introduction

We have recently completed our first run [21.1] with the Stony Brook scanning microscope at the National Synchrotron Light Source (NSLS), using a Fresnel zone plate fabricated at IBM as the principal focusing element. The brightness of this synchrotron radiation source coupled with our high-resolution optics have made a substantial improvement in resolution and throughput compared with our earlier work [21.2,3] which lacked these features.

The microscope images biological specimens by forming a submicron spot of monochromatic soft x rays which it scans in a raster fashion across the specimen, located in a one-atmosphere environment. The resolution is presently determined by the size of this spot. The transmitted radiation is detected by a flow proportional counter, the output of which is used to control a computer-generated real-time image. Since the specimen is in air, the inefficient optics are all located upstream of the specimen, and the detector is over 50% efficient, the radiation dose to unstained wet natural specimens is minimized. We believe [21.4] that the resolution limit for scanning microscopy of unfixed specimens will be set more by damage considerations than by optics; our scheme aims to minimize this damage.

This paper considers the components of the microscope in turn, starting from the monochromator and the optics. How we test our optics is considered in some detail as it illustrates the microscope's capabilities as well as the optical performance achieved to date. Some of our first images of biological specimens are then presented. We conclude with some remarks about the future of the microscope.

21.2 The Source for the Microscope

The microscope uses radiation monochromatized by a Toroidal Grating Monochromator (TGM) located on port U15 of the VUV ring at the NSLS [21.5]. The characteristics of the monochromator are summarized in Table 21.1.

The toroidal grating both disperses and focuses the radiation; there are no other reflections. The monochromator is the most grazing incidence TGM in existence and has performed remarkably well as a source for the microscope. Its efficiency seems at least as good as any design we have seen. In addition, the single reflection and straightforward scanning mechanism (a simple rotation driven by a sinebar-linked stepping motor) allow for a simple, easily aligned instrument with small vacuum enclosures.

The TGM accepts only about 1 mrad^2 of radiation. This does not limit the microscope's throughput as can be seen from the following argument. For

Table 21.1 TGM characteristics

Grating type	Holographic, ion milled, gold coating[1]
Grating grooves	600 1/mm, laminar, 120 A deep
Ruled area	15 x 50 mm[1]
Radii of curvature	0.19 m x 88 m
Plane of dispersion	Vertical
Total deviation	5.72°
Object distance	3.024 m
Image distance	6.061 m
Scan range	15 - 50 A
Magnification at 32 A (h x v)	2 x 0.9
Resolving power at 32 A	400, aberration-limited
Resolving power at 32 A	800, design source size-limited
Measured resolving power	70-200, depending on beam conditions
Measured efficiency at 32 A	~15%

diffraction-limited scanned imagery at wavelength λ, we can only accept a volume in phase space of about $(2.44\lambda)^2$ [21.6]. Even for a source of dimensions ~0.1 mm, this still corresponds to a solid angle two orders of magnitude less than our monochromator acceptance! So efficiency is much more important than solid angle collection in achieving acceptable scan times.

Since the monochromator uses the electron beam as an "entrance slit", its energy resolution is presently limited by the vertical size of the electron beam. During the period of our run the ring was plagued by various instabilities which produced a vertical size over 1 mm, giving the resolving power listed in the Table 21.1 and causing chromatic aberration in our zone plate to be the limit to our spatial resolution. Subsequent to the period of our run, the beam size has been reduced and the resolving power of the TGM is now routinely[2] 200 at 32 A.

The problem of carbon being cracked onto optical surfaces exposed to intense synchrotron radiation is well known [21.7]. This contamination is especially serious for imaging applications near the low transmission side of the carbon K edge, where our microscope operates. We have made every effort to keep a good hydrocarbon-free vacuum in the TGM chamber. Our base pressure is ~1 x 10^{-10} torr. After over a year of seeing beam, the grating has developed a barely visible brown streak down the central third, which is the only area that presently sees beam. By masking this area we can expose a fresh area to beam and thus delay the time when grating refurbishing becomes necessary.

21.3 The Optical Components of the Microscope

The function of the microscope's optics is to focus the beam emerging from the TGM into as small a spot as diffraction will allow. Magnification is accomplished electronically by varying the ratio between the mechanical scan dimensions and the image raster dimensions. The first optical element the beam sees after emerging from the TGM exit slit is a small (~1 cm²) fused quartz optical flat inclined at 2.86° to the beam. The function of the flat is twofold. First, it acts as a relay mirror to remove the vertical compo-

---

1) Astron Developments, Ltd.
2) B.X. Yang, private communication

nent of the beam's direction and thus allows the microscope to lie level. Second, because quartz has a reasonably sharp critical angle cutoff below about 20 A, the mirror removes the higher-order spectral contamination present in the TGM output ($\sim$15% at 30 A). This serves to improve the signal to background ratio when imaging thick specimens. The measured reflectivity of the mirror at 30 A is 60%, in agreement with previously published values [21.8].

After reflection from the mirror the beam is about 1 mm high and 3 mm wide with a vertical divergence of 0.5 mrad and a horizontal divergence of 1 mrad. Only a very small fraction of this radiation can be accepted and still produce an acceptably small image spot. A pinhole serves as the "source" for our zone plate demagnification. We can select a pinhole from a series ranging in diameter from 20 to 200 μm, thus trading resolution for flux. The zone plate then images this source onto the specimen through a thin window so that the specimen can sit outside the vacuum. The characteristics of the zone plate will be discussed more fully in section [21.7].

The specimen is mechanically scanned through this spot. The transmitted x rays are then detected by a proportional counter which need have no spatial resolution. The optical parameters are summarized in Table 21.2. The resolution will also be more fully discussed in section 21.7.

Table 21.2 Typical optical parameters

Source size	30 μm
Source distance	600 mm
Zone plate diameter	90 μm
Image distance	5 mm
Geometrical image size	0.25 μm
Theoretical resolution	0.2 μm
Measured resolution	0.3 μm

21.4 Mechanical Components of the Microscope

The function of the mechanical components is to: align the microscope to the beam; dampen vibrations; monitor the beam position and intensity; scan the specimen to the beam; and detect the transmitted radiation. The entire instrument, from source to alignment microscope, is rigidly mounted on a one meter optical bench which is vibration-isolated from the laboratory floor via a massive air table and from the rest of the beam line via welded bellows. We have not seen any vibration problems so far.

The sources are mounted on a two-dimensional mechanical feedthrough so as to get them into the brightest area of the beam. The area surrounding the sources is coated with P31 phosphor (visible through a window) which aids in rough alignment. (We have observed that this phosphor is particularly efficient in converting soft x rays into visible light.) The beam emerging from the source (typically 10^7 photons/s) is monitored by a transmission photodiode. This is a thin layer of cesium iodide evaporated onto $\sim$500 A of aluminum and unsupported over 2 mm. The transmission of the two foils is about 60% at 30 A. We detect the secondary electron current emitted from the downstream side of the CsI with an electrometer. This signal (typically 10^{-12} A) is monitored continuously and is used to normalize the image to the incident intensity which can change appreciably over the time necessary to make a high-resolution image.

The zone plate is mounted on a small housebuilt x,y,θ,φ manipulator at the end of the vacuum pipe. It also is surrounded by P31 phosphor visible through a mirror and window for a rough alignment. Translation of the zone plate is then optimized by observing the count rate with no specimen present. Tilting the zone plate is a little trickier and requires measurement of the astigmatism which will be discussed in section 21.7.

The specimen is mounted on a two-dimensional homebuilt scanning stage. The specimen is constrained to move along almost perfect straight lines by two sets of parallelogram phosphor-bronze "leaf springs" (Fig.21.1). These motions are driven by piezoelectric translators with a maximum range of 50 μm which are in turn controlled by high voltage operational amplifiers. The scanning stage motion is monitored by two linear variable differential transformers (LVDTs). The position monitoring is good to about 0.05 μm over the course of a picture and is limited mostly by electronic noise. The scanning stage is mounted on a conventional three-dimensional mechanical stage for alignment and focusing. These motions are driven by DC motors with integral optical encoders. If a field size larger than 50 μm is required for low magnification alignment these motors can also be used to scan the specimen.

Specimen change is accomplished in a few minutes. The specimen holder is attached magnetically to the scanning stage which is easily accessible and also visible through the alignment microscope. Rough focusing (to 100 μm) can be accomplished using the optical microscope. Fine focusing requires actual x-ray images. The specimen stage is under computer control which will be discussed in section 21.6. The specimen area is surrounded by a flexible clear plastic dust enclosure. When desired, this area can be flooded with

Fig.21.1 Scanning stage -- the specimen is at the top left

dry or humidified helium for increased x-ray transmission and /or moisture retention with wet specimens.

Our detector is a homebuilt flow proportional counter operated at 1 atmosphere of P10 gas (argon/methane). The counter body is made from 1/4" copper tubing, its entrance window is 1000 A of carbon supported on a coarse mesh, and it has a rear optically transparent window. We use an easily replaceable 12 μm gold-coated tungsten wire as an anode, which is located less than 3 mm from the entrance window. Typical operating voltages are only about 1200 V. Because of its small size and relatively low gas gain, count rate response is reasonably linear to $\sim 3 \times 10^5$ Hz although typical counting rates for diffraction-limited imaging have never exceeded 10^4 Hz. The detector has enough energy resolution at 30 A to do a fair job of discrimination against second and third order TGM radiation.

The final mechanical component is our alignment microscope. This is a 60x microscope with a working distance of 30 mm. Besides being used to focus and align the specimen, it can check the condition of the exit window as well as the counter's entrance window and central wire, all without disturbing the optical alignment.

21.5 Vacuum Considerations

As mentioned previously, the TGM operates at 10^{-10} torr. Since we bring the beam out of the vacuum through a thin window which must hold off one atmosphere, some leakage through the window is inevitable. In addition, easy access to the zone plate is desirable. Finally, other experiments sharing our beam line including contact x-ray microscopy involve exposing the microscope vacuum to organic x-ray resists and biological specimens. For all these reasons it is desirable to separate the UHV region (up the quartz mirror) from the rougher vacuum ($\sim 10^{-7}$ torr) where the zone plate and sources reside.

We isolate these regions by a "contamination barrier", a 1200 A thick IBM-fabricated foil of silicon nitride, unsupported over an area of 2 x 8 mm^2. These windows are made by chemical vapor deposition of Si_3N_4 on a silicon wafer which is then masked and anisotropically etched away. To take the heat load when the TGM is run in zero order, we overcoat both sides of the Si_3N_4 with ~ 400 A of aluminum. The transmission of the contamination barrier is over 50% at 32 A, but of course drops off considerably on the low transmission side of the nitrogen K edge. Our experience with this barrier indicates it can easily hold off a few torr pressure difference.

We have found Si_3N_4 to be an excellent thin window for x-ray work. It is extremely strong and brittle: if a foil is going to break, it usually does so upon initial pumpdown when no harm can come to the delicate optics. We use similar uncoated foils both for exit windows and zone plate substrates. Our exit window has been as large as a 250 μm square of Si_3N_4 which can hold off one atmosphere without breaking or leaking unacceptably.

Our vacuum safety system must contend with the possible catastrophic rupture of one of these windows and is comprised of a buffer volume, two interlocked valves and an accoustic delay line. The region between the contamination barrier and exit window contains a large unused buffer volume. Immediately downstream of the contamination barrier and upstream of the exit window are two interlocked pneumatically operated valves. In the event that the exit window should break both valves close and the buffer volume assures that the pressure rise will not threaten the contamination barrier. If the contamination barrier should break, only the valve near it closes. If all of this fails, a fast valve sensor immediately downstream of our exit slit will trigger the fast valve in the front end of our beam line to shut and our acoustic delay line upstream of the exit slit will slow the shock wave sufficiently so that the ring's vacuum is protected.

21.6 Control and Data Acquisition

The function of the control and data acquisition hardware and software is to automate the operation of the microscope and yet allow the operator easy control over those functions which require his intervention. In addition, the images must be stored for postprocessing and viewing. The hardware for the system is centralized in a dedicated PDP 11/23 microcomputer, CAMAC interface and color image processor.[3] The software is a set of FORTRAN and assembly language programs which are interactive, fast, and "user friendly".

The operator can specify the microscope parameters from an extensive menu so as to align, focus and quickly zoom in on the interesting areas of a specimen. He can specify the scan characteristics such as pixel size, field size, starting position, fast scan direction, and pixel dwell time. He can specify the image presentation in both color and black and white versions; the color scale translation into either linear or logarithmic false-color values; a real time graphics presentation of incident intensity, transmitted intensity, or stage position. He can invoke an "autofocus" feature whereby a series of scans are produced automatically at various specified axial distances. He can suspend a picture, interactively examine the acquired data, and then resume the picture with more appropriate parameters.

The inputs to the hardware consist of logic pulses from the detector's counting chain, analog current from the I_0 monitor, and analog voltages from the position transducers. These signals are all digitized and stored in computer memory. The computer controls the scanning stage through a closed-loop feedback system accurate to about 0.1 µm. The image is presented in real time on a color TV display (512 x 512 x 8 bits resolution), on a black and white TV display, and on graphics display on the operator's console. All of the above can be done in close to 1ms/pixel although present data rates are about two orders of magnitude slower than this. When faster optics and brighter sources become available, the software will be ready to handle the increased rates.

The images can be stored on a hard disk or diskette. A single image typically requires about 100 kilowords of mass storage. Simple image processing can be done on these stored images including changing contrast and color levels, regularizing the raster to correct for small irregularities in the motion, stitching multiple images into a mosaic, and simultaneous display of multiple images. Future software will allow for more sophisticated processing including resolution improvement by deconvoluting the image with the zone plate point-spread function, edge enhancement, digital filtering, etc.

21.7 The Fresnel Zone Plate

The zone plate is the most critical component of the microscope. So far only these optical elements have achieved submicron resolution with soft x rays.

---

[3] Peritek Corp.

The authors have embarked on a collaborative effort to produce and test what we hope will be a continuing series of ever improving zone plates. The first zone plates to be produced in this effort had an outer zone width of 2500 Å, followed soon thereafter by the 1500 Å zone plates used to produce the images shown in this paper. We soon hope to have zone plates of 1000 Å outer zone width.

The theory of how a zone plate focuses x rays is well known [21.9]. The specification of our zone plate is a little unusual in that a significant fraction of its area is blocked or "apodized". This is because most of the radiation transmitted by the zone plate is undiffracted (zeroth order) radiation. In order to block this radiation, a collimator is placed between the zone plate and specimen which is illuminated by the first-order spot in the shadow of the central stop as illustrated schematically in Fig.21.2. For mechanical reasons the collimator cannot be too close to the moving specimen and this requires that the apodized area covers an appreciable fraction of the zone plate.

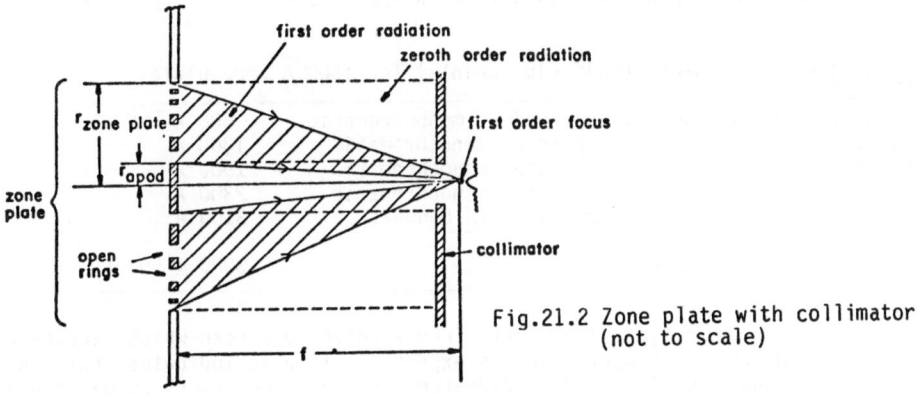

Fig.21.2 Zone plate with collimator (not to scale)

The geometry of our first zone plates is indicated in Table 21.3. Notice that the "opaque" zones need not be fully opaque to produce first-order diffraction efficiency near the ideal $1/\pi^2$ [21.10]. Thus the 10% transmission of the opaque zones is not of concern. However, the apodized area must be truly opaque in order not to swamp the focused signal with background.

Table 21.3 Zone plate geometry

Outer diameter	89 μm
Apodized diameter	48 μm
Total number of Fresnel zones	140
Number of open zones	50
Focal length at 32 Å	5 mm
Outer zone width	0.15 μm
Material of opaque zones	gold
Thickness of opaque zones	0.14 μm
Thickness of apodized area	0.3 μm
Specimen-collimator distance	1.2 mm
Diameter of rad. at collimator	22 μm

Our procedure for testing the resolution of our zone plates is a knife edge test: we scan a fine silver bar (part of a mesh with 5 μm wide bars) linearly through the focal spot and observe how far the edge must move to change the transmitted intensity from 25-75% of its total change. Repeating the test with an orthogonal scan defines the spot size rather well. But first we had to calculate what edge profiles a theoretically perfect zone plate would produce in this same test. Four factors unrelated to the zone plate quality could contribute to a broader measured profile than the theoretical Airy disk image of a point source:

1. the apodization has the effect of throwing flux into secondary lobes around the principal maximum [21.11]
2. our source is not a true point
3. our knife edge is not perfectly sharp
4. our illumination is not perfectly monochromatic.

The shape of the knife edge was measured in a scanning electron microscope. The size of the source was measured in an optical microscope. The results for our 1500 A zone plate are indicated in Table 21.4.

Table 21.4 Edge change (half-flux points) for 1500 A zone plates

Unapodized, point source, sharp edge, monochromatic	900 A
Apodized, point source, sharp edge, monochromatic	1300 A
Apodized, point source, fuzzy edge, monochromatic	1900 A
Apodized, finite source fuzzy edge, monochromatic	2300 A
Apodized, finite source fuzzy edge, 1% bandpass	3100 A
Measured edge change	3000 ± 500 A

Figure 21.3 shows typical raw data from a knife edge scan which indicate that the zone plate is performing as expected. It also indicates that the total background outside a 2 μm diameter is over two orders of magnitude less than the central flux.

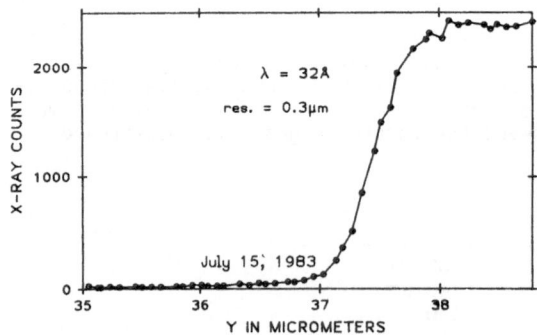

λ = 32A

res. = 0.3μm

July 15, 1983

Fig.21.3 Knife edge test of resolution

Because the zone plate has a low (f/50) numerical aperture and images on axis, we do not expect optical aberrations to be a severe problem. In fact, the spherical aberration correction was not.even taken into account in building our first generation zone plates. We have already mentioned the effect of chromatic aberration as a slight degradation of our spot size. If the zone plate has a total of n Fresnel zones (including any covered by a cen-

210

tral stop), then the illumination should be monochromatic to $1/2n$ or better in order not to introduce significant chromatic aberration. Note that this criterion is more stringent than the monochromaticity produced by a zone plate used as a spectrometer. In our application, a resolving power of 200 should produce a good focus.

The other significant aberration we encountered is astigmatism. In an electron-beam microfabricator controlled by Cartesian scanning axes, it is extremely difficult to produce perfectly circular rings. If the scan axes are not quite orthogonal or if the gains of the two axes are not perfectly matched, the zones will be elliptical. In fact, due to nonorthogonality in the scan axes, our first zone plates have a 2% difference in major and minor radii. If they are used on axis, they produce two line foci with 4% different focal lengths. The circle of least confusion would then have a diameter about twice the theoretical minimum. To correct for this astigmatism, we needed to tilt the zone plates about 12°. We have calculated that the additional aberrations produced by this tilt (chiefly coma) are negligible.

The alignment of the collimator with respect to the zone plate took some care. As can be seen from Table 21.3, the position of the collimator must be good to about 5 μm. If it is misaligned, then not only do we lose intensity; but, since the collimator is now the defining aperture, diffraction will further spread the beam.

We measured the absolute efficiency of the zone plate into first order. In order to do so, we mounted the zone plate out of the vacuum and replaced the knife edge with a 5 μm pinhole located at the first-order focus. By measuring the count rates before and after the zone plate was removed and knowing the ratio of the area of the pinhole to that of the zone plate, we determined the absolute efficiency to be 1%. This corresponds to a diffracting efficiency of 3% after accounting for the transmission of the zone plate substrate. This is a factor of three less than the diffracting efficiency of an ideal amplitude zone plate. Some of the loss of efficiency can be explained by the fact that our open zones were smaller than our opaque zones. Scattering from irregularities at the zone boundaries probably also contributed to the loss of efficiency.

We also measured the relative efficiency of diffraction into second and third orders by comparing the count rate through the 5 μm pinhole located at the first-, second- and third-order foci. Both second- and third-order count rates were a factor of ten less than first order. The fact that second order was present at all indicated that our open and opaque zones were of unequal widths.

The details of the fabrication of our zone plates are described elsewhere [21,12,13]. Briefly, the electron-beam microfabricator has a minimum spot size of 50 A and a design total distortion of less than 200 A over a high resolution field of about 250 μm. A specially built polar to Cartesian real-time converter enables the polar coordinate dataset to be rapidly written in the machine's coordinate system. Since the fabrication of fine lines with high aspect ratios is difficult, the central stop is thickened by a two step exposure process. A plating base is laid down over a Si_3N_4 substrate, followed by 2000 A of PMMA resist. The central stop is then written with a higher dose than the zone plate's opaque rings. The center is developed to the bottom and plated up with gold to about 1500 A. Then the rings are developed and an additional 1400 A of gold is plated. The result is a supported zone plate which is relatively robust. The disadvantage at present is the relatively low (∼30%) transmission of the substrate and plating base.

21.8 First Results

Almost all our first images were made at 32 A, on the high transmission side of the nitrogen K edge. We selected this wavelength for several reasons: the monochromator has its peak output there; wet organic specimens show good contrast; and our windows and air path are reasonably transparent at this wavelength. At 100 mA of beam current, we were able to get 10,000 photons/s into our detector using a 30 μm source pinhole. Our chromatic aberration-limited resolution was about 3000 A as discussed above. The depth of field was about 40 μm, more than adequate for our specimens.

Figure 21.4(a) is an example of the real-time display of an image of our test object, a mesh with 5 μm wide grid bars. The mapping from x-ray counts collected to pixel color is given by the color bars to the right of the picture. The bottom image of Fig.21.4(a) was made by selecting a smaller pixel size and shows that the mesh is not in focus, since the transitions take more than a single step of 0.65 μm. These images took only a few minutes to collect. Figure 21.4(b) is a lower magnification view of the same mesh and demonstrates the accuracy of the scanning feedback system over a field of 200 μm. (The object on the mesh is probably a piece of dust.)

In Fig.21.5(a), we see the ability of soft x rays to image internal structure in wet specimens. The specimen is an alga, *Phaeodactylum tricornutum*, which was mounted wet in the living state. Shortly before the exposure the specimen was drained and a thin carbon foil floated on top of the grid to keep it hydrated. The black area is the cell's nucleus; other cellular structures are visible. Absorptivity changes can be seen in one pixel step of 0.2 μm. This 100^2 pixel image took about one hour. The field of Fig.21.5(b) shows part of a neuron from a rat that was cultured on a thin substrate. The specimen was then fixed and dried. The cell body itself is rather thick and almost opaque, but two processes which communicate with a nearby cell (out of the field) are clearly visible.

21.9 Plans for the Future

A scanning microscope is ideally suited for differential absorption microanalysis [21.14]. In the near future we plan to implement a system to rock the grating on either side of an absorption edge and directly obtain differential absorption micrographs. The calcium L_{III} edge is both biologically interesting and easily accessible and will probably be the first element for which we search.

Improvements in resolution should also be forthcoming. With a smaller vertical beam size and our next generation zone plates, we hope to improve our resolution by a factor of two. Further improvements will involve more careful attention to minimizing the zone plate aberrations. Our ultimate hope is to attain ∿200 A resolution; this will require a better scanning stage and position feedback system as well as zone plates surpassing the present state of the art.

The main disadvantage of a scanning system compared to direct imaging is the longer times necessary to make an image. Brighter sources are necessary to image thick specimens with large fields at high resolution. Our focused intensity is sensitive only to the horizontal brightness of the source. Since the horizontal emittance of the VUV ring is already close to design, we expect about a factor of five improvement in the horizontal brightness when the ring achieves design current. With more efficient zone plates, we

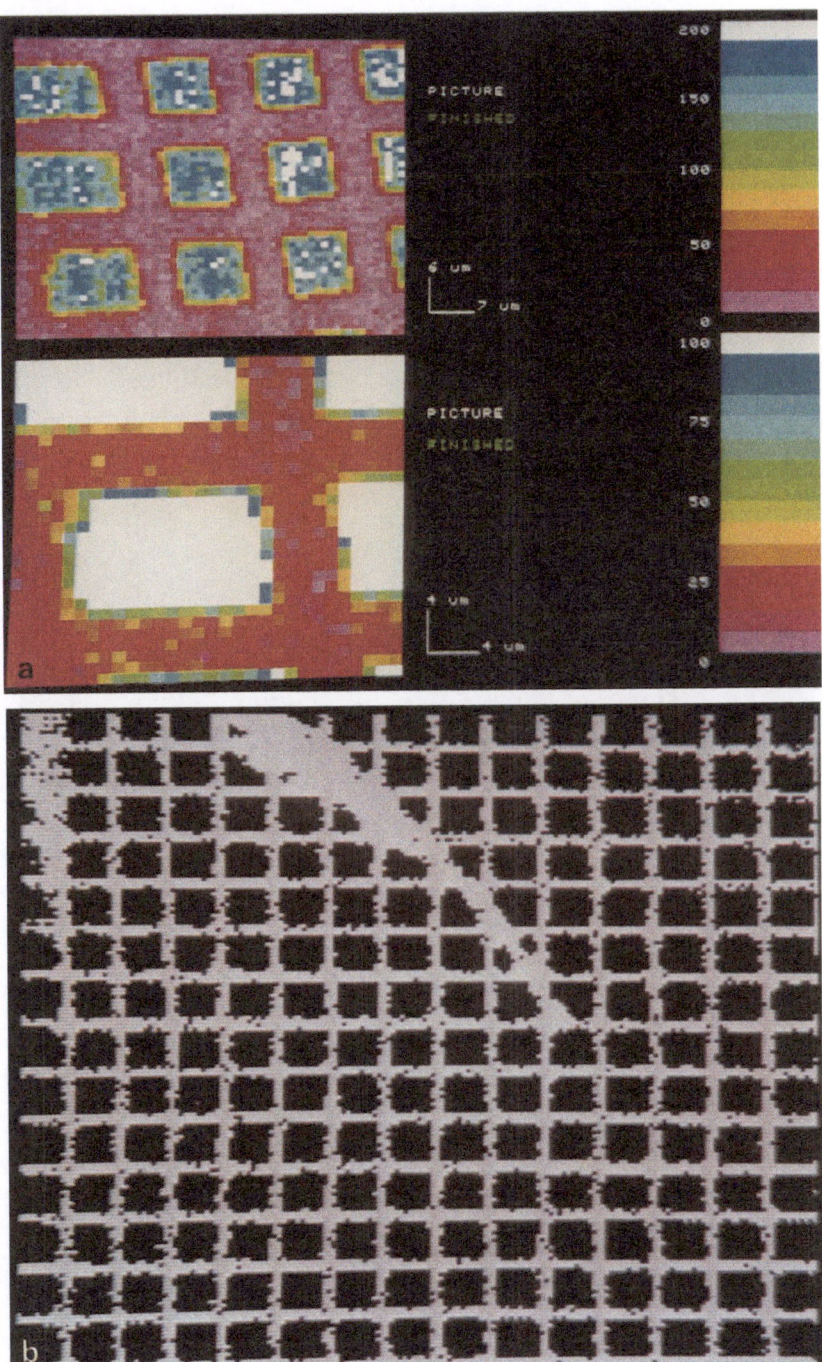

Fig.21.4(a) [Top] Real-time display of 1500 mesh image
Fig.21.4(b) [Bottom] Low magnification 1500 mesh image

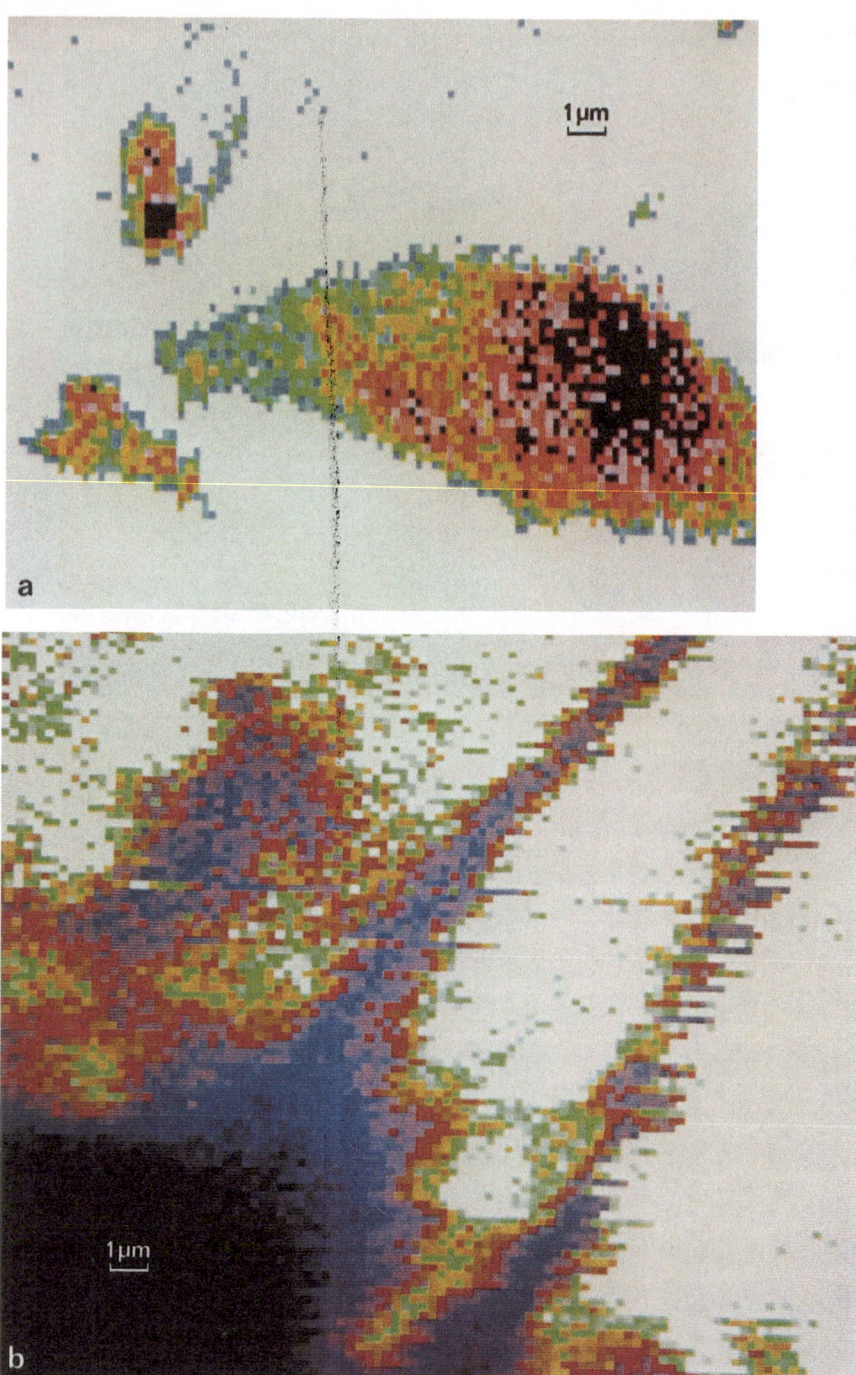

Fig.21.5(a) [Top] Image of a wet *Phaeodactylum alga*
Fig.21.5(b) [Bottom] Image of part of a neuron with processes

214

should then have an order of magnitude improvement in throughput. A significant improvement in the source brightness will become a reality when the soft x-ray undulator [21.6] is installed on the x-ray ring at the NSLS. At least three orders of magnitude more brightness should then make it possible to undertake dynamic studies of biological structure. We eagerly await this new source of soft x rays.

The microscope will be a useful instrument only once biologists become actively involved in using it as a research tool. We have already received specimens from scientists involved in neurophysiology, chromosome structure, and microchemical analysis of blood platelets. We have just begun to learn the art of dealing with wet specimens. The high-resolution imaging of initally living cell cultures is a unique ability of the x-ray microscope and will certainly involve biologist's abilities at the microscope site. We look forward to all of these scientific endeavors.

Acknowledgments

We are grateful to Drs. J. Pine and H. Rayburn of Caltech for providing the specimens of Fig.21.5(b). Components of the microscope were built in the Physics Department machine shop under the able direction of Louis Lenzi. Some of the thin windows were produced by Dan Riel of the Stony Brook Nuclear Structure Lab. Dan Tierney and Waiman Na provided valuable technical assistance. We are pleased to acknowledge the excellent support provided by the technical and operations staff of the NSLS.

The Stony Brook group is supported in part by the National Science Foundation. The NSLS is supported by the Department of Energy.

References

21.1 H. Rarback: Ph.D. Thesis, State University of New York at Stony Brook
21.2 J. Kirz, R. Burg and H. Rarback: "Plans for a Scanning Transmission X Ray Microscope" in vol. 342 of Ann. of NY Acad. of Sci. (1980) pp. 135-148
21.3 H. Rarback, J. Kenney, J. Kirz and X. Xie: "Scanning Soft X-Ray Microscopy--First Tests with Synchrotron Radiation" in Scanned Image Microscopy (Academic Press, London 1980), pp.449-455
21.4 D. Sayre, J. Kirz, R. Feder and E. Spiller: Ultramicroscopy 2, 337-349 (1977)
21.5 G. Williams and M. Howells: "Soft X-Ray and Vacuum Ultra-Violet Beamlines at the National Synchrotron Light Source 700 MeV Storage Ring" in 1982 Proc. of the Intl. Conf. on Synch. Rad. Instr. (to be published)
21.6 M. Howells, J. Kirz and S. Krinsky: "A Beam Line for Experiments with Coherent Soft X Rays", BNL report 32519, Dec. 1982
21.7 K. Boller et al.: Nuc. Instr. Meth. 208, 273-279 (1983)
21.8 B. Henke et al.: Atomic Data and Nuc. Data Tables 27, 118 (1982)
21.9 D. Rudolph and G. Schmahl: "High Power Zone Plates for a Soft X Ray Microscope" in vol. 342 of Ann. of NY Acad. of Sci. (1980), pp. 94-104
21.10 J. Kirz: "Thin Zone Plates for Soft X-Rays" in Proc. of VIIIth Intl. Conf. on X-Ray Optics and Microanalysis (1980), pp.268-269
21.11 M. Born and E. Wolf: Principles of Optics (Pergamon Press, Oxford 1980), p. 416
21.12 H. Smith: "Planar Techniques for Fabricating X-Ray Diffraction Grating and Zone Plates", this volume, no.7

21.13 D. Kern et al.: "Electron Beam Fabrication and Characterization Studies of Fresnel Zone Plates for Soft X-Ray Microscopy" in the SPIE Proc. of the Brookhaven Conf. on Advances in Soft X-Ray Science and Technology (to be published)
21.14 J. Kirz: "Mapping the Distribution of Particular Atomic Species" in vol. 342 of Ann. of NY Acad. of Sci. (1980), pp. 273-288

22. The Göttingen Scanning X-Ray Microscope

B. Niemann

Forschungsgruppe Röntgenmikroskopie, Universität Göttingen
Geismarlandstraße 11, D-3400 Göttingen, Fed. Rep. of Germany

22.1 Introduction

During the past years the capabilities of an imaging x-ray microscope using zone plate optics as focussing elements were demonstrated gradually [22.1,2]. In addition to this microscope we built a scanning x-ray microscope using zone plates. It has been developed because the radiation dose exposed to an object in such a system is reduced significantly. For biological applications in the resolution range below $\delta \lesssim 0.1\ \mu m$ this is important, as the dose transferred to an object increases drastically with increasing resolution [22.3].

The difference in radiation dose transferred to the object is explained briefly. In the imaging microscope the object is followed by the low-efficient amplitude zone plate and by the photographic film. The efficiency of the zone plate is in the range of 3 - 5 %, the detective quantum efficiency of films is in the range of 5 - 20 % [22.4]. Therefore roughly one percent of the radiation which illuminates the object and causes radiation damage contributes to the image.

In the scanning x-ray microscope the radiation is focussed by an x-ray optic to a small spot, which is the reduced image of the radiation source point. It is diffraction limited in the case of highest resolution. Through this spot the object is scanned, and all the transmitted radiation will be collected by a highly efficient x-ray detector, which in case of a gas proportional counter can have 70 % quantum efficiency. Thus the radiation dose transferred to the object in a scanning microscope can be 50 times less than in our imaging microscope.

The radiation dose transferred to the object in an imaging microscope could be reduced a) by increasing the efficiency of the zone plates, e.g. 15 - 20 % should be possible with phase zone plates [22.5], and b) using a more efficient detector; possibly nuclear emulsions, which have no protective layer on top of the emulsion and are fine grained, may increase the efficiency by a factor of two. However, the dosage transferred in a scanning system would still be 5 to 10 times less than in an imaging microscope.

22.2 The BESSY Source

The optical arrangement of the scanning microscope is adapted to the source at which it is used, otherwise we lose more flux than necessary. It is matched to the BESSY storage ring running at METRO optic conditions. This METRO optic mode of the storage ring is designed to deliver the best possible spectral brilliance. While the beam divergence is limited to $1/\gamma \approx 0.64$

mrad a source diameter of ≤ 0.5 mm FWHM should be possible [22.6] (γ is the electron-beam energy in units of m_0c^2, m_0 = electron mass). The number of photons is $N \simeq 1.2 \times 10^{13}$ photons/sec, 150mA, 1 % bandwidth, at 4.5 nm, 0.64 mrad horizontal and vertical beam divergence and 0.4-0.7 mm source diameter.

22.3 The Pinhole

We located a pinhole of 100 μm diameter in the vacuum chamber of the bending magnet, about 0.5 m away from the synchrotron radiation source point. We matched the optical layout to the pinhole diameter. Thus any changes in source diameter or source position do not affect the resolution of the microscope but only the transmitted flux through the pinhole, which can be measured easily.

The pinhole is made from OFHC copper. It is 1 mm in length and 0.1 mm in diameter at the small end. Its diameter increases towards the other end, as the bore is of conical shape. Therefore no significant shadowing will occur, if the axis of the pinhole bore is not exactly parallel to the direction of the radiation used. The pinhole is located at the end of a water-cooled, bellow sealed copper tube. It is adjustable from outside to ± 1.5 mm in height and can be removed from the synchrotron beam in horizontal direction, as this way of motion is 25 mm. Both motions are remote controlled. So it is possible to optimize the flux through the pinhole, provided the real electron beam orbit position is close enough to the ideal one.

22.4 The Beam Line

The beam line set-up follows next to the exit chamber, in which the pinhole is located. Within the first 3 m until the beam line passes the radiation shielding lead brick wall, there are a beam shutter, two ion getter pumps, an all-metal valve and a fast-acting shutter. The shutter protects the storage ring from sudden pressure increases from the connected x-ray microscope. The beam line set-up in the experimental area, outside the interlock area surrounded by the radiation shielding wall, is shown in detail in Fig.20.1 in [22.2] (this volume, no. 20). The experimental area can be used all the time, only a region directly at the lead brick wall, near valve V1 in Fig.20.1, must not be entered during electron injection to the ring. The line divides into two straight lines, beam lines 11.11 and 11.12. They are used for x-ray microscopy work only. At the beam line 11.12 the scanning x-ray microscope is located, 12 m away from the source point.

For safety reasons we introduced near the mirror chamber MC the valve V5 in the beam line, which is closed whenever the radiation is not used or the microscope is filled with air to open it. The latter can be necessary at intervals of a few minutes.

The vacuum system of the storage ring is protected against leaks and vacuum accidents at the beam line with two pressure sensors working in coincidence and located behind a vacuum delay line regarding the fast-acting shutter. Any significant pressure rise measured at the two sensors ($p \geq 10^{-5}$ mbar) will cause the fast-acting shutter to close within a few milliseconds.

The main seal plate of valve V5 contains a viton O-ring. Therefore, it can be often actuated without maintenance. Each time one opens or closes valve V5, it produces a small, short-lasting pressure rise. It was useful to install the valve V5 at some distance from the pressure sensors, because the pressure rise is bafflled by the beam line and the pumps and is less intense when it reaches the sensors. Thus it can be prevented from closing the fast-acting shutter unnessarily, it prolongs the life time of the shutter mechanism.

22.5 The Scanning X-Ray Microscope

The pressure in the microscope must only be a millitorr, so the absorption of photons is negligible. There is no optical element used in grazing incidence or reflection, all elements are used in transmission. Therefore contamination, which might occur at this poor vacuum, is no problem. The beam line with high vacuum, about 10^{-8} mbar, is connected to the microscope by a differential pumping system, consisting of a turbo pump and a one meter long vacuum-insulated tube type cold trap. It has a low conductance and works as a pump, as it is filled with liquid nitrogen.

A schematic of the arrangement of the scanning x-ray microscope is shown in Fig.22.2. The optical set-up within the microscope consists of two stages: a zone plate monochromator, in which a condenser zone plate focusses monochromatic light through a small diaphragm, and a micro-zone-plate, which images the diaphragm to the ultimate spot size of 50nm (resolution-limited).

As the zone plate has about 250 zones, or 125 zone pairs, we designed a zone plate monochromator, which has a spectral resolution of R= $\lambda/\Delta\lambda \simeq 250$. This resolution can be achieved if the diaphragm diameter d and the condenser zone plate diameter D fulfil the relation D/2d $\simeq$ 250.

The microscope to source distance is 12 m, the condenser zone plate accepts 0.2 mrad of the radiation and has 2.4 mm diameter; so for the dia-

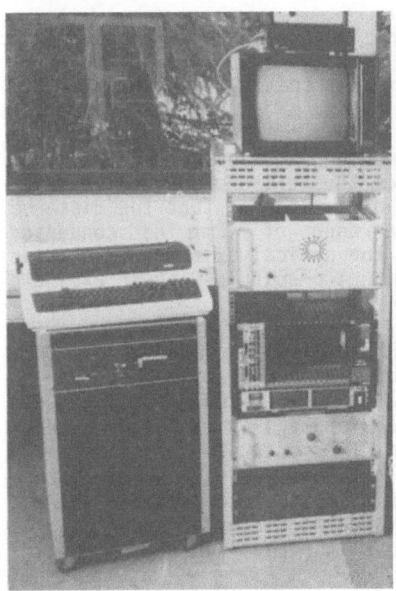

Fig.22.1 The scanning x-ray microscope

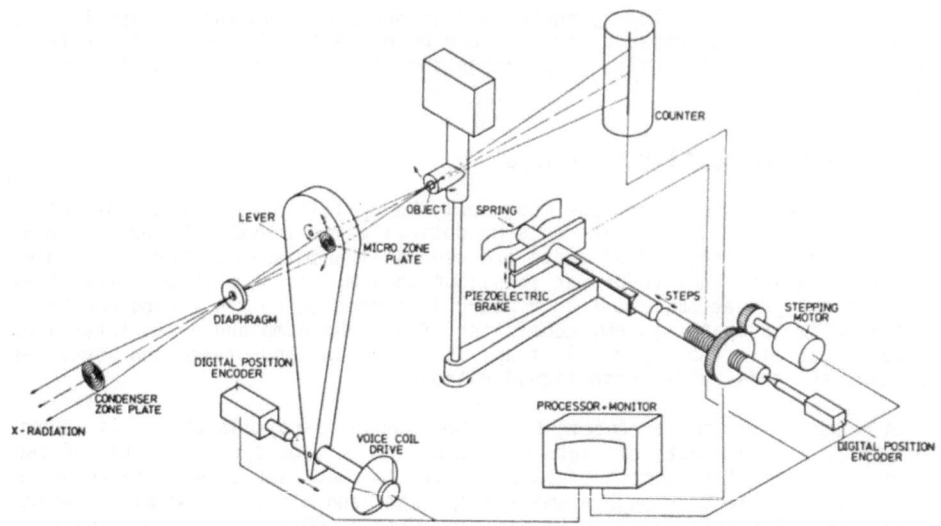

Fig.22.2 Schematic of the mechanical and optical arrangement in the microscope

phragm diameter D = 5 μm. Consequently the source pinhole has to be imaged to the monochromator diaphragm by the condenser zone plate with a demagnification scale of 20, and we need a condenser zone plate with 57 cm focal length. Solutions of two optical arrangements which generate two condenser zone plates KZP4 and KZP5 for 4.5 nm and 2.36 nm wavelength are presented in [22.7](this volume, no.10).

The micro-zone-plate demagnifies the monochromator diaphragm about 100 times to the 50 nm scan spot size. In this demagnification much light is lost. The micro-zone-plate accepts only the inner, spatial coherent part of the radiation from the monochromator.

Figure 22.3 shows another schematic of the scanning microscope. The synchrotron radiation enters the microscope from the left and is collected by the condenser zone plate. It is mounted on a ball slide so that the distance to the diaphragm can be adjusted within the range of 5 cm. The condenser zone plate can be adjusted perpendicular to the optical axis also. This is done from outside the evacuated microscope with two spindles, their axes are O-ring sealed.

The monochromator diaphragm is mounted on a ball-slide arrangement too. It is adjustable perpendicular to the optical axis from outside the vacuum chamber. The sequence of adjustments starts at the micro-zone-plate: it defines an optical axis in the microscope. The diaphragm is adjusted to this axis with polychromatic synchrotron radiation. The radiation passes the diaphragm. When it runs into the micro-zone-plate, it passes the micro-zone-plate and can be measured behind it. The radiation is absorbed, if it runs into the rim of the micro-zone-plate. By moving the diaphragm the location of the rim can be determined. Via this procedure the position of the optical axis can be figured out.

Next the condenser zone plate is centered to this diaphragm by optimizing the photon flux behind it.

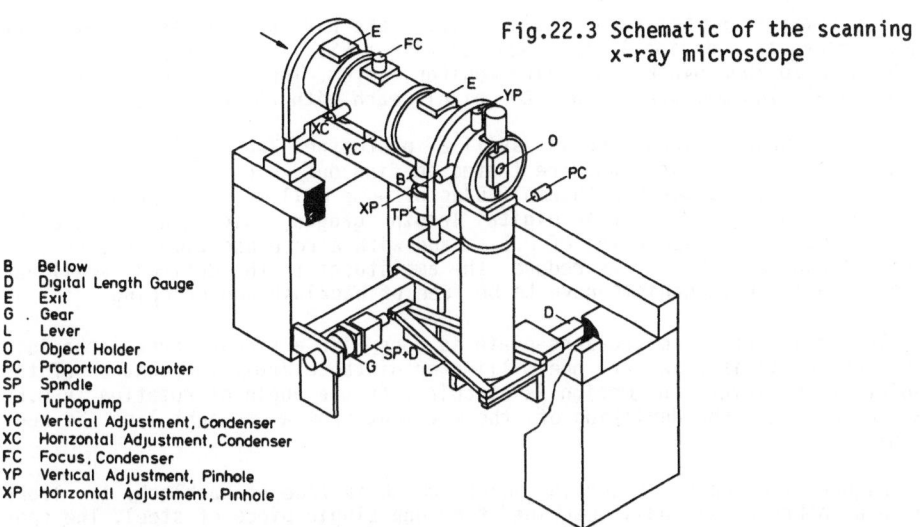

Fig.22.3 Schematic of the scanning
x-ray microscope

B   Bellow
D   Digital Length Gauge
E   Exit
G . Gear
L   Lever
O   Object Holder
PC   Proportional Counter
SP   Spindle
TP   Turbopump
YC   Vertical Adjustment, Condenser
XC   Horizontal Adjustment, Condenser
FC   Focus, Condenser
YP   Vertical Adjustment, Pinhole
XP   Horizontal Adjustment, Pinhole

The diaphragm is vacuum sealed with a thin polyimide foil, the adjustable base plate is sealed with a gliding O-ring. Therefore, the diaphragm separates the microscope into two vacuum regions; they can be pumped separately. The region containing the mechanical scanner, the micro-zone-plate, the object and the entrance window of the counter, can be filled with a gas of low absorption coefficient and held at a pressure of a few hundred millibar during operation.

In our microscope the object will also be irradiated with zero-order radiation of the micro-zone-plate. Therefore on the entrance window of the counter a small stop has to be located, which absorbs the zero-order photons having passed the object. Nevertheless, zero-order photons increase the radiation dose transferred to the object. For example, if micro-zone-plate MZP3 [22.8] (this volume, no. 9) with 56 µm diameter, with approximately 30-40% zero-order and 3% first-order efficiency is used and scans 40000 points with 50 nm spot size, then the zero-order radiation contributes 1/3 and the focused radiation contributes 2/3 to the dose in the scanned area. The additional dose from zero-order radiation will decrease below 1/3, if a micro-zone-plate with higher first-order efficiency is used.

For micro-zone-plates used at a given resolution the ratio of first-to zero-order radiation dose in the scanned area is proportional to the zone plate area.

22.6 The Scanning System

The scanning stage consists of two independent mechanical motions in perpendicular directions, each with a maximum amplitude of about 10 µm. This corresponds to an angular shift of the scan spot of some seconds of arc, when it is viewed from the centre of the zone plate. Thus, no aberration occurs. The micro-zone-plate scans the lines at a rapid speed. At the end of each line the zone plate reverses direction. In this moment the object is moved one step perpendicular, so that the micro-zone-plate scans the next line (see Fig.22.2). For a picture of 200 x 200 pixel elements the minimum scan time is 40 seconds corresponding to a pixel scan time of one millisecond and a scan time of 0.2 sec for one line. Therefore the time scale for the mechanical motions and the data processing is matched to one millisecond.

The computer system comprises a Camac crate with a dedicated controller and memory, an additional host computer with a LSI 11/23 microprocessor, and a DE ANZA ID 2000 black and white monitor system. The system can store approximately 100 pictures, 200 x 200 pixels each with 256 grey levels.

This scanning microscope is still a prototype. Two different types of mechanism do the motions and are tested to find out which work best. Levers are used. They are moved with amplitudes of some millimeters. The amplitudes are measured with Heidenhain digital length gauges, which can measure the amplitudes with a resolution of 0.1 μm and with a response time of less than one microsecond. The levers reduce the amplitudes to the desired small values. The lever mechanisms have to be free of backlash and creeping.

The motion of the micro-zone-plate is done with a single lever mechanism. The zone plate is fixed at one millimeter distance from the centre of the joint of the lever; the motion is circular. As the angle of rotation is 0.5° at the maximum, the deviation of the movement from a straight line is negligible.

Figure 22.4 shows the torsion joint, which is free of backlash. It looks like a spoked wheel and is machined from one single piece of steel. The spokes are strictly radial, thin and long. Therefore they can easily be bent and the inner ring can be twisted half a degree.

Figure 22.5 shows the joint with a lever arm in a test mounting. The lever is one meter long and its free end is continuously moved with a voice

Fig. 22.4

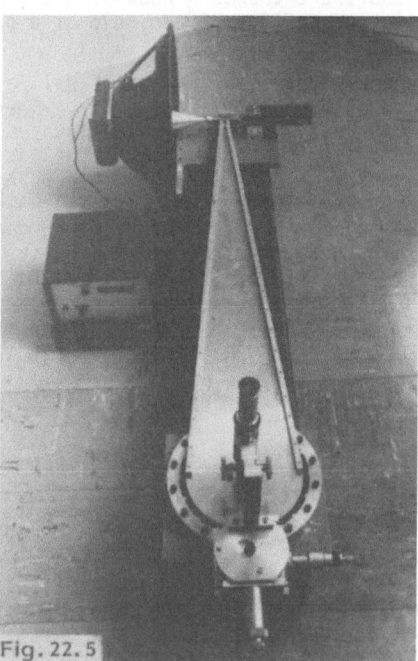
Fig. 22.5

Fig.22.4 Part of the mechanical scanner showing a torsion joint

Fig.22.5 The same as Fig.22.4, with a lever arm connected to the
joint and a loud speaker drive and a digital length gauge

coil of a loud speaker. The amplitude of the voice coil is measured with a digital length gauge. As the amplitude of motion is a thousand times smaller at the zone plate, a movement of 0.1 μm of the digital length gauge corresponds to a 0.1 nm step size of the zone plate and produces 1 pulse at the electronic output of the gauge.

The mechanism for moving the object includes a long steel rod, which is fixed with one end to the case of the microscope. At the other end a long lever arm is connected, located at the bottom in Fig.22.2. The lever can twist the rod. The angle of rotation decreases to zero at the fixed end. At a distance of only a few cm from the fixed end, a short lever is clamped to the rod. The short lever moves the object. The mechanism delivers a total reduction of the amplitude of about 1000 times.

A 10 nm step of the scan spot produces 100 pulses at the output of the digital length gauge; the pulses are counted. Presetting "100" to the counter, it produces interrupts in the Camac system every 10 nm during the scan. The Camac system subsequently counts x-ray photons between the interrupts and produces a pixel on the monitor of corresponding intensity and position. The correct horizontal position of the lines on the monitor is controlled by a reference mark in the digital length gauge.

During the scan time of each line, which is 0,2 seconds at the minimum, the stepping motor driven spindle of the other coordinate is prealigned to the next line position, with the help of another digital length gauge. Every time when one line scan is finished and the motion reverses, a piezoelectric brake, which is attached to the lever arm of the other coordinate, is inactivated for a few milliseconds. Hereby, a spring load pushes the lever arm to the new position of the prealigned spindle. Afterwards, within the next 0,2 seconds, the spindle is prealigned to the position of the next line and so on.

22.7 Estimated Photon Flux

As the system has not yet been connected to the beam line, only an estimation can be made to calculate the photon flux in the scanning spot and the detectable photon flux in the counter.

Much light is lost in the microscope. The following list shows where the losses occur:

	Loss factor
a) at the pinhole $(0.7 \text{ mm}/0.1 \text{ mm})^2$	= 50
b) at the condenser zone plate $(0.63 \text{ mrad}/0.2 \text{ mrad})^2$	= 10
c) 5 % efficiency of the condenser zone plate	= 20
d) 3.5 % efficiency of the micro-zone-plate	= 28
e) ω (zone plate monochromator)/ω (accepted by the micro-zone-plate, 50nm (100nm) scan spot)	= 20 (5)
f) 70 % transmitting foil on pinhole	= 1.4
g) 0.4 % bandwidth of monochromator/1% bandwidth when referring to the photons flux given above	= 2.5

The total loss in a scanning microscope, which achieves a 50 nm scan spot, is $2 \cdot 10^7$; this factor reduces the number of photons from BESSY given in section 22.2, resulting in $6 \cdot 10^5$ photons/sec in the focal spot.

Introducing an object on a foil with 70% transmission, and a proportional counter with an efficiency of 30-70%, the detected photon flux decreases by an additional factor of 2-3, resulting in $\leq 3\cdot10^5$ detectable photons/sec with a 50 nm scan spot diameter.

If the distance of the micro-zone-plate to the monochromator diaphragm is reduced to half its value, the scan spot is no longer resolution limited and increases to 100 nm diameter. The total loss factor is $5\cdot10^6$, corresponding to $\leq 2\cdot10^6$ photons/sec in the spot and to $\simeq 10^6$ detectable photons/sec in the counter.

Replacing condenser zone plate KZP4 by condenser zone plate KZP5, the monochromator works at 2.36 nm wavelength with the same resolution of $\lambda/\Delta\lambda$ =250 as KZP4 at 4.5 nm wavelength. The micro-zone-plate MZP3 can also be used at 2.36 nm wavelength, which is preferable for biological objects, as water is low absorbing at this wavelength. At 2.36 nm wavelength the micro-zone-plate delivers the same photon flux in the scan spot as at 4.5 nm wavelength, but the scan spot diameter doubles because the focal length doubles at 2.36 nm wavelength.

At 2.36 nm wavelength $6\cdot10^5$ photons/sec can be achieved in a scan spot 100 nm in diameter, corresponding to $3\cdot10^5$ detectable photons/sec.

A 50 nm scan spot diameter is achieved at 2.36 nm wavelength, if the distance of the micro-zone-plate to the diaphragm is doubled. Unfortunately the photon flux in the spot is reduced 4 times in this case.

22.8 Future Development

The future work will concentrate on testing the scanner, on changing the wavelength to 2.36 nm, on increasing the number of detectable photons and on increasing resolution. The photon flux can be increased manyfold: phase zone plates and micro-zone-plates with better f number (for example MZP4) have to the developed. A 10^2-fold increase of flux through the scan spot should be possible. A further increase of flux, possibly some orders of magnitude, can be obtained using a dedicated undulator, e.g. at the projected European Synchrotron Radiation Facility (ESRF).

An increased flux and micro-zone-plates with smaller outermost zone widths will allow to improve the scanning microscope: to run it at a higher resolution, in the 10 - 50 nm range, and at 2.36 nm wavelength.

Acknowledgements

This work was generously supported by the Stiftung Volkswagenwerk. I thank D. Rudolph and G. Schmahl for numerous fruitful discussions.

References

22.1 B. Niemann, D. Rudolph and G. Schmahl: Applied Optics, 15, 1883 (1976)
22.2 D. Rudolph, B. Niemann, G. Schmahl and O. Christ: "The Göttingen X-Ray Microscope and X-Ray Experiments at the BESSY Storage Ring", this volume, no. 20
22.3 J. Kirz: "Mapping the Distribution of Particular Atomic Species", Annals N.Y. Acad. Sci., 342, 273 (1980)

22.4 B. Niemann: "Detective Quantum Efficiency of some Film Materials", Annals N.Y. Acad. Sci., <u>342</u>, 230 (1980)

22.5 R. Tatchyn: "Optimum Zone Plate Theory and Design", this volume, no.6

22.6 G. Mühlhaupt: "The BESSY Soft X-Ray Source and Future Development", this volume, no. 1

22.7 J. Thieme: "Construction of Condenser Zone Plates for a Scanning X-Ray Microscope", this volume, no.10

22.8 P. Guttmann: "Construction of a Micro-Zone-Plate and Evaluation of Imaging Properties", this volume, no.9

23. A Scanning Soft X-Ray Microscope Using Normal Incidence Mirrors

E. Spiller

IBM T.J. Watson Research Center, P.O. Box 218
Yorktown Heights, NY 10598, USA

A scanning x-ray microscope has been installed at the National Synchrotron Light Source at Brookhaven. Normal incidence mirrors with multilayer coatings are used to form a small local spot and the specimen is mechanically scanned through this spot.

23.1 Introduction

Normal incidence optics for soft x rays can be realized by using multilayer coatings. Reflectivities between 10% and 25% have been obtained in the wavelength region λ = 45 - 100 A [23.1,2]. Normal incidence optics can be fabricated with considerably higher precision than the grazing incidence optics which has been predominantly used for x-ray telescopes and microscopes [23.3-5]. Phase measuring interferometry with visible light [23.6,7] can test the figure of a surface with an error of less than 10 A and promises that reflecting optical elements with a spatial resolution around 100 A can be fabricated in the future. This resolution capability is comparable to the most optimistic performance one can expect for focussing with Fresnel zone plates [23.8]. The long-term goal of our work is to approach the resolution limit possible with normal incidence mirrors and provide the experimental data required to compare normal incidence optics with zone plate optics.

23.2 Multilayer Coatings

Theoretical and experimental data on the performance of multilayer coatings for soft x rays have been presented before [23.1,2]. A recent compilation of the optical constants of all elements [23.9] makes it possible to calculate the theoretical performance of all material combinations and select the most suitable one [23.10]. It turns out that the combinations of the transition elements (V, Cr, Mn, Fe, Co, Ni) with carbon give theoretically higher reflectivities for wavelengths around 50 A than the most successful systems reported up to now: W-C or ReW-C.

In order to test the theoretical predictions we have fabricated multilayers of the transition elements with carbon and measured their reflectivities at the Stanford synchrotron [23.11]. Figure 23.1 is a comparison of the measured and calculated reflectivity curves of a NiC multilayer tested at various angles of incidence and photon energies. The measurements are in good agreement with theory.

Figure 23.2 is a plot of the measured peak reflectivity versus the corresponding photon energy for coatings made during the last few years. The data confirm two general theoretical predictions: 1) the increase in reflectivity

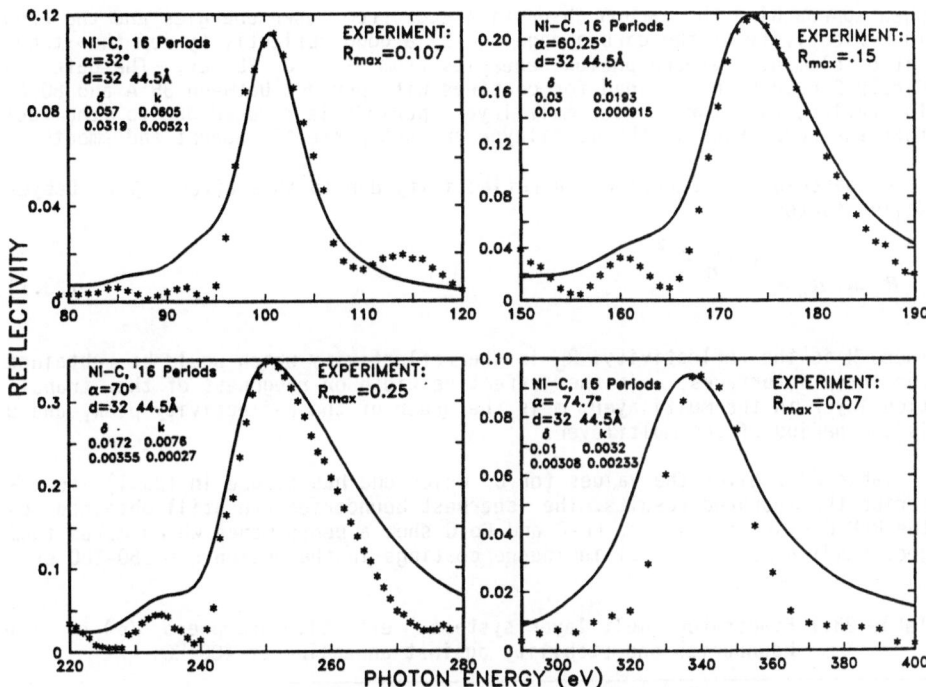

Fig.23.1 Calculated and measured performance of a Ni-C multilayer with a period 76.5 A and various incidence angles and energies. The experimental curves are scaled to the same maximum as the theoretical curves; the measured peak reflectivity is indicated for each curve (from [23.1])

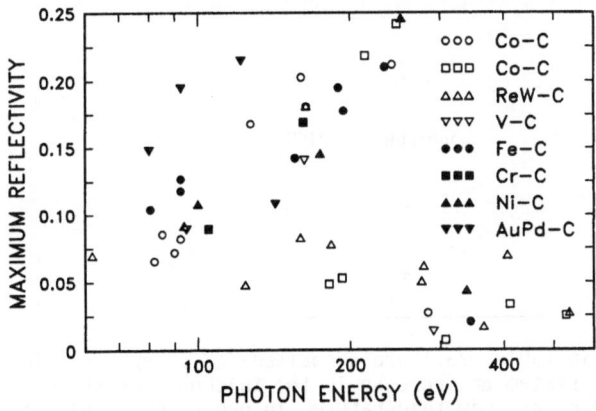

Fig.23.2 Measured maximum reflectivity versus photon energy for different multilayer systems. The periods of the multilayers are between 35 A and 78 A; the peak reflectivity is shifted to different energies by changing the angle of incidence

when approaching the carbon absorption edge from lower energies and the drop in reflectivity at the carbon edge and 2) a good reflectivity for the transition elements around photon energies from 150 to 280 eV. The data in Fig.23.2 have been obtained for coatings with periods between 35 A and 80 A. The reflectivity for shorter multilayer periods is reduced due to the fact that the boundaries in the multilayer are not perfectly abrupt and smooth.

We describe the reduction in reflectivity due to this effect by a Debye-Waller factor

$$R = R_o\, e^{-(\frac{2\pi\sigma\cdot m}{D})^2}, \tag{23.1}$$

where R is the reflectivity, R_0 is the reflectivity which would be obtained for sharp interfaces, σ is the effective width or roughness of the transition layer in the multilayer, m is the order of the reflectivity peak, and D is the period of the multilayer.

Table 23.1 gives the values for σ which one has to use in (23.1) to describe the measured results. The sharpest boundaries are still obtained by the ReW-C system; however, Fe-C and Co-C show a performance which makes them good candidates for normal incidence coatings in the region E = 150-280 eV.

Table 23.1 Evaporated multilayer systems, effective sharpness (σ) of the boundaries and stability against annealing at 400 $^{\circ}$C

	$\sigma[\text{Å}]$	Stable at 400°[C]
ReW-C	2-4	+
ReW-B	~ 7	
AuPd-C	~ 7	
Pt-C	< 5	
V-C	6-8	+
Cr-C	7-8	—
Mn-C	9	unstable at 20°[C]
Fe-C	3-5	—
Co-C	3-5	—
Ni-C	7	—
Ru-C	8	+

Most of the multilayers in Table 23.1 are deposited as amorphous films and we observed in the ReW-C system an increase in the thickness of the multilayer of about 0.5% per year at room temperature. In order to accelerate this aging process, we now subject some coatings of each run to several annealing steps (400 $^{\circ}$C for 4 hrs) in a vacuum furnace. We find that all coatings increase their period during annealing (about +5% after 10 steps = 40 hrs at 400 $^{\circ}$C). No drop in reflectivity is observed in the coatings marked (+) in Table 23.1. Coatings which lose their reflectivity are marked by (-); this loss in reflectivity is generally connected with the formation of microcrystallites in the metal film and an increase in the value of σ.

23.3 Microscope Design

The scanning x-ray microscope uses the outermost part of the radiation from the port U8 at the 700 MeV storage ring at Brookhaven. It shares the port with three other experiments (monochromators) and part of the geometry of the microscope was dictated by space restrictions on the experimental floor.

For diffraction-limited performance the microscope objective should be coherently illuminated, i.e., the coherence condition

$$d_s \sin \theta_a \lesssim \lambda , \tag{23.2}$$

where d_s is the source size and θ_a the divergence of the irradiating beam should be fulfilled. Present synchrotron sources have source sizes which are about a factor of 100 larger than the value given by condition (23.2).

We use two demagnification steps to obtain coherent illumination of the microscope. In the first step the synchrotron source is demagnified by a grazing incidence ellipsoidal mirror (grazing angle $\theta = 3^0$, focal length $f_1 = 5.70$ m, $f_2 = 0.70$ m). An adjustable pinhole is located at the image produced by the elliptical mirror and this pinhole serves as a secondary source and is further demagnified by a normal incidence Schwarzschild objective. The size of the pinhole (2 μm - 20 μm) is selected to fulfill the coherence condition (23.2), where θ_a is now the illumination angle of the first mirror of the Schwarzschild objective. The theoretical performance, assembly and test of the Schwarzschild objective has been described previously [23.12]. In our first set-up we are using an objective with an effective focal length f = 46 mm and a demagnification factor 64. The mirrors of the objective are coated with multilayers of ReW-C for a photon energy of 185 eV and we estimate their peak reflectivity between 10 and 15%, giving a throughput after 2 reflections of about 1% in a bandwidth of around 1%.

Figure 23.3 gives a top and side view of the microscope. The top view shows the mirror box with the elliptical mirror (5.7 m from the source) which deflects the beam by 6^0 and images the source onto the pinhole. The pinhole assembly and support can be seen in the side view. The chamber which houses the microscope consists of two parts, one for the mirrors and the other for the specimen and scan mechanism. Both chambers can be either evacuated together for operation in vacuum or they can be separated by a thin window (1-2.5 μm thick polyimide or Mylar) for observation of the specimen in air or helium. In the present set-up the specimen is moved through the focal spot by motor-driven micrometers and piezoelectric transducers. The photons transmitted through the sample are detected by a flow proportional counter with P10 gas (argon with 10% methane).

The parameters and photon fluxes of the microscope beam line are summarized in Table 23.2. The first column gives the expected flux for the case that the Brookhaven ring will achieve its design spot size (0.1 mm); however, we are assuming that the figuring of the elliptical mirror will determine the image size in front of the pinhole for this case. The second column uses typical source sizes (∼1 mm) now obtained at Brookhaven. The third column gives values obtained in a first preliminary test. We experienced unexpected delays in the fabrication of the first elliptical mirror and performed a first test with a diamond-turned (but not polished) paraboloid that was available at Brookhaven.

Probably mostly due to scattering in the parboloid the observed flux through the pinhole of about 1000 photons/sec in 1% bandwidth was considera-

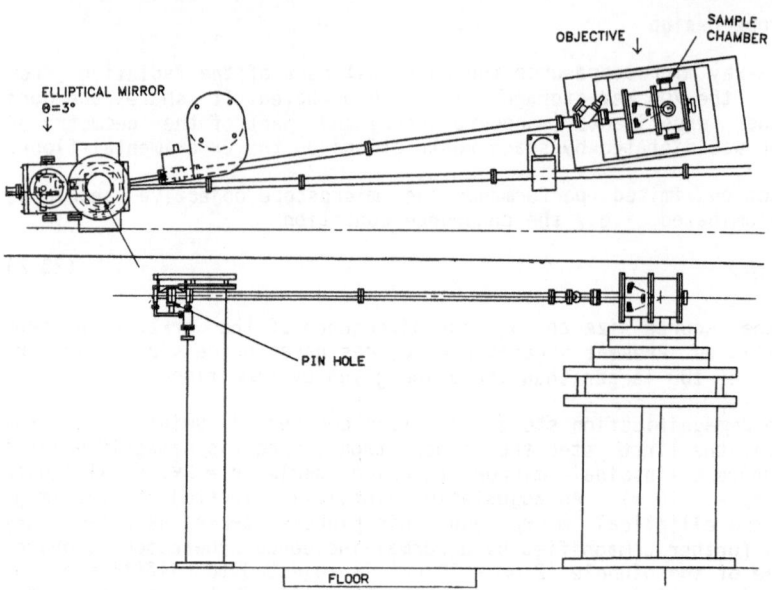

Fig.23.3 Top view and side view of the scanning x-ray microscope. The distance from pinhole to microscope chamber is about 3 m

Table 23.2 Parameters of microscope beam line

	Photon flux for 100ma current in 1% bandwidth		
	Ring Design	Present Ring Performance	Test with unpolished Paraboloid
Illumination of first mirror	10^{13}/sec.	10^{13}/sec	10^{13}/sec.
Demagnification	1:8	1:8	1:8
Image size (2σ)	50μm x 150μm	150μm x 300μm	5mm x 1mm
Flux through 2μm pinhole	5 x 10^9/sec.	10^9/sec.	10^8/sec. for 10 μm pinhole
Observed Flux	----	----	10^3/sec.
Demagnification of Microscope	64	64	64
Flux at focus of Microscope	3 x 10^7/sec.	5 x 10^6/sec.	----
Observed Flux	----	----	5/sec.

bly lower than expected and not high enough to test the resolution of the microscope. However, the measured throughput of the Schwarzschild objective was close to the expected value and we are optimistic to obtain fluxes close to 10^6 photons/sec once the elliptical mirror is installed.

References

23.1 E. Spiller: Low Energy X-Ray Diagnostics 1981, AIP Proc 75, eds. D.T. Attwood and B.L. Henke, p.124, (1981)
23.2 T. Barbee: Low Energy X-Ray Diagnostics 1981, AIP Proc. 75, eds. D.T. Attwood and B.L. Henke, p.131, (1981)
23.3 A. Franks: Sci. Progr. Oxf. 64, 371 (1977)
23.4 K. Silk: SPIE Vol. 106, X-Ray Imaging, eds. R.C. Chase and G.W. Kuswa (Society Photo-Opt. Instrum. Eng. Bellingham 1977) p.113
23.5 L.P. Van Speybroeck: SPIE, Space Optics, 184, 2 (1979)
23.6 C.W. Chen and P.W. Maymon: SPIE 316, 9 (1981)
23.7 G. Makosch and B. Solf: SPIE 316, 42 (1981)
23.8 G. Schmahl, D. Rudolph, and B. Niemann: SPIE 316, 100 (1981)
23.9 B.L. Henke,P. Lee, T.J. Tanaka, R.L. Shimabukuro, and B.K. Fujikawa: AIP Proc. 75, 340 (1981)
23.10 A.E. Rosenbluth: Thesis, U. of Rochester, 1982
23.11 R. Bartlett, D. Kania, W.J. Trela, E. Källne, P. Lee, E. Spiller: SPIE 447 (1983), to be published
23.12 I. Lovas, W. Santy, E. Spiller, T. Tibbetts, and J. Wilczynski: SPIE 316, 90 (1981)
23.13 Paraboloid with f = 1 m; grazing angle θ = 4°. We thank M. Howells for providing this mirror.

24. X-Ray Microscopy at the Daresbury Laboratory

P. J. Duke

SERC, Daresbury Laboratory, Warrington WA4 4AD, UK

24.1 Introduction

The 2 GeV electron storage ring (SRS) at the Daresbury Laboratory has been operating as a dedicated synchrotron radiation (SR) user facility since the summer of 1981. The operating parameters of the SRS (at least - those which are of primary interest to users of the facility) are shown in Table 24.1. The radiation spectrum itself is shown in Fig.24.1 for both the standard dipole magnets (magnetic field = 1.2 T) and the high-field super - conducting wiggler insertion (5.0 T). The source dimensions at present are listed in the table and cover a range of values, depending on the region of the electron beam orbit viewed by the experimenter.

Table 24.1 SRS parameters

BEAM ENERGY	2.0 GeV
ELECTRON CURRENT (maximum)	100 - 300 mA
BEAM LIFE TIME	8 - 10 hrs or greater
BEAM DIMENSIONS (fwhm) from	4.8×1.2 mm^2
to	12.9×0.6 mm^2
CHARACTERISTIC WAVELENGTH	
DIPOLE MAGNET (1.2 T)	0.388 nm (3.88 Å, 3.2 keV)
WIGGLER MAGNET (5 T)	0.093 nm (0.93 Å, 13.3 keV)
PHOTON FLUX AT 4.4 nm	4×10^{12} photons/(mrad.sec) into 0.1% BW at 100 mA

The intense flux from the SRS (see Fig.24.1), particularly in the x-ray region, means that in common with users at other SR sources there is an increasing emphasis on the development and application of x-ray imaging techniques. These have been reviewed recently by BOWEN [24.1], particularly in relation to materials science. Table 24.2 lists the imaging techniques at present in use or under development at the SRS. Recent results from several of these techniques are presented in [24.1]. It is outside the scope of this article to present these in detail but results in contact microscopy are

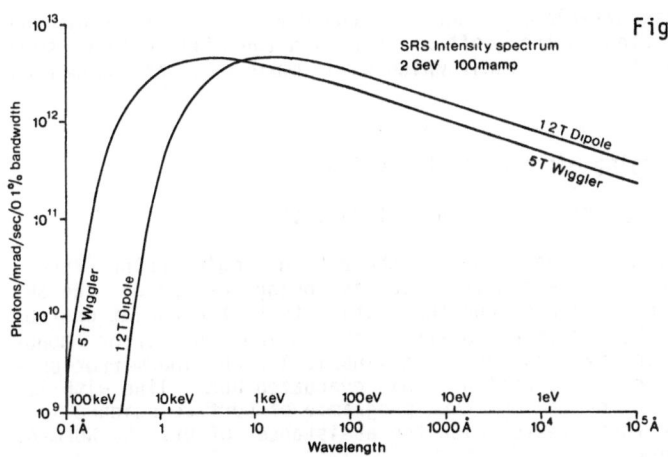

Fig. 24.1 Photon spectrum from the SRS

presented in [24.2]. Reference [24.1] gives several examples of the applications of x-ray topography and the combination of stroboscopy with x-ray topography is presented there and in WHATMORE [24.3]. First results using EXAFS contrast are described in BOWEN [24.1,4]. Both stroboscopic and EXAFS topography are particularly important because they illustrate the introduction of further imaging dimensions such as time and energy. The latter in particular could be used to distinguish different chemical bonding conditions within the imaged matrix. Finally the photoelectron spectromicroscope [24.5] can be used to provide direct recording of magnified x-ray images both by conversion of x rays to photoelectrons in a high–resolution photocathode and also by direct imaging of the photoelectrons produced in the sample by the x rays themselves. Both this technique and microbeam EXAFS are included in Table 24.2 because they are in process of development at the SRS.

METHOD	APPLICATION
CONTACT MICROSCOPY	HIGH RESOLUTION (1-50 nm) UNSTAINED SLICES EM EXAMINATION
TOPOGRAPHY	DEFECT IMAGING IN CRYSTALS RESOLUTION ~ 1 μm STROBOSCOPIC/DYNAMIC STUDIES
EXAFS CONTRAST	MAPPING OF VALENCE STATES AND CHEMICAL BONDING
X-RAY MICROSCOPY	POLYMER STUDIES BIOLOGICAL APPLICATIONS RESOLUTION > 0.1 μm
PHOTOELECTRON MICROSCOPY	SURFACE MICROANALYSIS MONOLAYER DEPTH RESOLUTION 1-10 μm LATERAL RESOLUTION
X-RAY MICROANALYSIS MICRO BEAM EXAFS	QUANTITATIVE ANALYSIS WITH MICROSTRUCTURE CORRELATION PROBE SIZE ~ 1 μm.

Table 24.2 Imaging techniques at the SRS

The remainder of this article will describe the development of an imaging x-ray microscope, operating in the soft x-ray region and will outline some of the developments of the SRS itself which will make a direct impact on this programme.

24.2   A Soft X-Ray Imaging Microscope at the SRS

24.2.1   Description of the Instrument and Initial Tests

An imaging x-ray microscope operating in the 2-5 nm region using Fresnel zone plates as the focussing elements which is being set up on the SRS [24.6] is described in Figs.24.2-4. For the initial tests the radiation from the SRS was conditioned by a combined diffraction grating and mirror mono-chromator and was focussed by a horizontal cylindrical focussing mirror operating in the grazing incidence region. This evacuated beam line with its optical system was set up for a long-term programme of surface science and was made available for these tests with the assistance of Drs. D. Norman, A. McDowell and J.B. West.

A condensing zone plate about 800 μm diameter was used to focus the radiation onto the object which was then viewed by a micro-zone-plate. This combination of condenser, object and micro-zone-plate was assembled "off-line" in the unit shown in Fig.24.3 and aligned using an optical microscope. The condenser unit was then assembled in the x-ray microscope (see Fig.24.2) and the microscope pumped down to a pressure of 10^{-6} torr. The region of the microscope containing the object, micro-zone-plate and camera unit (Fig.24.4) operated at a pressure of 10^{-4} - 10^{-5} torr.

A 150 μm diameter, 36 mm focal length micro-zone-plate fabricated by laser interference [24.7] has been used in an initial test and it is planned

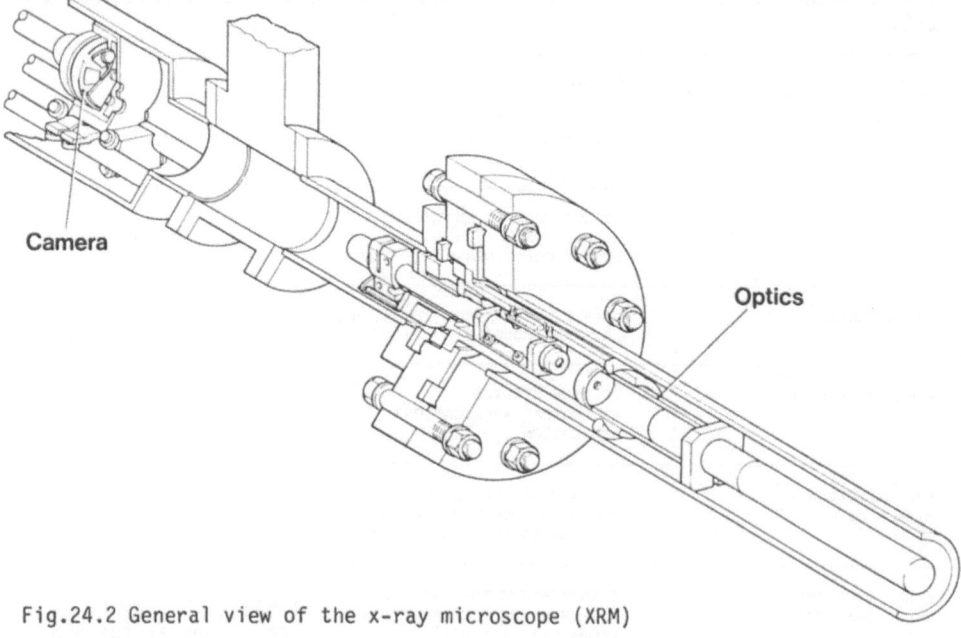

Camera

Optics

Fig.24.2 General view of the x-ray microscope (XRM)

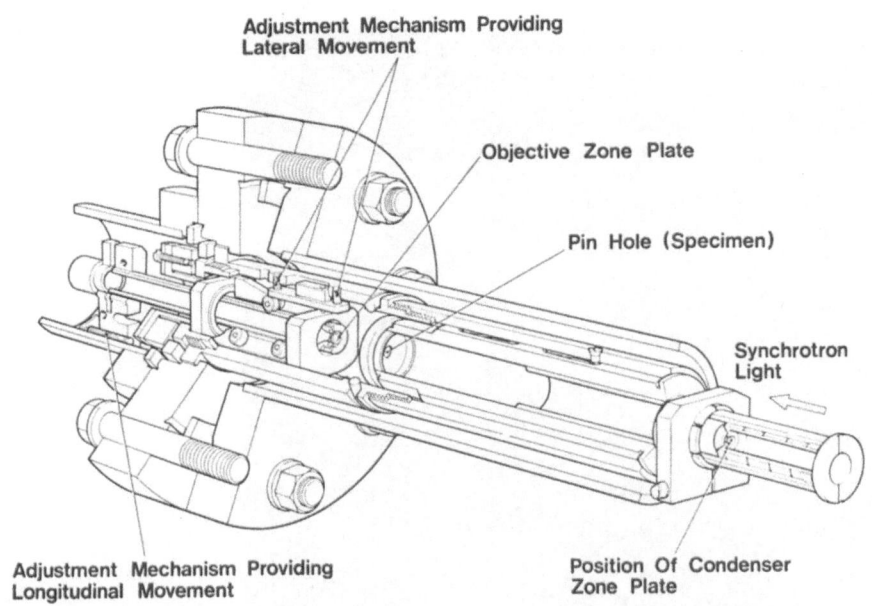

Fig.24.3 Close-up of the XRM condenser unit - the unit is 350 mm long

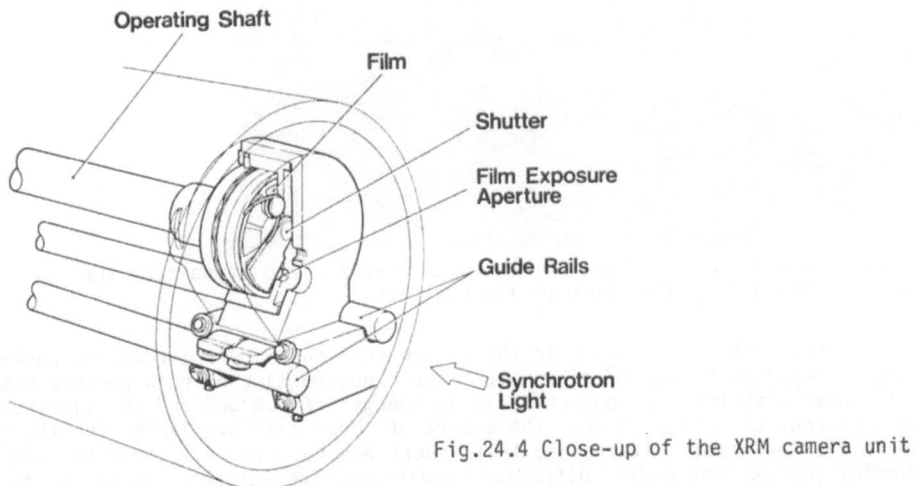

Fig.24.4 Close-up of the XRM camera unit

to use much higher resolution (50 nm) zone plates with a 250 μm focal length fabricated by STEM lithography [24.8] during further tests and commissioning which will take place during 1984. In the initial test the microscope has been aligned and synchrotron light recorded on the film. The commercially bought condenser zone plate was however damaged,so that instead monochromatic radiation produced at the diffraction grating was used to illuminate the object, which in this case was a gold wire grid of 120 μm periodicity supported on a 600 μm diameter pinhole. The first pictures taken with this microscope are shown in Fig.24.5. The camera unit (Fig.24.4) enables a series of up to 8 exposures to be made around a circular piece of film about 25 mm

235

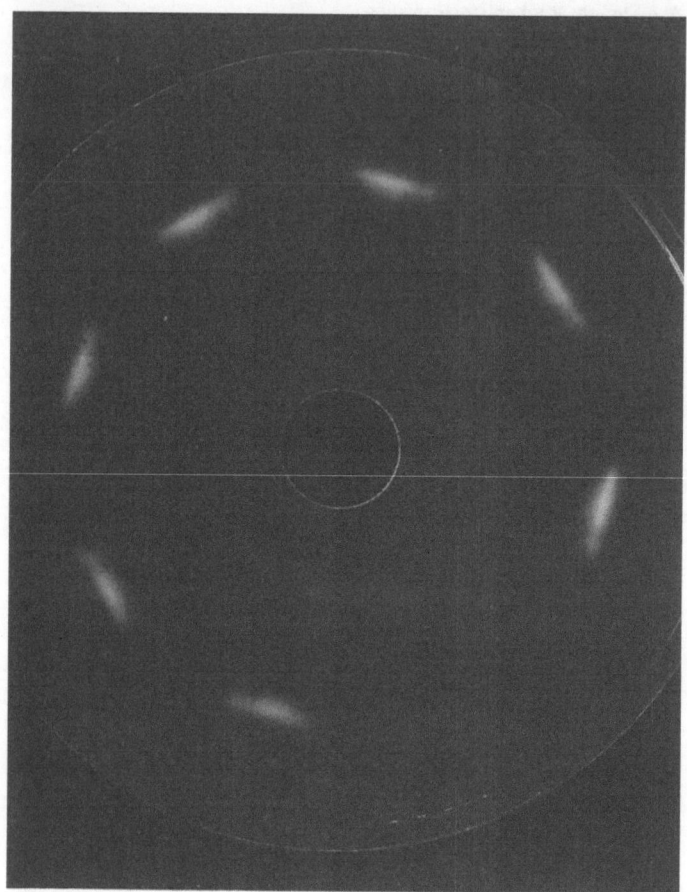

Fig.24.5 First photographs of monochromatic soft x radiation from the
SRS transmitted through the microscope

in diameter. The series shown in the figure were taken over a range of wavelengths supplied by the monochromator. The exposure time of each picture was
∿ 1 minute with the SRS operating at an energy of 1.8 GeV and an electron
beam current of about 60 mA. The amount of beam time available for these
tests was extremely limited so that there was no time available to take
further photographs under different conditions. The pictures shown in the
figure correspond to a series of broad wavelength bands between 3 and 4 nm.
The band pass of the monochromator was about 10% for each of these photographs so that a marked variation of image focusing with wavelength was not
to be expected. Each photograph shows a cigar-shaped image of the electron
beam projected through the 600 μm pinhole. Each of these photographs contains other structures but the absence of further scheduled beam time has
prevented the study and elucidation of these effects at the present time.

24.2.2 Longer Term Programme Development

The long-term aims of the soft x-ray microscopy programme at the SRS are
shown in Table 24.3. The first need is to commission the instrument and to

Table 24.3 Soft x-ray microscopy at the SRS

AIMS OF THE PROGRAMME	
IMAGING MICROSCOPE	
RESOLUTION ~ 100 nm	1984
SCANNING MICROSCOPE	
RESOLUTION ~ 100 nm	1985
RESOLUTION ~ 50 nm	
UNDULATOR	1985
HIGH BRIGHTNESS LATTICE	1986
STEREO CAPABILITY	
ELECTRON DETECTION	1987–1988
ENERGY DIMENSION	
TIME DIMENSION	
USER INSTRUMENT	1988
HIGH RESOLUTION (~ 10 nm) INSTRUMENT (ESRF?)	1992

do this using zone plates fabricated by the STEM lithography method [24.8,9]. The STEM lithography method provides a very flexible system for zone plate writing (or indeed for writing any coded aperture system) and it can be used to prepare modified Fresnel zone plates [24.9,10] which have the potential of providing high resolution in the diffraction limit with enhanced total diffracted intensity. Such an imaging microscope, with resolution better than 100 nm will make it possible to gain experience in the study of biological (and other) materials. This would take the form of comparison of specimens viewed with both x rays and electrons and extending the experience so gained to the study of wet, unstained specimens for which the electron microscope is fundamentally unsuitable. It will be important in this respect to make comparisons with results obtained from low-and medium-angle x-ray diffraction.

Although the commissioning and operation of an imaging microscope makes good sense in terms of the initial demonstration of the high–resolution zone plate focusing technique, the development of a scanning system has a much greater long-term potential. This is well recognized by all those working in this field. A scanning instrument can make use of high quantum efficiency detectors which leads to a lower x-ray dose (compared with,e.g.,x-ray film) and hence reduce damage for a given signal/noise ratio. The use of an electronic detector (e.g. ionization or proportional chamber) with digital signal processing means that one can take advantage of image processing techniques which are now used widely in electron and other forms of microscopy. In general applications of a scanning soft x-ray microscope can be listed as follows:

(i) 3D imaging. On the macroscopic scale the use of computerized tomography is now an established technique [24.18] and it is natural to apply these techniques on the microscopic scale once the possibility of examining 3D specimens (possible with x rays but not with electrons) becomes a reality. In principle a complete set of 2D projections is required to form a unique reconstruction of the object and although this is possible on the macroscopic scale, geometrical restrictions and the divergent x-ray beam at the scanning spot makes this ideal situation hard, if not impossible, to achieve on the microscopic scale. The problem is being studied by CHENG and MICHETTE [24.12] both from the theoretical and the experimental point of view, using a geometrically scaled optical analogue. This is a very promising area of development - 3D information on a cellular scale could generate a biological revolution comparable to the 3D solution of the structure of DNA by CRICK and WATSON [24.13].

(ii) Time dependence. Biological processes, by their very nature, are not static. Life implies movement and movememt implies the need to add the time dimension to the imaged picture. The importance of this in low-angle x-ray diffraction has been demonstrated by BORDAS et al. [24.14]. Processes such as muscle contraction take place on a very rapid time scale (fraction of a second) but processes such as cell division (mitosis), intake and expulsion of solute or particles (endocytosis, pinocytosis, exocytosis), intake of foreign particles (phagocytosis) whose assimilation may lead to loss of cell structure (lysis) and even to cell death take place on a much longer time scale (seconds or even minutes) and should be amenable to study in vivo with a scanning microscope. Other processes such as actin and tubulin assembly are already under study using diffraction techniques [24.14] and x-ray microscopy should provide useful complementary information.

(iii) Surface information. Many of the phenomena mentioned above take place on the surface of the cell and it will be necessary to extend the x-ray microscope so that it can, as it were, focus on the region where the living cell interacts with its external environment. The detection of photo-emitted electrons with both positional resolution using the x-ray beam as a micro-(if not a nano-) probe and energy resolution via an electron spectrometer in a standard way would provide photoelectron microscopy with surface and some depth sensitivity. Electron energy loss spectroscopy (EELS) is already becoming a standard technique for the characterization of biomaterials [24.15]. The use of the same technique but with x rays as the probe instead of electrons would eliminate the need for the subtraction of the direct background from the incident electron beam.

(iv) Chemical bonding information. The use of EXAFS (extended x-ray absorption fine structure) is now a well-established technique for the study of chemical bonding in biological and biochemical systems [24.16,17]. The application of EXAFS in an imaging context has already been mentioned [24.4]. A scanning x-ray microscope could be used to provide chemical bonding information as an additional dimension to the imaging process. The simplest way of doing this would be to scan the condenser zone plate in a direction parallel to the optic axis so that radiation of different wavelength would be focused on the defining aperture. The scanning system would be an oscillatory one locked in to the data acquisition system so that at each pixel an energy scan could be carried out if desired. This would not allow collection of data with sufficiently high quality to permit a full EXAFS or XANES analysis giving bond lengths, atomic coordination numbers, bond directional information, etc., but this could be collected in a conventional way using a wide-ranging series of model compounds. By this means a catalogue of well-characterized spectra would be built up which could then be compared with the results from the energy scanning microscope. In any case, a system such as the one outlined would be needed to obtain differential contrast across an absorption edge, although in that case the scanning range required would be considerably smaller.

The suggestions made in the previous section are somewhat speculative and although possible in principle are unlikely to be realized in the near future. What is required in the next 3-5 years is the construction of a user instrument which could be the prototype of a series of instruments at SR centres (or at other intense x-ray sources if these should become available). The user instrument should be as similar as possible to an electron microscope (at least as far as the user is concerned). It should have vertical geometry and ease of access to the specimen stage. Its operation should be computer controlled with a range of detection facilities. Standardisation is particularly important so that results from similar observations at different locations can be compared in a systematic way and arguments about the functioning properties of the instrument can be avoided.

24.3 Development of the SRS for X-Ray Microscopy

Table 24.4 shows a comparison of existing and proposed SR sources which are being or could be used for the development of x-ray microscopy. For each source the number of photons/(mm^2, mrad^2, sec) into 0.1% bandwidth is given at a wavelength of 4.4 nm (44 A, 310 eV) and with a stored electron beam current of 100 mA. This quantity is the photon brightness available to the designer of an optical system for the microscope. At present the SRS compares unfavourably with what is believed to be available either at the NSLS or at BESSY. This position is expected to change dramatically over the next three years with the installation of a soft x-ray undulator and the implementation of a major modifications to the storage ring itself known as the high brightness lattice (HBL) which will generate an order of magnitude improvement in source brightness compared with the present values.

Table 24.4 Figure of merit for soft x-ray microscopy at 4.4 nm

Storage Ring	Location	Approximate Date	Energy	FOM
SRS Dipole	Daresbury UK	Now	2 GeV	7×10^{11}
SRS Undulator	Daresbury UK	1985	2 GeV	6×10^{12}
SRS HBL Dipole	Daresbury UK	1986	2 GeV	6×10^{12}
SRS HBL Undulator	Daresbury UK	1986	2 GeV	2×10^{14}
BESSY Dipole	Berlin FRG	Now	800 MeV	2×10^{13}
NSLS Dipole	Brookhaven USA	Now	700 MeV	6×10^{12}
NSLS Undulator	Brookhaven USA	In preparation	2.5 GeV	6×10^{15}
ESRF Undulator		Proposed	800 MeV	4×10^{17}
ALS Undulator	Berkeley USA	Proposed	1.3 GeV	5×10^{17}
ESRF Undulator		Proposed	5 GeV	1×10^{18}

24.4 Future Prospects

An SR source for SXRM with a figure of merit of between 10^{13} and 10^{14} photons/(mm^2, mrad^2, sec) into 0.1% BW can deliver $\sim 10^6$ photons/sec into a scanning spot 10 nm in diameter [24.18]. According to the theoretical calculations of SAYRE et al. [24.19] 10^5 photons/pixel would be required to give a signal/noise ratio (S/N) of 5 at this level of resolution in a 1 μm thick sample of cellular protein in an aqueous matrix. It follows that at the present day intensity levels available a simple scan of 1 μm^2 at this S/N level would be completed in about 15 minutes. Relaxation of the resolution limit to, e.g., 50 nm would reduce this time for the same S/N to less than 10 sec so that at this level some of the ideas discussed in section 2 could be contemplated. Some improvements could be achieved by improved optical systems. However, really massive gains can come only from improved x-ray sources as shown in Table 24.4 where figures of merit $\sim 10^4$ times greater than those available today can, hopefully, be achieved [24.20,21,22].

24.5 Conclusion

A programme of SXRM development is underway at the Daresbury Laboratory in parallel with a corresponding programme of SRS development which will bring the x-ray source to a high standard for this type of work. However, many of the ideas for x-ray microscopy which are presented here and envisaged in this programme will have to await even stronger sources of radiation such as have been proposed and discussed elsewhere.

Acknowledgements

It is a pleasure to acknowledge the help of my colleagues especially at the Daresbury Laboratory, Queen Elizabeth College and the Imperial College of Science and Technology. I am indebted to the Akademie der Wissenschaften zu Göttingen, W. Germany, for the invitation to prepare this review and for the stimulating atmosphere of the x-ray microscopy symposium at which the manuscript was completed.

References

24.1 Application of Synchrotron Radiation to problems in materials science. D.K. Bowen (Ed). Daresbury Laboratory Report, DL/SCI/R19 (1983)

24.2 P.J. Duke: The application of synchrotron radiation to x-ray microscopy in Proceedings of Seminar on Microscopy-Techniques and Capabilities SPIE 368, 22 (1982)

24.3 R.W. Whatmore, P.A. Goddard, B.K. Tanner and G.F. Clark: Nature 299, 44 (1982); and B.K. Tanner: "x-ray diffraction using the time structure of the SRS" in DL/SCI/R19 (1983)

24.4 D.K. Bowen, S.R. Stock, S.T. Davies, E. Pantos, H.K. Birnbaum and H. H. Chen: DL Preprint DL/SCI/P399E, submitted to Nature

24.5 I.R. Plummer, H.Q. Porter, D.W. Turner, A.J. Dixon, K. Gehring and M. Keenlyside: Nature 303, 599 (1983)

24.6 M.T. Brown, R.E. Burge, P. Charalambous, P.J. Duke, A. Freake, A. McDowell, A. Michette, G. Nave, R. Rosser, M.J. Simpson, and W. Smith: 10th International Congress on X-Ray Optics and Microanalysis (ICXOM10), Toulouse (1983)

24.7 R. Rosser: Private Communication

24.8 M.T. Browne, P. Charalambous, R.E. Burge, P.J. Duke, A.G. Michette, and M.J. Simpson: 10th International Congress on X-Ray Optics and Microanalysis (ICXOM10), Toulouse (1983)

24.9 A.G. Michette, M.T. Browne, P. Charalambous, R.E. Burge, P.J. Duke, M.J. Simpson: This volume, no. 12

24.10 M.J. Simpson and A.G. Michette: Submitted to Optica Acta

24.11 G.T. Herman: "Image reconstruction form projection - the fundamentals of computerised tomography", Academic Press, 1980

24.12 L.M. Cheng and A.G. Michette: 10th International Congress on X-Ray Optics and Microanalysis (ICXOM10), Toulouse (1983)

24.13 "Molecular Biology comes of age". Nature 248, 765 (1974)

24.14 J. Bordas and E. Mandelkow: "Time resolved x-ray scattering from solutions using synchroton radiation", published in Fast methods in Physical Biochemistry and Cell Biology. R.I. Sha'ati and S.M. Fernandes (Eds). Elsevier (1983)

24.15 M. Isaacson: 10th International Congress on X-Ray Optics and Microanalysis (ICXOM10), Toulouse (1983)

24.16 C.D. Garner: in "Applications of Synchrotron Radiation to the Study of large molecules" DL/SCI/R13 (1979)

24.17 EXAFS and near edge structure, A. Bianconi, L. Inoccia and S. Stipcich (Eds), Springer-Verlag (1983)

24.18 C. Kunz and G. Schmahl: ESRP PI 5/83

24.19 D. Sayre, J. Kirz, R. Feder, D.M. Kim and E. Spiller: Ultramicroscopy 2, 337 (1977)

24.20 European Synchrotron Radiation Facility - The feasibility study. European Science Foundation, Y. Farge (1979)

24.21 An optimized vacuum ultraviolet storage ring. European Science Foundation, G.V. Marr and D.J. Thompson (Ed) (1981)

24.22 National Center for Advanced Materials - Conceptual Design Report. Lawrence Berkeley Laboratory, University of California, PUB-5084 (1983)

25. X-Ray Microscopy at Imperial College

R. J. Rosser

Imperial College of Science and Technology, Prince Consort Road
London SW7 2BZ, UK

Some theoretical aspects of using line radiation sources for soft x-ray microscopy are discussed. High-resolution, absolutely calibrated spectra of a gas-puff z-pinch source are presented.

A qualitative discussion of radiation damage mechanisms and the time scales involved indicate that any picture taken in less than a millisecond may be less damaged by the radiation. Results of a first attempt to use a 100J, nanosecond laser to produce a carbon plasma as a source for very short time scale microscopy are presented.

25.1 Introduction

At Imperial College we have become involved in x-ray microscopy using zone plates as part of a collaboration with Queen Elisabeth College, London University, and Daresbury Laboratory. This collaboration is well advanced in plans to set up a microscopy station at the Synchrotron Radiation Source, Daresbury [25.1]. Our part has been to provide facilities for producing holographic zone plates to act initially as micro-zone-plates and at a more advanced stage, as condenser units.

In addition, we have been looking at possible sources for doing laboratory x-ray microscopy and doing microscopy on very short time scales. The reason for wanting to do the work in the laboratory is self-evident. The most promising source for this work is a gas-puff z-pinch, which produces line spectra for plasmas. We have taken spectra of one such source and found that there is sufficient light to do single shot microscopy to magnifications of at least X100. The reason for wanting to do microscopy on a very short time scale is not as obvious. In assessing the potential of soft x-ray microscopy it is usually assumed, based upon the theoretical calculations of SAYRE et al. [25.2], that a fundamental limit to resolution will be set at about 10 nm by damage to the specimen. One method of going beyond this limit to the theoretical Rayleigh resolution of about 3 nm set by the wavelength of light being used is to obtain the data in a very short time [25.3]. This relies on the specimen not manifesting the damage in the time it takes to obtain the picture; there is no question of reduced dosage. It would be a single shot process as the specimen will be destroyed. The most promising source for this type of work are laser-produced plasmas, which give subnanosecond x-ray light sources. A very recent trial run using the Vulcan laser at Rutherford failed to produce convincing images but does give some encouragement in terms of spectra and flux.

25.2 High Resolution and Damage

In order to see how short-pulse microscopy might be of use in obviating the damage limit to resolution, it is necessary to consider how that damage occurs. The chemistry is complex and not well understood. Table 25.1 shows a time table of events after HENLEY et al. [25.4].

Table 25.1 A simplified time table of events after absorption of x rays

Time, sec	Event
10^{-18}	1 Mev electron traverses a molecule
10^{-18}	Excitation or ionization
10^{-16}	10 eV electron traverses a molecule
10^{-14}	Period of molecular vibration
	Dissociation of excited molecules
	Time for an ion-molecule reaction
10^{-13}	Collision time in liquids
	Internal conversion between electronically excited states
	Secondary (subexcitation) electrons reduced to thermal energy
10^{-12}	Radical moves one jump if diffusion coefficient is $5 \times 10^{-5} cm^2/sec$
10^{-11}	Dielectric relaxation time for dipoles
10^{-10}	Collision time in gas
10^{-10} to 10^{-8}	Radical-radical (diffusion controlled) reactions nearly complete in spurs and a-track
10^{-9} to 10^{-8}	Fluorescence
	Radiative lifetime of singlet excited state
10^{-7} to 10^{-5}	Lifetime of hydrated electron
10^{-7} to 10^{-5}	Reaction time for radical with solute (depends on activation energy)
10^{-4} to 10^{-3}	Time for radical to diffuse, ~ 5000 A (distance between primary MeV electron events)
10^{-3}	Radiative lifetime of triplet state

Figure 25.1 shows the comparative diffusion times for electrons, ions and a protein sphere of diameter 2.3 nm, molecular weight 10^4, in water.

A simple picture of the damage process is as follows. The initial photon absorption takes place in 10^{-17} sec, resulting in the production of ions and fast secondary electrons. These electrons are thermalized by further electron production, or captured to form negative ions in about 10^{-13} sec. These electrons and ions result in free radicals; that is highly reactive chemicals which are responsible for the actual bond breaking and damage to proteins. The free radicals persist for up to 10^{-3} sec. According to SAMMUAL and MAGEE [25.5] they are formed in clusters within 2 nm of the original ionisation. From here they diffuse out with a diffusion constant $D \sim 2 \cdot 10^{-2}$ $mm^2 sec^{-1}$. So in a nanosecond they have reached a diameter of ≈ 6 nm.

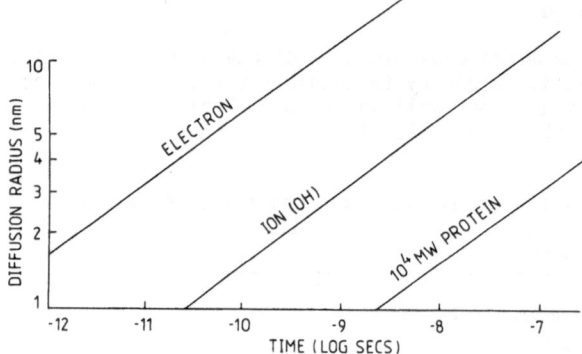

Fig.25.1 Comparative diffusion times for electrons, ions and a protein sphere of diameter 2.3 nm (molecular wt 10^4) in water

From this simple picture it is clear that a considerable amount of the damage will have been done in less than a nanosecond, but that the total structural damage will not have been completed until about a msec. So any way of getting the picture in under a millisecond will show the specimen before the full damage appears and so gain an apparent reduction in the close limit. The shorter the pulse, the greater the apparent reduction. It must be stressed that it is only an apparent reduction. The specimen will not survive and the very high resolution microscopy will be available only as a still picture.

One concern about such high resolution is the effect of heating. As about 10^3 absorbed photons per resolution element will be necessary for image formation. This would result in an enormous temperature rise if it were confined to a sphere of 2.3 nm diameter. If the energy is spread over a sphere of 10 nm diameter, the temperature rise is about 300 °C. For a sphere of 15 nm diameter this falls to less than 100 °C.

As electron diffusion would carry a great deal of the thermal energy, and would diffuse to a sphere of 15 nm diameter in the nanosecond time scales of the light source, it seems plausible to assume that heating will stay within reasonable limits.

25.3 Line Sources

Both the gas-puff z-pinch and the laser plasma produce line sources. In both there is the possibility of doing away with the monochromator and relying on the monochromatic source itself. There are three conditions that must be fulfilled for this to be possible.

1. The source must be sufficiently monochromatic. That is the width of the particular spectral line being used must satisfy the inequality

$$\frac{\lambda}{\delta\lambda} > 2 \, n$$

(25.1)

where n = number of zones in the zone plate.

2. The continuum background must be sufficiently low. If the range of intensities in the image is 50:1, the intensity of the spurious image must be at least one-fiftieth of the principal image. In line source soft x-

ray microscopy this means that the continuum intensity in the immediate vicinity of the spectral line being used must be less than one–fiftieth of the intensity of the line.

3. The spectral line being used must be separated from nearby lines by a sufficient amount. In order to satisfy the same intensity rule of thumb as in 2, the inequality

$$\frac{\lambda}{\Delta\lambda} < \frac{2\ n}{7} \qquad\qquad (25.2)$$

must be fulfilled.

Inequalities (25.1) and (25.2) can both be derived from the paraxial zone plate equation

$$\lambda f = \frac{r_n^2}{n} \quad . \qquad\qquad (25.3)$$

On differentiation and resubstitution this leads to

$$\frac{d\lambda}{\lambda} = \frac{df}{f} \quad . \qquad\qquad (25.4)$$

For inequality (25.1) we require that the change in focal length of the zone plate due to a change in wavelength is less than the depth of focus, i.e.

$$\delta f \ < \ \text{depth of focus} \ . \qquad\qquad (25.5)$$

Inequality (25.2) is found from the requirement that the change in wavelength from one line to the next is sufficiently large that the blurred, out of focus image due to a second line is less than one-fiftieth the intensity of the image due to the principal line. Because of the square law factor, this leads to the requirement of a displacement of the second image by at least $\sqrt{50}\cdot$ (depth of focus) or approximately 7 times, i.e.

$$\Delta f \ > \ 7\cdot(\text{depth of focus}) \ . \qquad\qquad (25.6)$$

So apart from the factor of 7 and the sign of the inequality, the quantities involved are the same.

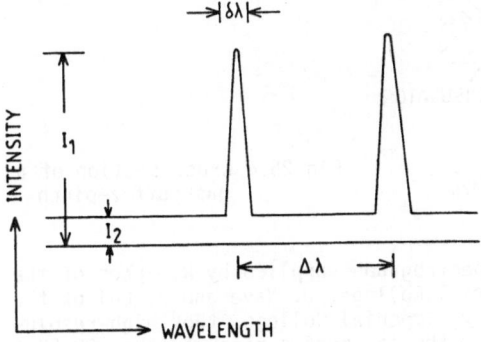

Fig.25.2 Nomenclature for inequalities governing line-source imaging

At 30 A, with a zone plate of 100 zones

if

$$\delta\lambda \ < \ 0.15 \ A$$

$$\Delta\lambda \ > \ 1.00 \ A$$

imaging can be attempted without a condenser, as shown schematically in Fig.25.3.

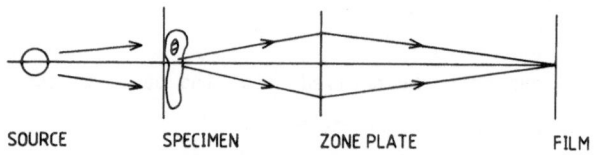

SOURCE   SPECIMEN   ZONE PLATE   FILM

Fig.25.3 A schematic drawing of an x-ray imaging system which does not use a condenser or monochromator

25.4 A Gas-Puff Z-Pinch Source

Under A.E. Dangor, a gas-puff z-pinch source has been set up in the Plasma Group at the Blackett Laboratory, Imperial College, by C. Challis. The device is shown schematically in Fig.25.4. A cylindrical shell of gas 50 mm in diameter and about 20 mm deep is formed by a change of gas flowing at Mach 10 into a vacuum. The change is injected in by an electromagnetically operated fast valve. This shell is energized by a 9 µF capacitor, charged to 25-35 kV. The resulting plasma collapses as a z-pinch, to a diameter of the order of 100 µm. The soft x-ray emitting region of the plasma is about 0.5 mm by 5 mm.

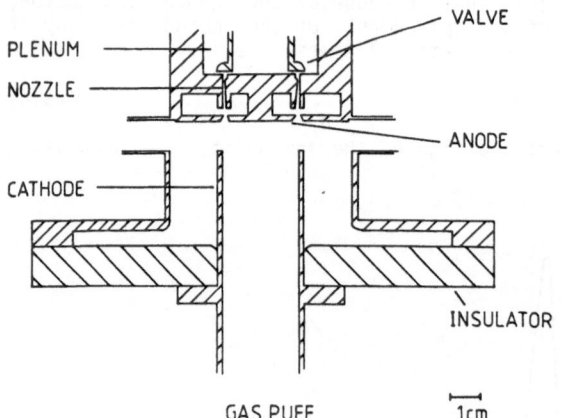

GAS PUFF   1cm

Fig.25.4 Cross section of a gas-puff z-pinch

Using a 5M grazing incidence GML spectrograph supplied by R. Speer of the soft x-ray Spectroscopy Group at Imperial College, J. Nave and J. Lui of the Spectroscopy Group, Blackett Laboratory, Imperial College, took high-resolution spectra of the light emitted from the gas-puff z-pinch in the 20-50 A region. The spectra for Ar, CO_2, C_5H_8 and N_2 are shown in Fig.25.5-7.

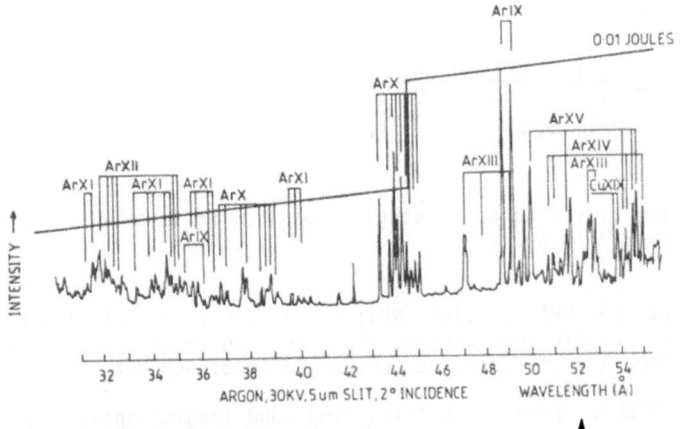

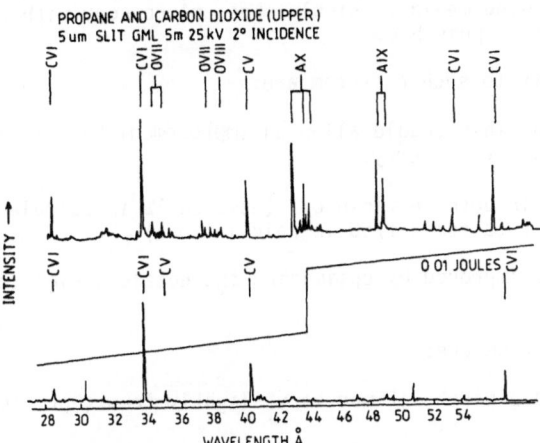

Fig.25.5 High-resolution spectrum of gas-puff z-pinch produced plasma, charged with argon, taken with a GML 5 metre spectrograph

Fig.25.6 As for Fig.25.5, but with carbon dioxide and a propane charge

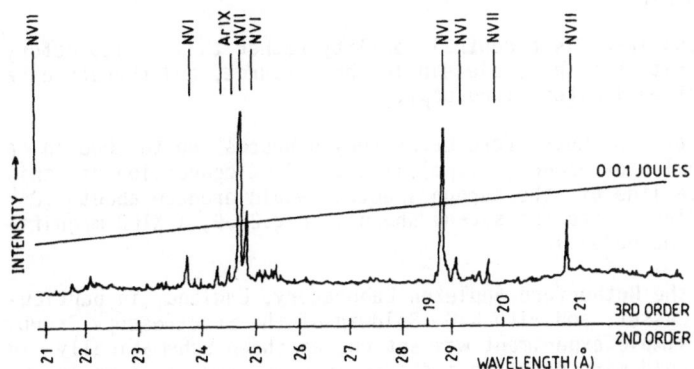

Fig.25.7 As for Fig.25.5, but with a nitrogen charge

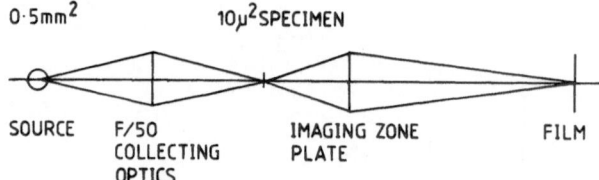

SOURCE   F/50   IMAGING ZONE   FILM
          COLLECTING   PLATE
          OPTICS

Fig.25.8 Arrangement for calculations on single shot imaging using a
         gas-puff z-pinch line source

For x-ray microscopy, the NVI (28.1A), NVII (24.8A), CVI (33.7A) and CV (40.3A) lines are particularly interesting. They contain about 0.02J in a single line, emitted over 4π steradian, or about 10^{15} photons/shot.

Using currently available f/50 x-ray collecting and imaging optics and photographic film as the recording medium, single shot microscopy with a magnification of X10 should just be possible.

The obvious possible improvements to such a system are:

a) Microchannel plate detection. This should allow an improvement by a factor of a hundred in collection efficiency.

b) Zone plates should improve in both efficiency (assumed 1% in calculations) and numerical aperture (assumed f/50 in calculations).

c) The gas puff itself could be improved by optimizing the nozzle design to the gas being used.

Possible disadvantages of the device are:

a) The source position. Being a z-pinch, it may vary slightly from shot to shot.

b) The repetition rate would be limited to about a minute by the vacuum pumps needing to clear the previous discharge.

25.5 Laser-Produced Plasma

Although a large glass laser is a central facility rather than a laboratory piece of equipment, it is a possible single shot source, and therefore a candidate for resolution-limited microscopy.

The Vulcan laser at the Rutherford Laboratory produces up to 100J in a single, nanosecond pulse of green (0.53μm) light. A 0.1% conversion of this energy into a single line of the carbon spectra would produce about 10^{16} photons in 4π steradians. With the set-up shown in Fig.25.9, a X100 magnification set-up should be possible.

With the help of the Rutherford Appleton Laboratory, England, in particular R. Eason and D. Basset, and with K.G. Baldwin of the Spectroscopy Group, Imperial College, a simple experiment was set up, as shown schematically in Fig.25.9. The 50 μm gold wire provided a shadow in which the image could be formed. A gold pointer indicated the centre of the zone plate as shown in Fig.25.10. Figures 25.11 and 25.12 show two results from the experiment.

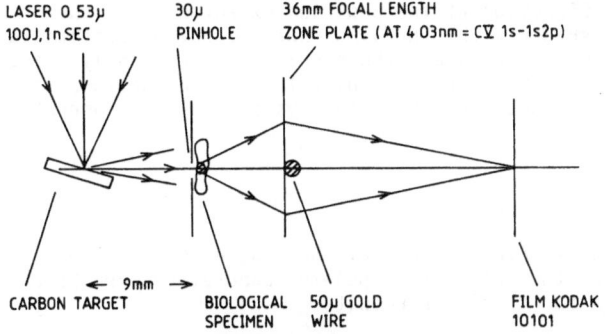

LASER 0 53µ   30µ   36mm FOCAL LENGTH
100J,1n SEC   PINHOLE   ZONE PLATE (AT 4 03nm = C $\overline{\text{V}}$ 1s–1s2p)

← 9mm →
CARBON TARGET   BIOLOGICAL   50µ GOLD   FILM KODAK
                 SPECIMEN   WIRE   10101

Fig.25.9 Schematic diagram of an experimental arrangement for doing microscopy with a laser-produced plasma

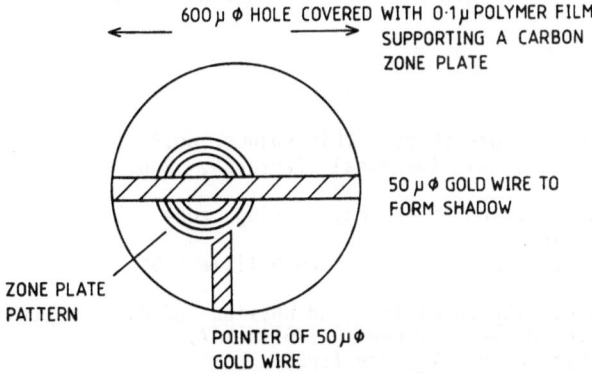

600µ ⌀ HOLE COVERED WITH 0·1µ POLYMER FILM
SUPPORTING A CARBON
ZONE PLATE

50µ ⌀ GOLD WIRE TO
FORM SHADOW

ZONE PLATE
PATTERN

POINTER OF 50µ⌀
GOLD WIRE

Fig.25.10 The apodizing arrangement for eliminating the zero-order light
through the centre of the zone plate

Fig. 25.11   Fig. 25.12

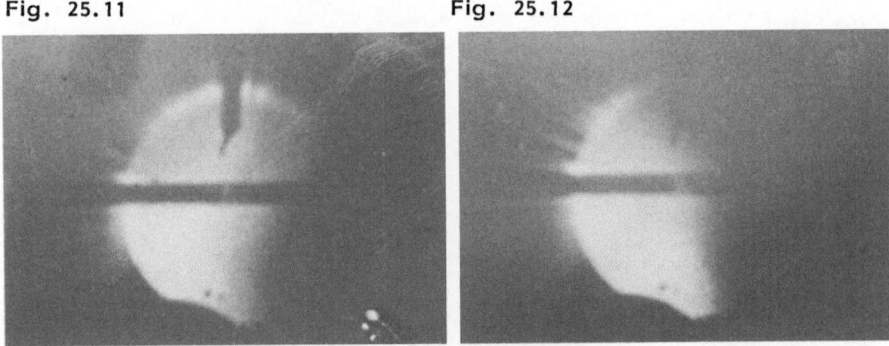

Fig.25.11 Single shot photographs taken using the arrangements shown in
and   Fig.25.9 and Fig.25.10. Although there is no evidence of imaging,
Fig.25.12 there is diffraction into the horizontal shadow near the zone
plate centre (pointed at by the vertical shadow)

There is some evidence of diffraction at the position expected, i.e. the intersection of the pointer shadow with the line shadow, but no conclusive evidence of imaging. The spectra taken and the results are currently being analysed to determine the reason for a null result. Initial indications are that there were sufficient photons, but that the system was not correctly positioned.

Acknowledgements

The author would like to thank A.E. Dangor, C. Challis, J. Nave and J. Lui for the gas-puff spectra, and the Rutherford Appleton Laboratory, Rob Eason, Dave Basset, and Ken Baldwin for the laser results.

The SERC is thanked both for the fellowship which has enabled this work to be undertaken by myself and for the provision, via one means and another, of all the equipment used.

The Stiftung Volkswagenwerk are thanked for sponsoring this excellent meeting, so ably organized by Professor Schmahl and his colleagues at Göttingen.

References

25.1 P. Duke: "X-ray microscopy at Daresbury", this volume, no.24
25.2 D. Sayre, R. Feder: "Additional technical factors", Ann. N.Y. Acd. Sci., $\underline{342}$, 268 (1980)
25.3 R. Rosser: "Soft x-ray imaging microscopy using zone plates and non-synchrotron sources" in SPIE Vol. 368 - Microscopy - Techniques and Capabilities (Sira), 17-20 (1982)
25.4 E.J. Henley, E.R. Johnson: The chemistry and physics of high energy reactions (Washington D.C. University Press, 1969 p 147)
25.5 J.L. Magee, Ann. Rev. Phys. Chem., $\underline{12}$, 389 (1961)

26. Photoelectron X-Ray Microscopy: Recent Developments

F. Polack

L.U.R.E., Bat. 209 C, Université Paris-Sud, F-91405 Orsay, France

S. Lowenthal

Institut d'Optique, B.P. 43, F-91406 Orsay, France

26.1 Introduction

Using absorption edges for contrast enhancement or microchemical analysis has been often claimed as one of the main advantages of the soft x-ray microscopy; the K edges of carbon to aluminium and L_{23} edges of phosphorus to bromine are located in the 0,8 - 9 nm wavelength range which is very favourable to image biological objects. But in practice, this opportunity has been very seldom used [26.1]. As a matter of fact, none of the common techniques of x-ray microscopy is really suitable for this purpose.

Imaging using focussing gratings like zone plates suffers from a very strong chromatic aberration, thus focus and magnification are altered by wavelength tuning [26.2]. The efficiency and the numerical aperture of these lenses also steeply decrease at shorter wavelength. With scanning x-ray microscopes, fluorescence analysis can be done around 1 nm wavelength, but should be difficult to exploit for the low Z elements [26.3]. Presently, resolutions in the order of 10 nm only are expected for such instruments [26.3-5].

On the contrary, contact microscopy has already produced analytical results, using photoresists as image detectors [26.1]. The resolution can be as good as a few nanometers and the quantum efficiency is excellent. The process mostly suffers from the detector saturation and nonlinearity which strongly limit the sensitivity. This point can be overcome by replacing the photochemical detector by a photocathode layer used in transmission mode and followed by an emission electron microscope. A few attempts have been made in this area [26.6,7,3], and the former instrument of this type has been constructed by HUANG and MÖLLENSTEDT, who, in 1955, using a special built-in x-ray tube, obtained a 120 nm resolution.

Synchrotron radiation has now opened the way to new developments. We examine here what characteristics can be expected from photoelectron x-ray microscopy, and discuss the means of improving the resolution and efficiency. We describe the instrument which is presently under realization at the Institut d'Optique (Orsay) and make a few comments on the solutions given to the main technical problems.

26.2 Technical Aspects of Photoelectron X-Ray Microscopy

26.2.1 Principles

As said before, photoelectron x-ray microscopy is nothing else than contact microscopy with an image-converting detector. The object is fixed on a sup-

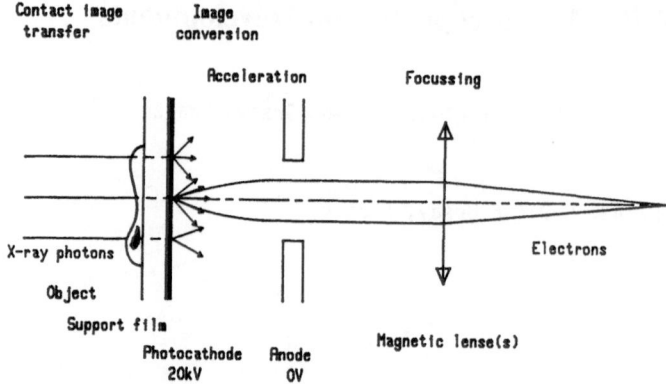

Fig.26.1 Photoelectron magnification of x-ray images

port film (0.1 to 0.5 μm thick) which is coated on the other side by a very thin layer of a photoemissive material, as shown on Fig.26.1. An x-ray beam of small aperture projects a transmission image of the object on the cathode layer. The x-ray photons which are absorbed in this layer generate photoelectrons of various energies. Most of them undergo multiple interactions and therefore exit as low-energy secondary electrons; with usual photocathodes, from 50% to 90% of the electrons have an initial energy lower than 10 eV [26.8] . This is actually a very profitable situation since, as it will be shown later, the electrons of very low initial velocity only can be focussed accurately to produce an enlarged image of the emission plane.

The photoelectron must be at first collected from a 2π solid angle in order to match the aperture of the collected beam with the 10^{-4} -10^{-3} aperture of an electron microscope. It is usually done by a strong electrostatic field. Only electrostatic acceleration will be considered here, but collection by a strong magnetic field [26.9] or by combined electrostatic and magnetic fields [26.10] has been also described.

A large amount of data can be found in the literature on emission microscopes of this type (see for instance the review by P. GRIVET [26.11]). However, most of them have been designed for thermoemission in the case of which the electron energy distribution is much more narrow than what we expect. Some adaptations are therefore needed.

For the evaluation of its fundamental characteristics such as resolution and quantum efficiency, the instrument can be divided in two separated parts: the object and photocathode assembly where a unit magnification image is generated by contact x-ray projection and photoemission; the emission electron microscope which enlarges this electron image.

26.2.2 Formation of the Contact Electron Image

In this part of the instrument the resolution-limiting factors are those of contact imaging: Fresnel diffraction, penumbral blurring and the intrinsic resolution of the detector. The presence of the support film maintains a distance d between the object and the photocathode, and therefore Fresnel diffraction limits the size of the smallest visible detail around $\delta = \sqrt{\lambda d}$. If the film is thin enough, penumbra will remain negligible so long as the aperture of the x-ray beam is lower than $\theta = \sqrt{\lambda/d}$ (for instance, with

$\lambda = 5$ nm and $d = 0.5$ μm, then $\delta = 50$ nm and $\theta = 0.1$ rd). The intrinsic resolution of the cathode layer is limited by the diffusion of the photoelec-
trons and is rather difficult to evaluate because the various interactions
and the mean free paths of the generated particles are not known precisely.

To give an order of magnitude to this effect, we shall develop a very
simple model of the photocathode. A more exact model is developed by HENKE
et al. in [26.12]. However, the small differences between the predictions
which can be obtained do not justify here this increase of complexity.

The fraction of the x-ray photons which are absorbed in a layer of thickness dx at a depth x in the photocathode is

$$\mu(E) \exp\left[-\mu(E) x\right] dx \qquad (26.1)$$

where $\mu(E)$ is the absorption coefficient of the cathode material for x rays
of energy E. We assume firstly that the number of generated secondary electrons is proportional to the energy deposition, secondly that they are uniformly absorbed along their path to the surface. The yield of a photocathode
layer of thickness t in transmission mode, then, is given by

$$Y(t,E) = K \int_0^t E \,\mu(E) \,\exp\left[-\mu(E) x\right] \exp\left[-\nu(t-x)\right] dx \qquad (26.2)$$

where ν is the mean absorption coefficient of the cathode material for the
secondaries (usually $\nu \gg \mu$). Hence we obtain the following expression of the
yield

$$Y(t,E) = K \frac{E \,\mu(E)}{\nu - \mu(E)} \left[\exp\left[-\mu(E) t\right] - \exp\left[-\nu t\right]\right] \qquad (26.3)$$

which presents a maximum value for

$$t = \frac{1}{\nu - \mu} \text{Log} \,(\nu/\mu) . \qquad (26.4)$$

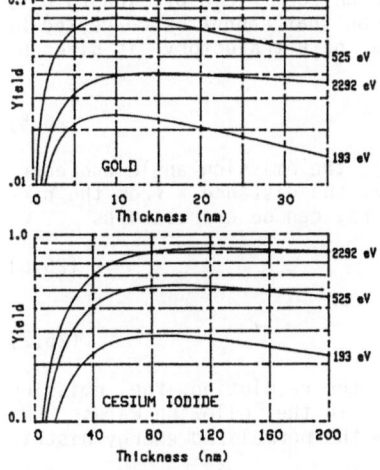
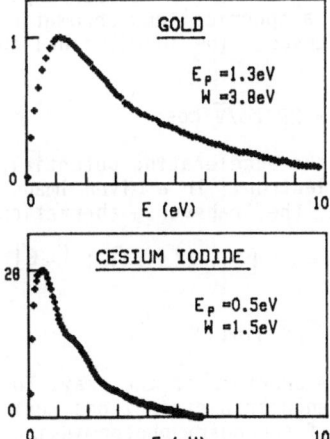

Fig.26.2 Secondary electron yield for
Au and CsI photocathodes in transmission mode, versus cathode thickness,
after HENKE et al. [26.12]

Fig.26.3 Secondary electron energy
distribution for Au and CsI photocathodes, excited by C-K$_\alpha$ (277 eV)
x rays, data from HENKE et al.
[26.12]

The above expression does not differ significantly from the results of HENKE et al. [26.12], and from the shape of the curves Y(E) on Fig.26.2. We can get an estimation of the mean free path $\lambda = 1/\nu$ of the secondary electrons. This value of λ is also an estimation of the radius of the point spread function. We therefore predict an intrinsic resolution around 5 nm for gold cathodes and around 50 nm for CsI.

In defining the conversion efficiency great care must be taken of the fact that photoelectrons are emitted by groups of a few elements that we call emission events. The quantum efficiency η is the number of events per incident photon and differs from the secondary yield by a factor G that we call gain factor and which is the mean number of electrons per event, $Y = \eta \cdot G$. Photoemission yield and gain are strongly dependent on the cathode material and the x-ray energy. In agreement with HENKE [26.12] we have found a yield variation proportional to $E \cdot \mu(E)$. According to our model the $\mu(E)$ factor should be attributed to η and the E factor to G. Values of η and G can be obtained from the measurement of [26.13]; for instance at $\lambda = 8.3$ A, $\eta = 4\%$, $G \simeq 5$ for gold and $\eta \simeq 35\%$, $G \simeq 20$ for CsI.

26.2.3 Resolution of the Emission Microscope

The resolution which is required from this part of the instrument is quite moderate for an electron microscope and is not difficult to obtain except in the regions where the aperture of the beam and the relative energy disper- sion of the electrons are large. In a typical cathodic field of 50 kV/cm the energy dispersion is reduced by a factor of 100 and the aperture angle passes from $\pi/2$ to 0.1 in the first 0.1 mm. Therefore, as long as the spherical and chromatic aberrations are concerned, the accelerating lens (often called immersion lens or cathode lens) can be considered as the superposition of the uniform accelerating field which exists near the cathode, with an aber- ration free-lens.

The image given of an emissive plane by a uniform field of length D is affected by a spherical and chromatic aberration that cannot be corrected by the other lenses. The longitudinal aberration of a trajectory is given by [26.14,11]

$$\ell(\alpha,\phi) = 2D \sqrt{\phi/V} \cos\alpha \qquad (26.5)$$

where V is the accelerating potential; α and ϕ, the emission angle and ener- gy of the electron. In a given image plane at the distance z from the par- axial plane, the transverse aberration of the ray can be expressed as

$$F_z(\alpha,\phi) = \alpha' \left[z + \ell(\alpha,\phi) \right] ; \; F_z \in] -\infty ,+\infty [\qquad (26.6)$$

where $\alpha' = \sqrt{\phi/V} \sin\alpha$ \qquad (26.7)

is the image aperture of the ray. To evaluate the resolution, the emission energy and angular distributions must be known. In the following we use HEN- KE's model of secondary photoemission in which the normalized energy distri- bution is

$$N(\phi) = 6W \phi (\phi + W)^{-4} \qquad (26.8)$$

where W is the work function of the cathode material, and the angular distribution is assumed to be Lambertian. Figure 26.3 shows the energy distribution for Au and CsI cathodes, and indicates the value of the work function which fits best with Henke's model.

Following TSYGANENKO [26.15], we can now write the point spread function as

$$f_z(r) = \frac{1}{2\pi r} \int_0^S \int_0^{\pi/2} N(\phi) \sin2\alpha \; \delta(r^2 - F_z^2(\alpha,\phi)) \; d\alpha \; d\phi \qquad (26.9)$$

in which $\delta(r^2 - F_z^2(\alpha,\phi))$ denotes the following Dirac distribution

$$\delta(r - F_z(\alpha,\phi)) + \delta(r + F_z(\alpha,\phi)) \; ; \; r > 0 \; . \qquad (26.10)$$

The expression (26.9) however cannot be directly evaluated and anyway would not be of great help since, as in any case of pure geometrical aberration, the illumination is infinite at the centre of the point spread function. It is therefore convenient to calculate the Modulation Transfer Function (M.T.F.) which is the 2-dimensional Fourier transform of $f(r)$. Both are rotationaly symmetric and hence, the M.T.F. is given by the following Hankel transform

$$\tilde{f}_z(\nu) = 2\pi \int_0^\infty f_z(\pi) \; J_0(2\pi \nu r) \; r \; dr \qquad (26.11)$$

using (26.9), (26.10) and the parity of J_0 we then have

$$\tilde{f}_z(\nu) = \int_0^\infty \int_0^{\pi/2} N(\phi) \sin2\alpha \; J_0(2\pi \nu \; F_z(\alpha,\phi)) \; d\alpha \; d\phi \; . \qquad (26.12)$$

We obtain here an expression of the FTM (similar to that of [26.15]) which is easy to compute numerically with the help of (26.5) to (26.8). In trying to obtain (26.12) we have supposed that all the natural emission laws were not modified by instrumental factors, and that all the electrons were imaged by the microscope. This does not happen in practice and the most aberrant rays are suppressed by two ways. The first way is to limit the image aperture $\alpha' < \alpha_m$ by means of a small diaphragm in the cross-over plane; it performs a filtering of the radial component of the emission energy

$$Er = \phi\sin^2\alpha < V\alpha_m'^2 = \phi\sin^2 \alpha_m(\phi) \; . \qquad (26.13)$$

The second way, which is more unusual, is to limit the maximum emission energy by means of a dispersive energy filter. These filterings are easily taken into account by modifying the integration limits in (26.12) so that

$$\tilde{f}_z(\nu) = \int_0^{\phi_m} d\phi \int_0^{\alpha_m(\phi)} N(\phi) \sin2\alpha \; J_0(2\pi\nu \; F_z(\alpha,\phi)) \; d\alpha \; . \qquad (26.14)$$

26.2.4 Optimization of the Characteristics

The influence of the filtering parameters is illustrated by Fig.26.4, which shows the M.T.F. curves, computed at the best focus, for different conditions. The value of $\tilde{f}(0)$ given by (26.14) is the transmittance of the optics. To make comparisons easier, the F.T.M. has been normalized so that

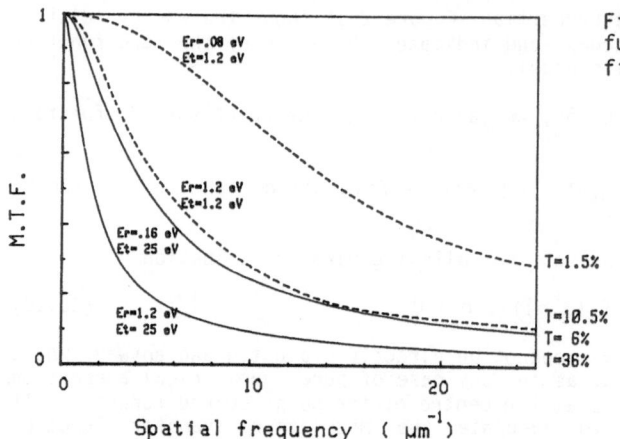

Fig.26.4 Modulation transfer function in different filtering conditions

M.T.F.

Er=.08 eV
Et=1.2 eV

Er=1.2 eV
Et=1.2 eV

Er=.16 eV
Et= 25 eV

Er=1.2 eV
Et= 25 eV

T=1.5%

T=10.5%
T= 6%
T=36%

Spatial frequency (μm^{-1})

Photocathode work function: W=5 eV ; cathodic field: 50 kV / cm
Er: Maximum radial energy (aperture limitation)
Et: Maximum total energy (dispersive filter)

$\tilde{f}(0)$ = 1 and the optics transmittance T is quoted in the margin. Curves in full line stand for an instrument without energy filter and illustrate the influence of aperture filtering on resolution and also on efficiency. Dotted curves show how the energy filtering (total energy < 1.2 eV) can enhance the resolving power.

The data of Fig.26.4 apply to a model of gold cathode with a work function W = 5eV and a reasonable field ε = 50 kV/cm, but these curves became universal by using W as the energy unit and W/ε as distance unit. As ε cannot easily be increased, low work functions as for CsI should be interesting, and thus a gain of overall efficiency χ (cathode + optics) can be expected. χ is roughly given for low optic transmittance T by χ = $Y \cdot T$, where Y is the total secondary yield and by χ = η for a high T. Then for 50 nm resolution the expected overall quantum efficiency χ is of the order of 1% for a F.T.M. = 20% and 2% for a F.T.M. value of 10%. With CsI, χ could reach 40 or 50% for a F.T.M. = 20%.

26.3   Design of a Photoelectron X-Ray Microscope for ACO Storage Ring

26.3.1 General Description

The general structure of our instrument is shown on Fig.26.5. It directly proceeds from the above considerations, and thus we have the following essential elements:
- The emission lens which has been designed in order to offer the maximum cathodic field for a given accelerating potential as well as minimum field aberrations.
- The magnetic prism and electrostatic mirror of the Castaing-Henry filter which will be used for total energy filtering.

Between them, a group of magnetic lenses makes the necessary conjugations. It must also give a real magnified image of the cross-over for lateral energy filtering. Behind the prism, a second group of lenses produces the final magnification.

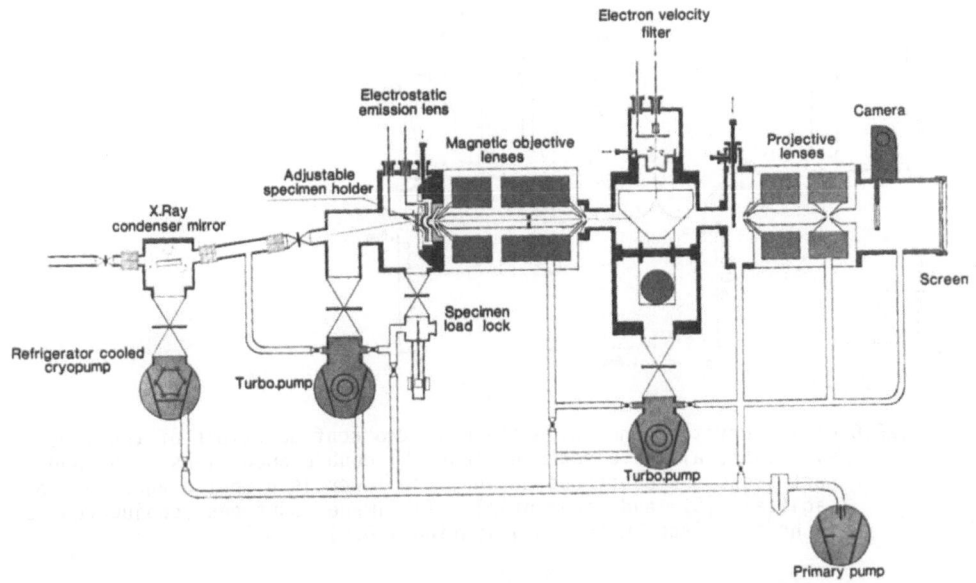

Fig.26.5 Diagrammatic drawing of the photoelectron x-ray microscope

The object and photocathode are mounted on a very thin polyimide foil ($\sim$ 0.2 μm) and are inserted in a specimen holder which is part of the immersion lens and can be moved in its plane. They are illuminated with soft x-ray synchrotron radiation by means of a grazing incidence toroidal mirror (focal length 1 m; aperture 15 mrad) to ensure a sufficient flux at high magnifications. It also acts as a low-pass filter removing the hard x-ray component approximately above 800 eV.

26.3.2 The Immersion Lens

A uniform accelerating field can be approximated in practice by the design of Fig.26.6a. A plane cathode at potential 0 and, at the distance D, an annular anode at potential V produce near the cathode a quasi-uniform field ε = V/D which should give an image at the distance D behind the cathode. However, a field increase through an aperture has a converging action when a field decrease is divergent. The small hole of the anode can therefore be approximated as a thin lens of focal length 4D. To obtain a positive objective, two solutions have been used. In the most usual, a Wehnelt electrode at a lower potential is inserted between cathode and anode to provide a strong focussing action, but it also reduces the cathodic field by a factor 5 to 10. The second solution has been suggested by Fert and Simon, who have shown that a combined objective, made of a negative immersion lens followed by a magnetic lens, could be used with some advantages [26.16].

The ray tracing in this lens has shown important field aberrations (Fig.26.6a). The computation of various configurations proves that these field aberrations can be significantly reduced with the 3-electrode design of Fig.26.6b. The lower part of this figure compares the sagittal and tangen-

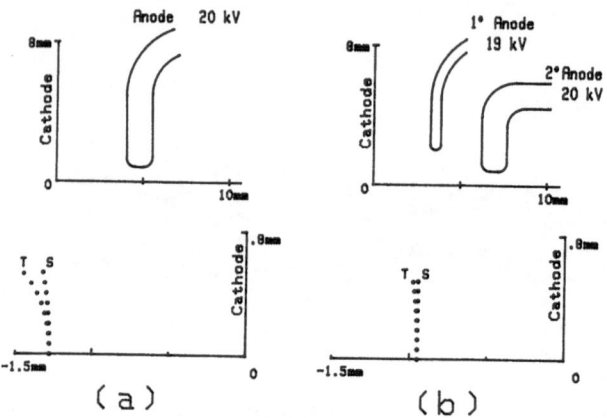

Fig.26.6 Field curvature and astigmatism in two configurations of the immersion lens; a) two electrode lens, b) double anode lens. The upper
part shows the electrode design; the lower is a spot diagram of the
sagittal (S) and tangential (T) image surfaces (conjugated to
cathode surface, with magnification ∿ 0.6)

tial image surfaces in two configurations of equal field ε = 50 kV/cm. This
new lens gives a slightly demagnified image of the cathode plane (0.6 magnification). The cathode and cross-over image plane are respectively located
at 1 and 10 mm behind the cathode.

26.3.3 The Magnetic Prism and the Electron Optics

The dispersive element of the energy filter is a magnetic prism in which the
electron beam follows;on 90°, a circular path of radius R = 40 mm. The electrons are then reflected from the electrostatic mirror and, going back are
again deflected the opposite way. This property of the filter ensures that
the incoming and outcoming beams are aligned while the dispersion is doubled
(10 μm/V at 20 kV). The prism which is fixed on a horizontal pivot can be
easily rotated away for direct imaging and initial alignments. The filter has
only two couples of stigmatic points and the optical parameters of the electrostatic mirror are chosen in such way that they are distributed symmetrically at approximately 3R and R/2 of the prism centre [26.16].

The remaining electron optics is completely determined by the properties
of the prism and of the immersion lens (Fig.26.7). A very small image of
the cross-over must be located in the real stigmatic point and its diameter limits the energy resolution (1 eV resolution at 20 kV for 10 μm diameter). The length of the electron beam is then set by the maximum size
of the intermediate image which ensures pupil stigmatism (∼ 800 μm) and
therefore the intermediate magnification is imposed by the energy and angular filtering conditions (from X20 to X50). An intermediate lens of 30 mm
focal length, which gives an image magnification of 3, produces the magnified image of the cross-over. An objective lens of 3 to 10 mm focal length
ensures the remaining conjugations and convenient magnification (X10-X30).

Fig.26.7 Diagram of the electron optics

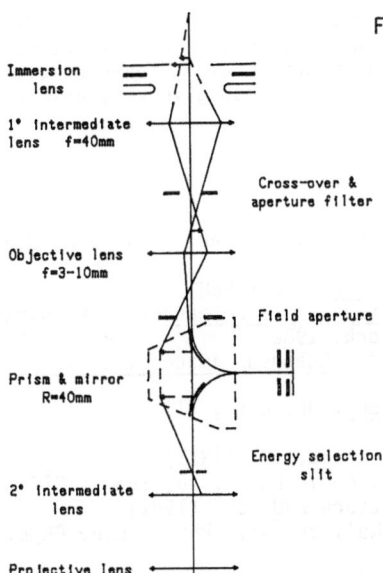

Immersion
lens

1° intermediate
lens   f=40mm

Cross-over &
aperture filter

Objective lens
f=3-10mm

Field aperture

Prism & mirror
R=40mm

Energy selection
slit

2° intermediate
lens

Projective lens

26.3.4 Designing a Horizontal Microscope

In our application, the synchrotron radiation illumination has only one drawback: the beam axis is horizontal. Many problems are encountered in de-signing a horizontal microscope that would not appear with a vertical col-umn. For instance, before evacuating a horizontal column, all the elements must be clamped tight to each other. But the same elements must be also held on some support, and the multiplicity of mechanical links can be harmful to stability.

Our instrument must therefore be laid on a granite table, each element being supported by air bearings with very small play ($\sim$ 10 μm) and the me-chanical axis of each element will be carefully aligned once for all during the construction. This, of course, does not ensure the alignment of the op-tical axis, and deflection coils and adjustable pole pieces are provided for this purpose.

26.4 Conclusion

We have established, in the first part, the theoretical characteristics which can be expected from a soft x-ray photoelectron microscope. It has been shown that with the combined use of an aperture and an energy filter, the resolution could be brought to 50 nm or even better. The efficiency of the instrument, around 1% at worst, will be quite similar to that of the existing soft x-ray microscope. On these principles we have designed a microscope, presently under realization, which will be installed on one of the beam lines of ACO storage ring (L.U.R.E., Orsay). However, the instrument will be tested, in a first stage, without the magnetic filter and should not offer a better resolution than 100 nm.

The main advantage of photoelectron x-ray microscopy is that its quantum efficiency is not very sensitive to the illuminating wavelength. This advantage should become significant when, as it appears to become practicable, the condenser mirror will be replaced by a monochromator. We are of course thinking of absorption analysis, but the illumination by a tunable wavelength, even if sharp lines are not available, should also give very helpful variations of contrast in soft x-ray images.

References

26.1 J.W. McGowan and M.J. Malachowski: Ann. N.Y. Acad. Sci. $\underline{342}$, 288 (1980)
26.2 G. Schmahl et al.: Ann. N.Y. Acad. Sci. $\underline{342}$, 368 (1980)
26.3 J. Kirz and D. Sayre: in Synchrotron radiation research, H. Winnick and S. Doniach eds., 277 (Plenum, New-York, 1980)
26.4 G. Schmahl, D. Rudolph, B. Niemann: in Scanned image microscopy, A. Ash ed., 393 (Academic Press, 1980)
26.5 R.P. Haelbich: in Scanned image microscopy, Op. Lit., 413
26.6 L.Y. Huang, Z. Phys., $\underline{149}$, 225 (1957)
26.7 F. Polack, S. Lowenthal: Rev. Sci. Instr., $\underline{52}$, 207 (1981)
26.8 B.L. Henke, J.A. Smith, D.T. Attwood: J. Appl. Phys., $\underline{48}$, 1852 (1977)
26.9 G. Beamson, H.Q. Porter, D.W. Turner: Nature $\underline{290}$, 556 (1981)
26.10 N.A. Smirnov, M.A. Monastyrkii, Yu. V. Kulikov: Sov. Phys. Tech. Phys. $\underline{24}$, 1462 (1979)
26.11 P. Grivet: Electron Optics (Pergamon, Oxford, 1965)
26.12 B.L. Henke, J.P. Knauer, K. Premaratne: J. Appl. Phys., $\underline{52}$, 1509 (1981)
26.13 L.G. Eliseenko, V.N. Shchenelev, M.A. Rumsh: Sov. Phys. Tech. Phys. $\underline{13}$, 122 (1968)
26.14 G. Slodzian: in Adv. Electron. Electron. Phys. Sup. 13 B (Academic Press, London, 1980) 1
26.15 V.V. Tsyganenko, R.A. Lachasvili, I.A. Bobrovskiy: Sov. J. Opt. Techn. $\underline{39}$, 733 (1972)
26.16 C. Fert: C.R. Acad. Sci. $\underline{243}$, 1300 (1956)
26.17 L. Henry: Thèse, Paris (1964)

Part V

Applications of X-Ray Microscopy

27. Prospects and Problems in X-Ray Microscopy

J. Kirz[1]

Physics Department, State University of New York
Stony Brook, NY 11794, USA

D. Sayre

IBM T.J. Watson Research Center, P.O. Box 218
Yorktown Heights, NY 10598, USA

27.1 Introduction

As various forms of x-ray microscopy develop from "state of the art" technical accomplishment to user-oriented tools of biological investigation, it
may be useful to pause and assess where we are, where we ought to be and how
we might get there. Most of the contributions to this meeting address the
first of these topics.

Here we propose to bring up the other subjects we should examine. We will
pose several questions, propose only a few answers, but hope to provoke a
lively debate from which more answers will emerge. We propose to make an
attempt at predicting in which areas soft x-ray microscopy will have a major
impact - fully aware that of all predictions the ones most likely to be far
off the mark are the ones pertaining to the future.

Once we have carved out our domain in biological research and have established the principal topics of the next symposium, we shall examine the
problems, the prospects and the limitations. We will discuss radiation damage, admit our ignorance, and look at ways to minimize both. We shall dwell
in particular on the problems of thick specimens, and the need to obtain
three-dimensional images.

27.2 The Future of Soft X-Ray Microscopy

In the study of biological structures a clear gap exists between two well-
developed instruments, the optical and the electron microscope. It is our
role to fill this gap. The limitation of optical microscopy in biology is
almost entirely in its resolution,[2] given that staining, and modern
contrast-enhancing optics have overcome many of the problems related to the
poor intrinsic contrast of most specimens.

The limitations of electron microscopy are more subtle. For thin,
stained, stabilized specimens images of very high resolution are routinely
obtained. We cannot and should not try to compete in the imaging of these
objects.

---

1) Supported in part by NSF Grant #PCM8109358.
2) In the study of specimens of nonbiological nature there is also another
   limitation in optical microscopy: opacity. For this reason x-ray microscopy may have important applications in areas such as metallurgy,
   ceramics, geology, and thin-film technology. Maybe we should pay more
   attention to their needs!

Soft x rays have major contributions to make where the specimen is thick, wet, or radiation sensitive.

In fact, from the perspective of the electron microscopist, most biological specimens start thick, wet, and radiation sensitive, and a host of tricks and techniques have been developed to overcome these "problems". Let us examine them briefly.

The standard treatment for thick specimens is to cut them into thin sections. To be able to visualize the entire specimen, large numbers of serial sections are examined. To study the organization of synaptic inputs to a single neuron, for example, DR. HAMOS [27.1] examines 1000 sections, and then reconstructs the neuron from over 5000 electron micrographs. He could in fact use fewer and thicker sections but each micrograph would become more complex, and the interpretation more difficult. A three-dimensional imaging method capable of dealing with thick specimens would clearly speed up the work, and make it possible to study many more similar systems!

Thick and wet specimens pose additional problems. Attempts at electron microscopy in the thick, wet state has not met with much success due to multiple scattering of the electron beam in the specimen even in high voltage instruments. Progress has been made in freezing specimens, then cutting and viewing them in a frozen hydrated state. This is a promising but difficult approach, requiring particularly clean instruments to avoid freezing out too much contamination onto the specimen. Though it is unclear just how well one could use this method to reconstruct an entire cell from serial sections, we could do away with all the problems of cryo sectioning, by providing a full three-dimensional image of the thick, wet specimen.

The electron microscopist handles the radiation damage problem in a variety of ways:

- The specimen is stained with high Z elements to increase the contrast. (The radiation dose required to form an image is inversely proportional to the square of the contrast, and unstained specimens are generally of very low contrast in the electron microscope.)
- Fixation, dehydration, embedding in plastic and negative staining all help make the specimen more radiation resistant. They do not necessarily preserve all features of the specimen with equal fidelity.
- Specimen damage is reduced if during exposure the specimen is kept at cryogenic temperatures.

Though in x-ray microscopy we can do away with most of these preparatory steps, we would keep in mind that cooling, fixation and staining may have a role in our efforts to minimize damage as well.

Let us look into the crystal ball then, and view our future. We see ourselves engaged in the study of uncut, wet, thick specimens (1-10 μm thick), with imaging in three dimensions. These specimens may be entire cells in culture, nuclei or other organelles, tissue sections, or unicellular organisms. In most studies we will be looking for a resolution of 300A-1000A, though higher resolution can be attained, at least for the thinner specimens. In many cases we will also map some major elemental constituents. In all of these studies we will be working with biologists who in fact identify the problems to attack, prepare the specimens, and interpret the results. In many cases they will even do the imaging for themselves. When that becomes routine, we can rest on our laurels or attack a different problem entirely.

27.3 Radiation Damage

Before embarking on our discussion on how we will attain the above-mentioned scenario, we need to face the problems of radiation damage [27.2]. While we know how to calculate radiation doses required to form a given image [27.3], we are unaware of experimental work to relate x-ray dose to damage in the exposure range of interest.

Ignorance shall not deter us from looking at the problem from several points of view. We shall admit at the outset that after the exposure the specimen, unless fixed and stabilized,will most likely be destroyed. Our aim is to preserve it during exposure only.

27.3.1 The Temporal Dimension

Most of the work in x-ray microscopy to date involves exposures which range from seconds to hours. The exception is the work with flash sources by FEDER et al. [27.4]. The idea in the use of flash sources involves at least two considerations. First, the flash is used as in every day photography, to eliminate blurring due to the motion of the subject. This is of particular importance for initially live specimens. Second, one might hope to eliminate blurring due to radiation damage in the same way, at the same time.

An interesting and careful analysis of the time dimension of the latter problem has been carried out by SOLEM [27.5]. He points out that if one uses a flash much shorter than that used by Feder et al. ($\sim 10^{-7}$ sec), the high rate of energy deposition in the specimen leads to rapid thermal expansion, decomposition, and finally a shock wave reminiscent of an explosion leaving but a plasma in the place of the specimen. Though this intermediate time scale is unsuitable for high-resolution imaging, an even shorter flash in the 10^{-12} sec range will fix the image before the components of the specimen have a chance to move very far.

On the other end of the time scale, it is known that living organisms in many cases repair the damage done by radiation, given enough time. Unfortunately,if we were patient enough to supply a low enough dose rate, the recovering specimen would walk away while we tried to take its picture.

27.3.2 The Thermal Dimension

Another way to freeze the motion is to freeze the specimen. Given a modest dose rate that does not cause too much local heating, low temperatures also promise to offer a significant amount of radiation protection. Studies in electron microscopy [27.6] indicate that while the specimen is kept cold it maintains its morphology to much higher doses than at room temperature. And the colder the better! The best results are obtained at 4 K, but even 4 $^{\circ}$C is better than not cooling at all. We hasten to add that careful attention must be paid to avoid the formation of ice crystals in the specimen and the condensation of contaminants on its cold surfaces, or else cooling may do more harm than good.

27.3.3 The Choice of X-Ray Energy

Since the radiation dose required to form the image depends so strongly on the contrast, it is important to select the beam energy that provides the

264

best contrast in a given specimen. This consideration was an important component of the calculations by SAYRE et al. [27.3]. It has been emphasized recently, especially by SOLEM and CHAPLINE [27.7], that in some materials large absorption maxima occur near absorption edges. Using a beam energy tuned to such a maximum will enhance the contrast for the particular material. Though large resonances have not been found in materials of biological interest within the energy range of interest to us, there are significant peaks which are worthy of attention.

Though we know little about the relationship between radiation dose and damage to the specimen, for soft x rays there used to be general agreement at least on the one point that inner-shell ionization is particularly damaging [27.8]. Since it is inner shell ionization that provides the contrast that we use for image formation, we expect that this will involve a significant penalty. There is general agreement concerning the process first emphasized by Platzmann, that inner-shell ionization almost invariably leads to the loss of several electrons from the outer shell of the atom. The resulting Coulomb forces tend to break up the molecule to which the atom belongs. In fact studies in $C_4H_{12}Pb$ indicate that subsequent to L-shell ionization of lead the molecule tends to break not just into a few fragments, but indeed into its atomic constituents [27.9]. Studies of damage to specimens by hard x rays bears out the damaging nature of inner-shell ionization [27.2].

Recent work by GOODHEAD [27.10] calls into question the relative importance of inner-shell ionization. He finds that for a given amount of energy absorbed in the specimen C_K x rays, with an energy too low to cause any inner-shell ionization, are much more damaging than harder x rays. Goodhead's criteria for damage have to do with mutation and survival. His dose rates are low enough to allow repair to play a role. The results have nevertheless called into question whatever consensus there existed concerning the nature of radiation damage for soft x rays.

Of course there is no substitute for hard facts. A realistic assessment of the damage problem in soft x-ray imaging must await serious experimental work. Different specimens will pose different problems and will require different solutions. There is a sizeable challenge ahead of us!

27.4 Thick Specimens, and the Third Dimension

Although two papers [27.11,12] at this Symposium are devoted to imaging methods which are inherently three-dimensional, most of our work so far has been concentrated on forming two-dimensional images. Here we shall examine to what extent these two-dimensional methods can evolve and provide three-dimensional information on thick specimens.

The recording of stereo pairs of images is compatible with just about every instrument in use today. In fact already at the 1979 conference FEDER and SAYRE showed stereo pairs made by the contact technique [27.13]. At the same conference SAYRE [27.14] reviewed the other possibilities as well. He pointed out that with the large depth-of-field optics available already at that time, tomographic techniques could generate a full three-dimensional image. In fact a step in that direction has been made by ELLIOTT and DOVER who generated a two-dimensional view from 54 one-dimensional projections with a resolution of 15 μm [27.15]. In addition GRODZINS [27.16] examined ways to optimize the exposure. Clearly much further progress could and should be made in the near future both with the zone plate imaging microscope of Prof. Schmahl's group [27.17] and with the scanning instruments described by several speakers [27.18].

As optical elements of shorter depths of field become available, the option of focusing on the specimen layer-by-layer opens up. Though the resolution in depth remains inferior to the transverse resolution by a factor that corresponds to the f number of the focusing element, the charm of this method is that it is direct and does not require any reconstruction algorithms.

One has a similar problem with depth resolution in holography, [27.12] but of all three-dimensional imaging methods, this is the one that has the potential to provide the most information with the least exposure. Only the method of diffraction pattern analysis [27.11] has the potential to provide an image with a resolution of $\lambda/2$ in all three dimensions.

Once a three-dimensional image is recorded the pleasures and problems of displaying it must be addressed. This is probably best done by computer. In this area we have a lot to learn from radiologists who have been facing this question for some years now. The work of Dr. Herman's group stands as a fine example of what can be done in displaying connected elements from a variety of angles [27.19].

While the image resides in the computer one can also consider the possibilities of deblurring it. When optics have been employed in forming the image, the point-spread function can be determined without much difficulty and correcting for it should produce an improved representation of the specimen.

27.5 Summary and Conclusions

In moving toward the study of thick, wet, radiation-sensitive specimens we must choose our approach with some care. The use of the flash to circumvent the damage problem is compatible with holography but not with procedures requiring multiple exposures to generate the three-dimensional image.

The use of low temperatures, and beam energy optimization are of universal applicability.

Finally, one should keep in mind the use of stains as another way to improve contrast. Every form of biological microscopy has benefitted from staining. In many cases these stains may be nonlethal, may be introduced via the specimen's metabolism through modified nutrients, but they will often contain elements (such as boron or fluorine) which are not normally major constituents of the specimen. Small particles of high opacity have already been used successfully by PANESSA-WARREN and WARREN [27.20]. At lower resolution, FLETCHER et al. [27.21] succeeded in delineating neural processes by heavy metal stains. A systematic study of the best techniques should pay handsome dividends!

References

27.1 J. Hamos: SUNY, Stony Brook, Dept. of Neurobiology private comm.
27.2 For detailed review see A. Halpern: "Damage to Biomolecules and Cells by Low-energy X-Rays and Vacuum Ultraviolet Light", in Uses of Synchrotron Radiation in Biology (H.B. Stuhrman, ed. Academic Press, London 1982 P 255)

27.3 D. Sayre et al.: Ultramicroscopy $\underline{2}$, 337 (1977)

27.4 R. Feder et al.: "Recent Developments in X-Ray Contact Microscopy" this volume, no.29

27.5 J.C. Solem: High Intensity X-Ray Imaging of Biological Specimens, Ultramicroscopy (in press)

27.6 V.E. Cosslett: J. Microsc. $\underline{113}$, 113 (1978)
      S.B. Hayward, and R.M Glaeser: Ultramicroscopy $\underline{4}$, 201 (1979)
      R.F. Egerton: J. Microsc. $\underline{126}$, 95 (1982)
      J. Dubochet et al: Ultramicroscopy $\underline{6}$, 77 (1981)

27.7 J.C. Solem, and G. Chapline: X-Ray Biomicroholography, Optical Engineering (in press)

27.8 A. Ore: "The Role of Multiple Ionization in Radiation Action", in $\underline{Radiation\ Reserarch,\ 1966}$ (Silini, ed. North Holland, Amsterdam 1967) P 54
      J. Durup and P.L. Platzmann: Int. J. Radiat. Phys.
      and Chem. $\underline{7}$, 121 (1975)
      M.S. Issaacson: "Specimen Damage in the Electron Microscope", in $\underline{Principles\ and\ Techniques\ of\ Electron\ Microscopy}$ (M.A. Hayat, ed. Van Nostrand, New York 1976) Vol 1, P 1

27.9 T.A. Carlson and R.M. White: J. Chem. Phys. $\underline{48}$, 5191 (1968)

27.10 D.T. Goodhead: Radiation Reserarch $\underline{91}$, 45 (1982)
      D.T. Goodhead et al: Int. J. Radiat. Biol. $\underline{36}$, 101 (1979)

27.11 D. Sayre: "On the Possibility of Imaging Structures by Soft X-Ray Diffraction Pattern Analysis", this volume, no.35

27.12 M. Howells: "Possibilities for X-Ray Holography using Synchrotron Radiation", this volume, no.36

27.13 R. Feder and D. Sayre: Ann. NY Acad. Sci. $\underline{342}$, 213 (1980)

27.14 D. Sayre, ibid. p. 387

27.15 J.C. Elliott and S.D. Dover: J. Microsc. $\underline{126}$, 211 (1982)

27.16 L. Grodzins: Nucl. Inst. and Meth. $\underline{206}$, 541 (1983)

27.17 D. Rudolph et al: "The Göttingen X-Ray Microscope and X-Ray Microscopy Experiments at the BESSY Storage Ring", this volume, no.20

27.18 B. Niemann:"The Göttingen Scanning X-Ray Microsope",this volume, no.22
      H. Rarback et al.: "Recent Results from the Stony Brook Scanning Microscope", this volume, no.21
      E. Spiller: "A Scanning Soft X-Ray Microscope using Normal Incidence Mirrors", this volume, no.23

27.19 G. Herman et al: "Two Dimensional Display of Three Dimensional Data: Foundations and Applications" Report MIPG 71, Univ. of PA Dept. of Radiology (1982)

27.20 B.J. Banessa-Warren, and J.B. Warren: Ann. NY Acad. Sci. $\underline{342}$, 350 (1980)

27.21 S. Fletscher et al.: J. Neurosci. Methods $\underline{7}$, 19 (1983)

28. Biological Applications of X-Ray Contact Microscopy

B. J. Panessa-Warren

Department of Anatomical Sciences, Health Sciences Center
School of Medicine, State University of New York
Stony Brook, NY 11794, USA

28.1 Introduction

In their living state biological specimens are hydrated, composed of low atomic number molecules (predominately C, O, N, H) and exhibit some form of movement. All of these inherent characteristics make morphological examination of the tissue in it's natural state difficult by conventional light and electron microscopy. To withstand the rigors of a light or electron beam (and vacuum) and provide adequate contrast for cellular or intracellular delineation, the biological sample must be chemically and/or physically treated. Chemical and cryofixation are commonly used to stabilize the cells and membranes of tissues so that the specimen may be carbon or metal coated, dehydrated and dried, or dehydrated and infiltrated with a supportive matrix such as plastic or wax to facilitate sectioning of the tissue prior to viewing. All of these procedures carry the risk of altering the elemental composition or ultrastructure of the tissue, and the question of whether data is real or artifactual must always be addressed. Each chemical treatment for tissue preparation risks the introduction of exogenous materials into the specimen, as well as the leaching and translocation of diffusible substances from specific cells.

In 1952, WOLTER [28.1] and later others [28.2-4] proposed that a natural contrast mechanism for biological specimens could be exploited by imaging samples with x rays between 2.3 to 4.4 nm wavelength. In this region there is an order of magnitude difference between the absorption coefficients for water and protein. This is not the case for routine light microscopy and electron microscopy where contrast must be artificially produced [28.5]. Because biological specimens inherently do not have adequate contrast for discrimination of cellular structures when examined by light and electron microscopy, the tissue must be chemically treated with heavy metals or organic stains, or coated with a thin layer of metal. Again this may alter the chemical and morphological nature of the specimen and necessitate the use of controls and additional experiments to prove the validity of the research results.

Soft x-ray contact microscopy has provided a new means for morphologically and elementally analyzing biological specimens. This relatively new method eliminates many of the problems of specimen preparation and tissue processing artifacts [28.5,10],thereby permitting the examination of hydrated and living tissue, as well as unique types of analysis on intact cells [28.7] or cryosections.

28.2 Advantages of Using X Rays for Microscopy

X rays have wavelengths on the order of angstroms (10^{-8}cm) whereas visible light rays have considerably longer wavelengths $0.4 - 0.7 \times 10^{-4}$cm. Since the resolving power of an optical system is directly proportional to the wavelength of the incident illumination, the chief advantage of using x rays for microscopy are increased penetration of the specimen and higher resolving power [28.8]. Because of their short wavelength, x rays can penetrate biological specimens that are opaque to visible light, making possible the morphological and elemental analysis of electron-dense biological structures that could not be morphologically imaged by conventional methods [28.10]. The choice of an incident x-ray wavelength for imaging is critical to the resolution of the final image. For imaging unstained biological specimens 1 micrometer thick or less, softer x rays must be used to take advantage of the contrast [28.4] and resolution [28.8] potential. For the examination of thicker specimens (up to several micrometers thick) higher energy x rays must be employed [28.8].

By closely examining the absorption and transmission of incident x rays of a known wavelength with a biological sample, it is possible to obtain information about the elemental composition of the tissue [28.8-10]. When a biological sample is placed on an x-ray sensitive substrate (photoresist) such as polymethyl methacrylate (PMMA) and exposed to x rays of a known wavelength, the incident x rays will be differentially absorbed according to the atomic number of the elements within the tissue sample. Areas of higher atomic number absorb or stop the incident x rays from penetrating the sample and exposing the underlying photoresist (Fig.28.1). Regions that are primarily aqueous or composed of low atomic number materials permit the incident x rays to pass through the specimen, exposing the photoresist below [28.9]. The degree of exposure of the photoresist (PMMA) is directly proportional to the elemental composition and thickness of the tissue [28.7,10]. After x-ray exposure, the biological sample is removed and the photoresist chemically developed to reveal a contact replica of the original tissue which shows areas of high atomic number or increased specimen thickness in high relief, and areas of low atomic number or decreased specimen thickness in low relief [28.9,10]. These contact replicas can be further analyzed if the general chemical composition of the specimen is known, to reveal the distribution of specific electrolytes in tissues and cells [28.7].

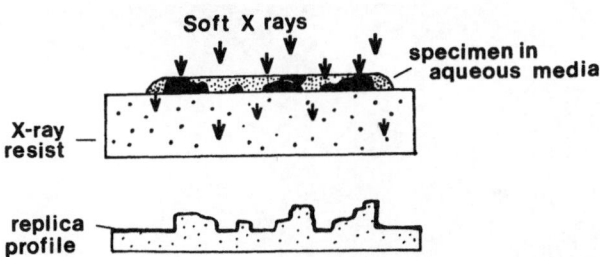

Fig.28.1 Schematic representation of soft x-ray (black arrows) exposure of a biological specimen on photoresist and the resultant x-ray replica following development in isopropanol/MIBK

Not only can x rays be used to obtain morphological and elemental information from tissues and produce good contrast images of thick and thin biological specimens without necessitating the extensive specimen preparation procedures essential for light and electron microscopy, but x rays can be used to examine fragile specimens that could not withstand 1) the rigors of a vacuum; 2) bombardment by an electron beam or 3) specimen heating. Although x rays can potentially be damaging to biological tissues and cause extensive covalent bond breakage, the dose necessary for x-ray contact microscopy can be delivered very rapidly (in nanoseconds) in a predetermined controlled amount minimizing radiation damage but still providing an adequate dose for optimal exposure. The ability of x rays easily to penetrate matter permits the specimen to be maintained at atmospheric pressure under aqueous conditions during x-ray exposure [28.6,7,11]. This opens an entirely new realm of possibilities for the clear visualization of intracellular details in living or hydrated fragile biological specimens which is now available by light, scanning or transmission (TEM or STEM) electron microscopy [28.6,7,9,11,12]. With the development of more x-ray sources capable of producing specific ranges of soft x rays and synchrotrons for general research purposes, it will be possible for biologists to expose living cells or hydrated fragile specimens that could not be analyzed by conventional methods, for a minimal amount of time (nanoseconds) to an x-ray beam tuned to a desired wavelength. In this way an image could be formed (before the sample had a chance to deteriorate) which would potentially have information about ultrastructure and elemental composition at spatial resolutions approaching those of scanning electron microscopy (the limiting factor being the x-ray resist or copying medium to record the image).

28.3 X-Ray Contact Microscopy of Opaque Specimens

Some biological specimens are too opaque to be analyzed by conventional light and electron microscopy. Because x rays have superior penetration capabilities to electrons, and photons (in the visible range) at doses that are relatively nondestructive to the sample, x-ray contact microscopy can be used to study morphological or compositional characteristics of totally electron-dense samples. The spindle-shaped pigment granules of the retina of the frog *Rana catesbiana* are extremely electron dense. Figure 28.2 shows an electron micrograph of unstained, plastic embedded retina fixed in 3% glutaraldehyde (osmium was not used as a fixative). The melanin pigment granules

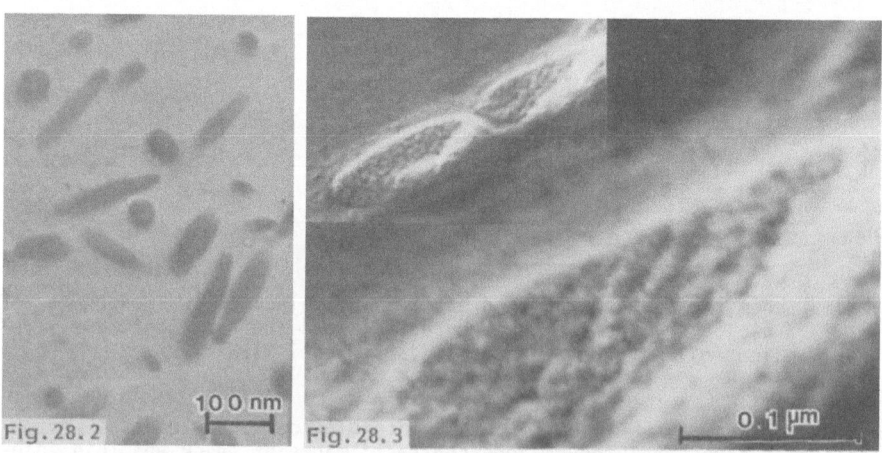

Fig.28.2 Transmission electron micrograph of unstained, plastic embedded thin section of frog retinal melanin granules

Fig.28.3 X-ray contact replica of the same melanin granules as in Fig.28.2 shows a periodicity in the granule not seen in the surrounding background (low magnification insert). Synchrotron radiation from DESY 3.0-4.4 nm wavelength

are quite electron dense and no intracellular details can be seen even in unstained ultrathin sections. However, when an identical plastic embedded section is placed on photoresist and exposed to soft x rays (synchrotron radiation DESY in the region of 3.0 - 4.4 nm) the contact replica clearly shows a periodicity of 5 to 7 nm globules arranged in rows within the elliptical melanin granule (Fig.28.3) [28.13]. This periodicity was not seen in the plastic surrounding the pigment granules, and therefore it may be assumed that these globules are representative of the internal ultrastructure, or the internal groupings of the calcium, barium, copper, zinc and chromium ions characteristically associated with these granules [28.14,15].

28.4 X-Ray Contact Microscopy of Fragile Specimens

Many biological specimens are extremely delicate and become easily damaged when exposed to an electron beam. Even when the specimen is kept at cold temperatures as low as $-126^{o}C$, each pass of the electron beam can severely alter proteins due to mass loss [28.6]. For many years, proteoglycans (the molecular building blocks of cartilage) have been visually examined by pretreating these macromolecules with a series of chemicals followed by metal shadow coating on a TEM grid to make them visible and durable enough to withstand the vacuum and electron bombardment of the electron microscope [28.16]. In actuality using the aforementioned technique, one is in reality not looking at the native proteoglycan aggregate, but at a vacuum dried, stain-and-metal coating of the proteoglycan resulting from treatment with ammonium acetate, cytochrome C, uranyl acetate (and/or phosphotungstic acid) and a platinum-carbon metal coating. Although this method has been used for a number of years, these images do not provide a very accurate idea of what these proteoglycan aggregates and monomers are like in their natural state (Fig.28.4a).

When isolated proteoglycans resuspended in distilled water were placed on silicon wafers uniformly coated with 1-2 μm thick PMMA and exposed for 16 hrs. to carbon x rays (4.4 nm) using a stationary target x-ray source $(10^4 J/cm^2)$, clear images of aggregated proteoglycans were obtainable. The exposed photoresist was sequentially developed in a 5:1 mixture of isopropanol/methyl isobutyl ketone and the contact replica coated with 1-4 nm of gold palladium [28.17]. For the first time it was possible to see proteoglycans without resorting to extensive staining and specimen preparation procedures. Figure 28.4 shows an x-ray contact replica of proteoglycan aggregates which had been extracted at $4^{o}C$ with 4M guanidinium chloride and reaggregated after equilibrium on a cesium chloride density gradient followed by velocity sedimentation on a cesium chloride gradient. Figure 28.5 shows an x-ray replica of the same proteoglycan aggregates after pretreatment with cytochrome C, routinely used to coat the glycosaminoglycans and prevent the collapse of the proteoglycan aggregates for TEM observation. To see the difference in the proteoglycan structure after the addition of metallic cations, similar proteoglycan aggregates were pretreated with $FeCl_3$ and imaged by x-ray contact microscopy (Fig.28.6). Figure 28.7 shows an x-ray replica of proteoglycan aggregates prepared directly by rate zonal sedimentation rather than by the conventional cesium chloride density gradient equilibration. The hyaluronic acid backbone and chondroitin and keratan sulfate side chains (see schematic diagram Fig.28.4a) are apparent in the conventional preparation of unstained aggregates (Fig.28.4). When these are compared to the cytochrome C treated (Fig.28.5) samples and $FeCl_3$ preparation (Fig.28.6) there is a marked difference in proteoglycan structure. The cytochrome C x-ray images show large clusters of aggregates that have numerous thickened side chains with a prominent hyaluronic acid backbone. Although it is difficult to determine

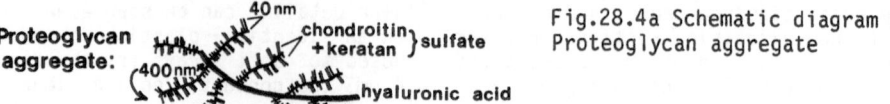

Proteoglycan aggregate: 40 nm, chondroitin + keratan } sulfate, 400 nm, hyaluronic acid, 1200 nm

Fig.28.4a Schematic diagram of Proteoglycan aggregate

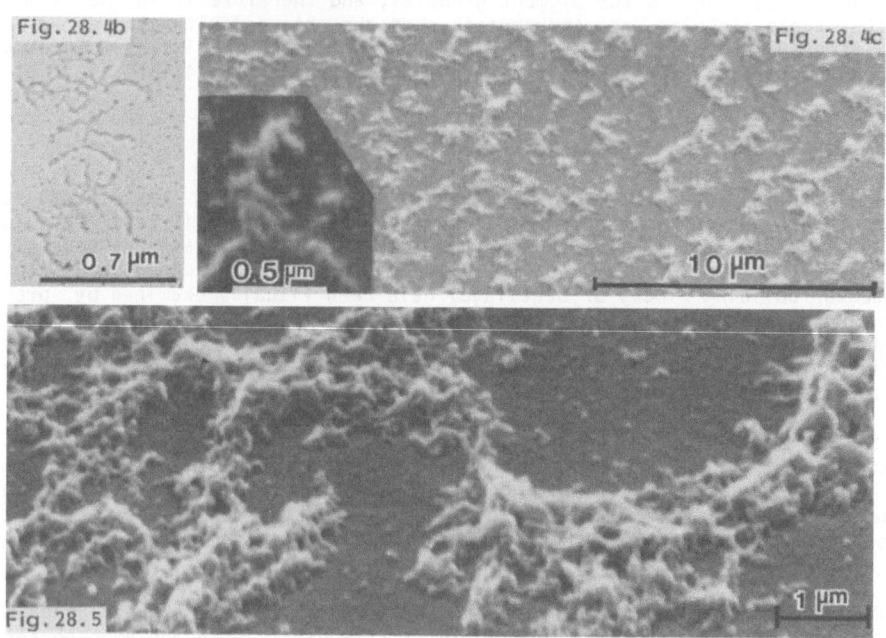

Fig. 28.4b 0.7 μm

0.5 μm 10 μm

Fig. 28.4c

Fig. 28.5 1 μm

Fig.28.4b Proteoglycan aggregate routinely prepared for TEM
Fig.28.4c Low and high magnification (insert) of x-ray replicas of freshly
        prepared proteoglycan aggregates resuspended in acetate buffer
        (flash x-ray source, wet cell, 60 nanosecond exposure)
Fig.28.5 X-ray replica of cytochrome C treated proteoglycan aggregates
        (16 hour exposure, stationary x-ray source, carbon K x rays)

from these x-ray replicas that the iron in the $FeCl_3$ treated preparation has
caused the collapse of the proteoglycan aggregates, it is obvious that
Fig.28.6 has aggregates with knob-like components instead of filamentous
side chains and a central backbone. The iron from the $FeCl_3$ may have caused
the glycosaminoglycans to collapse and coalesce causing the knob-like bare
chains. Figure 28.7 shows yet a different image of the proteoglycan
aggregates with intricate clusters of globular decorated chains. Using rate
zonal sedimentation there is a higher content of noncovalent protein
associated with the proteoglycan aggregates than would be found using the
conventional cesium chloride density gradient preparation represented in
Fig.28.4. Therefore, the globular-decorated aggregates of Fig.28.7 are probably representative of the higher noncovalent protein incorporated in the
aggregate. By conventional electron microscopy such a change in protein
structure would be difficult if not impossible to see. X-ray contact microscopy provides a way to image these delicate macromolecules quickly without
extensive specimen preparation. Although the x-ray contact replicas are not
always easy to interpret they offer a superior way to image changes in pro-

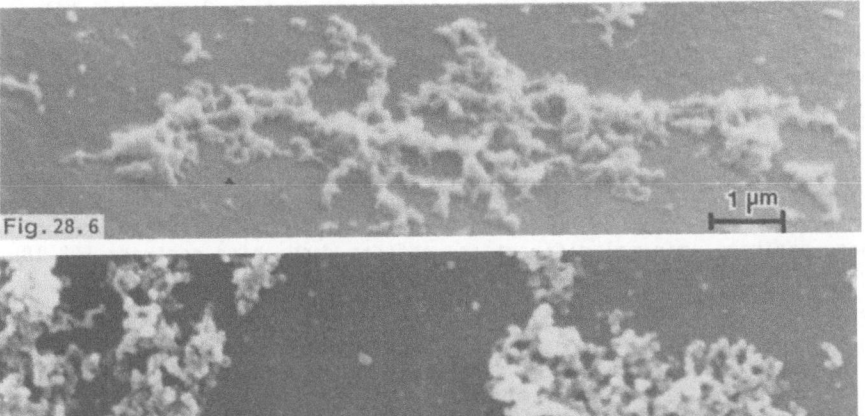

Fig.28.6　X-ray replica of $FeCl_3$ treated proteoglycan aggregates (16 hour exposure, stationary x-ray source, carbon K x rays)

Fig.28.7　X-ray replica of proteoglycan aggregates prepared by rate zonal sedimentation, which retains more bound protein than the conventional cesium chloride method (16 hour exposure, carbon K x rays)

tein morphology with unstained samples. The high relief in the x-ray replicas of the proteoglycan aggregates treated with cytochrome C and $FeCl_3$ (Fig.28.5 and 28.6) may also be due to the incorporation of iron into the aggregate proper. Since the topography of the x-ray replica is dependent on the thickness and atomic number of the specimen, the areas of the proteoglycan aggregate with bound Fe would appear in higher relief than those areas devoid of metals. Therefore, this method of imaging offers not only morphological but compositional information.

28.5 Examination of Hydrated and Living Biological Samples

To the biologist one of the most important advantages of using soft x-ray microscopy is the ability to examine hydrated specimens with sufficient contrast to image structures at better than 25 nm resolution [28.18]. A number of environmental and wet chambers have been developed for microscopy which can maintain a wet specimen at atmospheric pressure during x-ray exposure [28.18-20]. Using synchrotron radiation $\sim$ 4.5 nm, SCHMAHL et al. [28.19] successfully made x-ray images of living tissue culture cells cultivated on polyimide or polycarbonate foils and placed in an air-filled object chamber. MCGOWAN and MALACHOWSKI [28.20] have also developed an environmental chamber but due to the period of time necessary for an adequate x-ray exposure, cell movement often prevented clear imaging of the cells. FEDER et al. [28.21] report a technique whereby a biological specimen on a TEM grid could be placed on a silicon nitride (Si_3N_4) window sandwiched on top of another Si_3N_4 window with 400 to 800 nm of x-ray resist. X-ray exposures were made with a stationary x-ray source, and the x-ray resist was developed to a

273

maximal thickness of less than 400 nm, coated with a thin layer of metal and viewed by transmission electron microscopy. Although this biological Si_3N_4 sandwich chamber could be used for exposures of hydrated cells, the relatively long exposure time required using the x-ray source available was felt to rule out the viability of the cells and the production of reliable x-ray replicas [28.21].

In 1979, McCORKLE developed a flash soft x-ray tube [28.22,23] and wet cell that could rapidly (60 nanoseconds) deliver an intense burst of soft x rays over a band from 2.3 to 4.4 nm that is optimum for high-resolution imaging, exhibits minimal diffraction effects, and high contrast [28.18]. Although the total exposure time was approximately 60 nanoseconds, the actual image formation period was probably during the first 20-40 nanoseconds, which may account for the lack of problems with cell movement or Brownian motion. Using the pulsed plasma x-ray source and this wet cell it was possible to image totally hydrated proteoglycan aggregates and the smaller (300-400 nm) more delicate proteoglycan monomers resuspended in low salt buffers (0.01M sodium acetate or 0.01M sodium chloride) [28.6].

More recently we have examined hydrated invertebrate isolated myosin, actin and paramyosin paracrystal preparations using the same wet cell and pulsed plasma source [28.12]. Paramyosin paracrystals and myosin filaments were isolated from telson muscle of the horseshoe crab *Limulus polyphemus* L., and resuspended in a Tris buffer for dialysis overnight against a buffer containing EDTA at 4°C. Tissue from the white adductor muscle of the clam, *Mercenaria mercenaria* was similarly prepared and paracrystals of the paramyosin were formed by dialysis against a phosphate buffer with 100mM KCl. Aliquots of muscle filaments were diluted and resuspended in a relax solution (100mM KCl, 2mM $MgCl_2$, 10mM Tris, 5mM EGTA, 5mM ATP pH 7.4), and 1.5 microliter droplets of the mixture were placed on PMMA coated (1 μm thick) silicon chips and loaded into the wet cell. To see the effect in the filaments of reducing the KCl content, similar muscle preparations were diluted and resuspended with relax solution with 10mM KCl. Using the pulsed plasma source, all samples were exposed for less than 60 nanoseconds. Following x-ray exposure the tissue was rinsed away in distilled water and the resist sequentially developed in isopropanol/MIBK. For scanning electron microscopy the x-ray replicas were coated with 3-7 nm gold palladium.

By routine scanning electron microscopy of a fixed critical point dried, AuPd coated Limulus muscle preparation, the fine slender actin (a) filaments, larger and longer myosin (m) filaments and very large paramyosin (p) paracrystals are difficult to see clearly (Fig.28.8). Transmission electron microscopy povides a much clearer image of the isolated myosin and actin filaments but necessitates the use of negative staining and frequent fixation in glutaraldehyde to stabilize the specimen to withstand better the electron beam and increase contrast sufficiently for viewing (Fig.28.8). The drying process necessary for scanning or transmission electron microscopic examination may be responsible for some distortion or inaccurate measurement of the true length and morphological subunits of these specimens. To date x-ray contact microscopy has been the only means ultrastructurally to examine these filaments in their hydrated environment at the resolution of the electron microscope. Figure 28.9 shows an x-ray contact micrograph of actin and myosin filaments. A preparation taken from a different band of the concentration gradient (filaments were segregated by differential centrifigation) yielded a richer filament concentration with far less extraneous protein and actin filaments (Fig.28.10). Upon closer examination of the myosin filaments (Fig.28.11) a repeating period of 19-25 nm (black arrows) was often observed in the x-ray contact micrographs. This period seems to

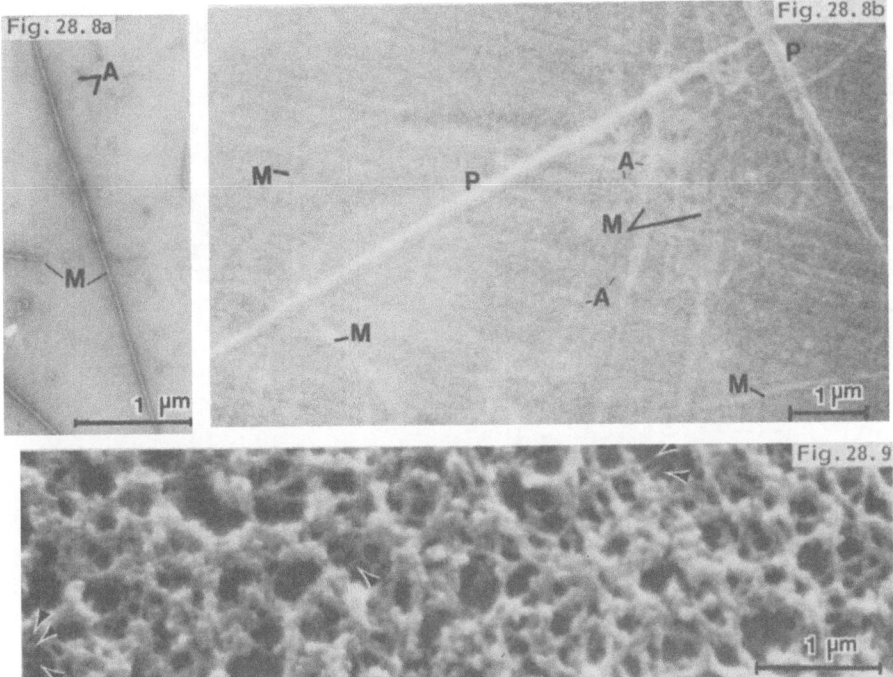

Fig.28.8a Transmission electron micrograph of negatively stained *Limulus* myosin and actin filaments;

b scanning electron micrograph of *Limulus* paramyosin, myosin and actin

Fig.28.9 X-ray replica of actin (arrows) and myosin

correspond to the cross striations measuring 14.5 nm seen in negatively stained myosin preparations. The variation between the TEM and x-ray replica period size may be the result of the metal coating and 45° tilt angle needed to view the x-ray replica, as well as the difference between viewing a hydrated filament and a dried filament outlined with a heavy metal stain. When the myosin filament preparation in relax solution (with only 10mM KCl) was examined by soft x-ray contact microscopy using the wet cell, the x-ray replica exhibited strands and pools of material (Fig.28.12) which upon higher magnification showed either no evidence of actin or myosin filaments or filaments in the process of coming apart. *Mercenaria* filaments were also easily imaged by wet-cell contact x-ray microscopy (Fig.28.13). This work dramatically illustrates the reliability and usefulness of x-ray contact microscopy as a high-resolution imaging technique for studying phenomena of hydrated organelles and cell fragments with excellent image contrast.

Summary

"Soft x-ray photons interact with matter chiefly through absorption" [28.24]. When this phenomenon is exploited by placing a biological specimen on an x ray or photo resist, "the variation of the number of absorption events from point to point on the specimen provides the contrast mechanisms

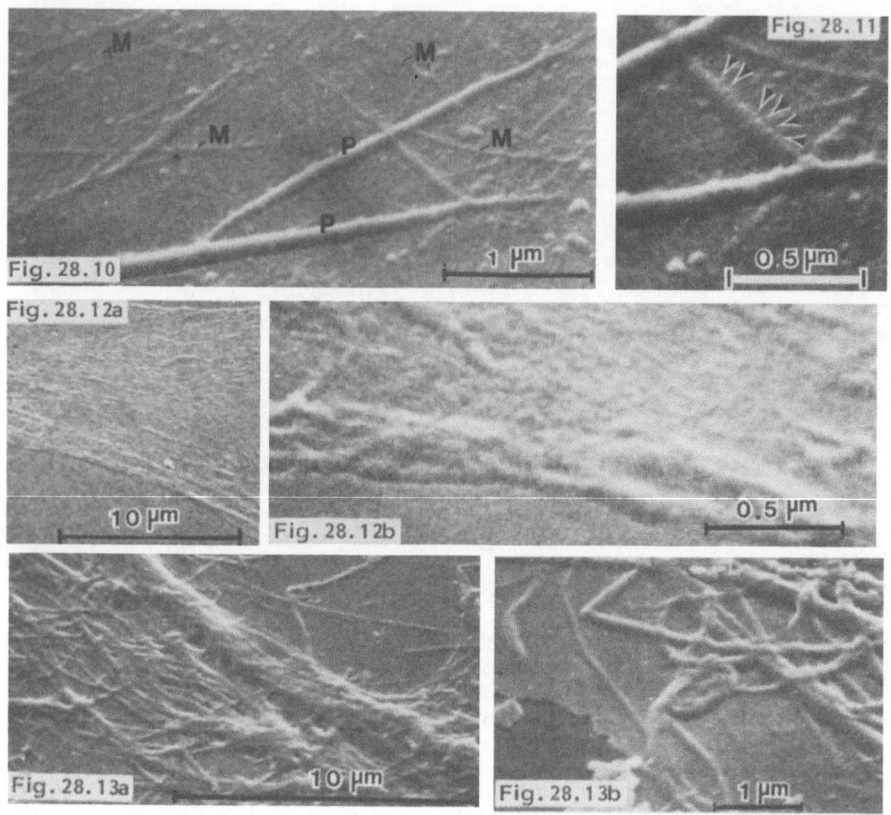

Fig.28.10 X-ray replica of hydrated *Limulus* paramyosin and myosin filaments
Fig.28.11 Higher magnification of Fig.28.10 showing periodicity in myosin filaments

Fig.28.12a X-ray replica of muscle filaments in low KCl relax solution;
b at higher magnification only amorphous material and remnants of filaments were visible.

Fig.28.13a X-ray replica of hydrated *Mercenaria* filaments at low and high magnification (b).

of soft x-ray microscopy" [28.24]. Therefore, samples do not need to be fixed, stained or pretreated to permit visualization of gross as well as fine structure. The well-documented artifacts attributed to specimen fixation, staining and drying can be circumvented completely by studying living cells and hydrated biological materials. The penetration capability of x rays makes it possible to image ultrastructure, or elemental composition of very dense (electron-dense) biological specimens, as well as image these specimens in air or various aqueous media. The ability to expose a specimen to a single rapid (less than a few hundred nanoseconds) dose of soft x rays using a flash x-ray source or synchrotron radiation permits the analysis of very fragile biological samples that would be altered or destroyed during electron or visible light microscopy. The newly developed x-ray sources and increasing beam time at regional synchrotrons will hopefully allow biologists to make x-ray exposures of tissues using monochromatic x rays of a specific wavelength above and below an absorption edge to produce x-ray replicas with specific topographic information about the elemental content and distribution within living cells [28.7,24]. Soft x-ray contact microscopy of hydrated biological specimens offers one of the most important methods in biological research since the development of the electron microscope. For the first time it is now possible to study not only cell ultrastructure of living, hydrated biological samples, but more importantly it is now within our grasp to determine elemental localization and content in living cells without having to labor with the very difficult and painstaking quench freezing ultracryomicrotomy and frozen-hydrated x-ray microanalysis methods now used. How exciting to be able to take a living sample and subject it to x-ray microscopy without any further specimen preparation than placing it in a wet-cell and aligning it with an x-ray beam port. There are so many new things to be learned by analyzing hydrated specimens, and x-ray contact microscopy offers the biologist a new threshold to cellular and analytical biology.

Acknowledgment

The author would like to express deep appreciation to Dr. P. Hoffman for providing the proteoglycans; Dr. M. Dewey, Dr. B. Gaylinn and Mr. D. Colflesh for collaboration on the muscle research and providing the isolated filaments; Mr. R. Feder for kindly providing advice and making the stationary x-ray source exposures; and to Dr. R.A. McCorkle for his encouragement, collaboration and use of the flash x-ray source and wet cell.

References

28.1 H. Wolter: Ann. Phys. Leipzig. 10, 94 (1952)

28.2 B.L. Henke: Monochromatic Sources of Ultrasoft X-Radiation for Quantitative Microradiographic Analysis in X-Ray Microscopy and Microradiography (Academic Press, New York 1957)

28.3 B.L. Henke: Advances in X-Ray Analysis Vol. 2 (Plenum Press, New York 1960)

28.4 D. Sayre, J. Kirz, R. Feder, D. Kim and E. Spiller: Ultramicroscopy. 2, 337 (1977)

28.5 P.J. Duke: "X-Ray Microscopy: Recent Developments and Future Prospects", in 1981 Proc. Royal Microscopy Society, pp. 186-192

28.6 B.J. Panessa, R. McCorkle, P. Hoffman, J.B. Warren and G. Coleman: Ultramicroscopy 6, 139 (1981)

28.7 J.W. McGowan, B. Borwein, J. Medeiros, T. Beveridge, J.D. Brown, E. Spiller, R. Feder, J. Topalian and W. Gudat: Journal of Cell Biology 80, (1979)

28.8 V.E. Cosslett: X-ray Microscopy and Microradiography (Academic Press, New York 1957)

28.9 B.J. Panessa-Warren, J.B. Warren: Ann. NY. Acad. of Sci. 342, 350 (1980)

28.10 B.J. Panessa, J.B. Warren, R. Feder, D. Sayre and P. Hoffman: Scanning Electron Microscopy/1980. 2, 106 (1980)

28.11 B.J. Panessa, G. Colemann, R.A. McCorkle and J.B. Warren: Scanning. 4, 21 (1981)

28.12 B.J. Panessa, M.M. Dewey, P. Brink, B. Gaylinn, R.A. McCorkle, G. Coleman and J.B. Warren: "High Resolution X-ray Contact Microscopy of Hydrated Muscle Filaments", in 1981 Proc. 39th Elec. Microsc. Soc. Ameri., pp. 512

28.13 R. Feder, E. Spiller, J. Topalian, A.N. Broers, W. Gudat, B.J. Panessa, J.A. Zadunaisky and J. Sedat: Science. 197, 259 (1977)

28.14 B.J. Panessa and J.A. Zadunaisky: Exp. Eye Res. $\underline{32}$, 593 (1981)

28.15 M. Burns, D. File, K. Brown and D. Flaming: Brain Res. $\underline{220}$, 173 (1981)

28.16 L. Rosenberg, L. Hellman and A. Kleinschmidt: J. Biol. Chem. $\underline{750}$, 1877 (1975)

28.17 B.J. Panessa, J.B. Warren, P. Hoffman and R. Feder: Ultramicroscopy. $\underline{5}$, 267 (1980)

28.18 B.J. Panessa, G. Coleman, R.A. McCorkle and J.B. Warren: Scanning. $\underline{4}$, 21 (1981)

18.19 G. Schmahl, D. Rudolph and B. Niemann: "X-ray Microscopy of Native Material", in 1980 Proc. European Congress on Electron Microscopy, pp. 668-669

28.20 J.W. McGowan and M.J. Malachowski: Ann. NY Acad. of Sci., $\underline{342}$, 228 (1980)

28.21 R. Feder, J. Costa, P. Chaudhari and D. Sayre: Science, $\underline{212}$, 1398 (1981)

28.22 R.A. McCorkle: J. Phys. B.: Atom Molec. Phys., $\underline{11}$, L407 (1978)

28.23 R.A. McCorkle, J. Angilello, G. Coleman, R. Feder and S. LaPlaca:

28.24 J. Kirz and D. Sayre: "Soft X-Ray Microscopy of Biological Specimens" in Synchroton Radiation Research (Plenum Press, New York 1980)

29. Recent Developments in X-Ray Contact Microscopy

R. Feder, V. Mayne-Banton, and D. Sayre   IBM Th.J. Watson Research Center
P.O. Box 218, Yorktown Heights, NY 10598, USA
J. Costa National Institute of Mental Health, Bethesda, MD 20205, USA
B. K. Kim and M. G. Baldini Deaconess Hospital and Harvard Medical School
Boston, MA 02215, USA
P. C. Cheng Department of Anatomy, University of Illinois at Chicago
Chicago, IL 60612, USA

29.1 Introduction

Contact imaging was briefly reviewed two years ago [29.1]. Important developments since then have occurred in two main areas. New submicrosecond
flash soft x-ray sources have provided the first suboptical images of a
living cell; and improved techniques of specimen preparation are yielding
higher quality images of biological material than were previously obtainable. We will give examples from both these areas.

In addition, a start has been made in the use of synchrotron radiation in
the identification of the atomic composition of image features. We will not
report on this topic here, but some recent results in this area are being
presented in another paper at this symposium [29.2].

Finally, the submicrosecond flash x-ray sources mentioned above, of
which the first moderate-cost commercial versions are now beginning to appear, may provide at least a partial solution to the problem of providing a
high-performance soft x-ray source which can be installed in the biology
laboratory. We give some details on these sources.

29.2 Image of a Living Cell with a Sub-Microsecond Flash X-Ray Source

Specimen preparation for x-ray imaging is a challenge which is probably almost as important to the ultimate success of the field as,e.g. , improved x-
ray optics. X-ray imaging typically calls for stabilization of structures in
three dimensions, because of its thick-specimen capability. It also calls
for stabilization with a minimum of added or subtracted material, since such
addition or subtraction is not needed for purposes of enhancing contrast.
(At the same time, site-specific labelling is a powerful tool for x-ray as
well as electron imaging, with x rays offering interesting possibilities for
the sensing of low-Z labels.)

Of all the possible stabilization techniques, the most desirable would be
to accomplish the imaging in an extremely short time, thereby removing the
need for any intervention with the structure other than that of the imaging
particles themselves. Recently this has been accomplished using a 200 ns
flash of soft x rays produced by a Maxwell Laboratories "Lexis II" soft. x-
ray source. The cells which were imaged were human blood platelet cells
which were alive (and in the case of the cell we show apparently in the act
of putting out pseudopods) at the time of the flash.

The image is shown in Fig.29.1 and shows excellent resolution of features
less than 100A in diameter, showing that flash imaging does not degrade re-

solution in the contact technique, and does effectively stabilize structures to at least that level. Details of the experimental procedure, and of the information contained in the images, will be given in a later publication [29.3].

29.3 Specimen Preparation Techniques for Improved Image Quality with Normal Exposure Techniques

In Figs.29.2-4 we show specimens prepared by fixation with glutaraldehyde (no heavy-metal postfixing), followed by critical-point drying. The material is *Xenopus laevis* muscle cell, cultured directly on the surface of the 1500A Formvar film used as support during the x-ray exposure. The Formvar was coated with approximately 100A of carbon to ensure cell attachment. Further details are given in the figure legends. Figure 29.2, taken with carbon K radiation (44A), shows the excellent x-ray imaging quality achievable. Figure 29.3, on the other hand, shows the difficulty of imaging the same lightly stabilized three-dimensional structure with electrons. Here the structure undergoes visible change during even minimal-dose exposure.

(Not shown here, but shown in the talk, is a digitized color display of Fig.29.1, showing the cellular structures very clearly in false color.)

Figure 29.4 is a carbon K stereo view of a specimen similar to that in Figs.29.2 and 29.3, showing the three-dimensional nature of the preparation.

In Figs.29.5-8 we show a *Xenopus laevis* fibroblast-like cell and muscle cell which have received heavy-metal fixation (glutaraldehyde plus post-fixation in OsO_4) before the critical-point drying. In this case the images are in vanadium L radiation (24A) and again the critical-point drying gives excellent image quality.

Freezing is another type of stabilization technique to be considered. A limited amount of work has been done on x-ray imaging of freeze-dried and freeze-substituted specimens [29.5].

In conclusion we stress that mild fixation or fast freezing, without subsequent water removal, has not been tried as yet as far as we know, and could prove superior to any methods (including critical-point) involving the removal of water.

Fig.29.1 High-speed (200 ns) image of a living human blood platelet, taken by contact soft x-ray microscopy. The platelet was apparently putting out pseudopods at the moment of imaging. In contrast to the subsequent pictures in this paper, the x-ray image was read out by SEM rather than TEM, since the x-ray resist here was supported on a thick substrate. Features of diameter less than 100A can be seen

Fig.29.2 *Xenopus laevis* muscle cell, cultured on the surface of Formvar film (∿1500A), fixed in 1% glutaraldehyde, dehydrated in EtOH, and critical-point dried in CO_2. For the method of cell culture see PENG et al. [29.4]. Carbon K (44A) imaging on 1.2μm photoresist (PMMA) on Si_3N_4. TEM viewing condition (also for Fig. 29.3): JEOL 100CX with 100 kV, LaB_6 gun. Key: (Y) yolk granule, (N) nucleus, (white arrow) nucleolus, (L) lipid granule (removed during dehydration)

Fig.29.3 Electron image of same specimen

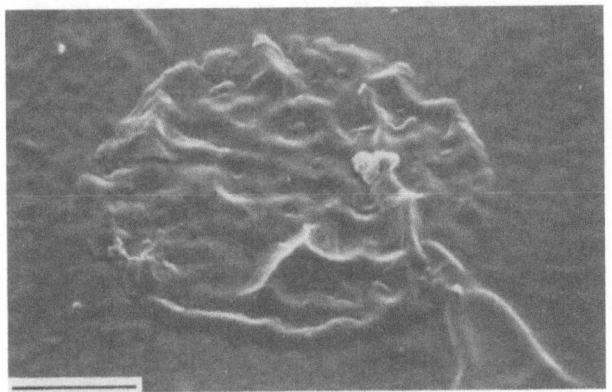

Fig.29.1. Caption
see opposite page

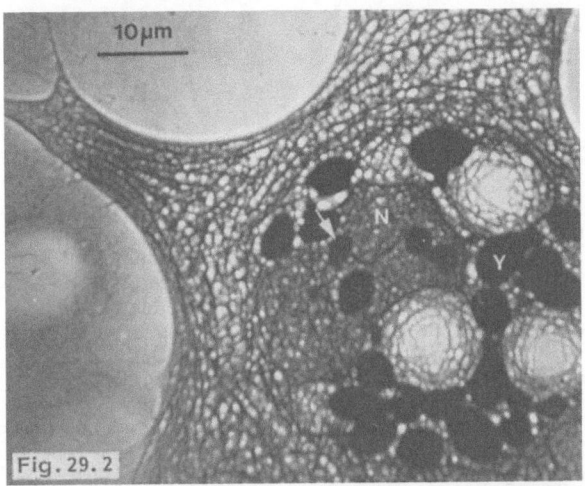

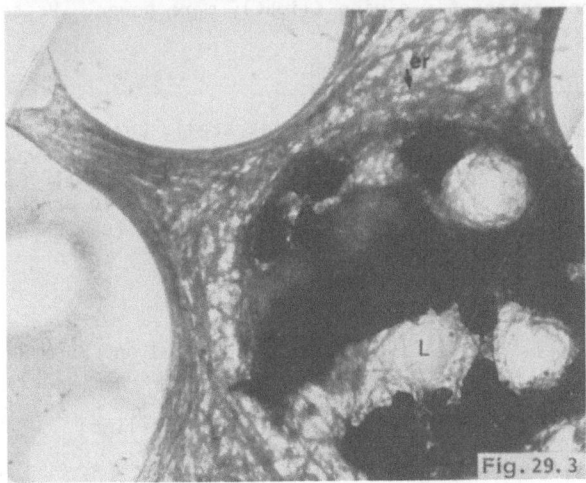

Figs.29.2,3. Caption
see opposite page

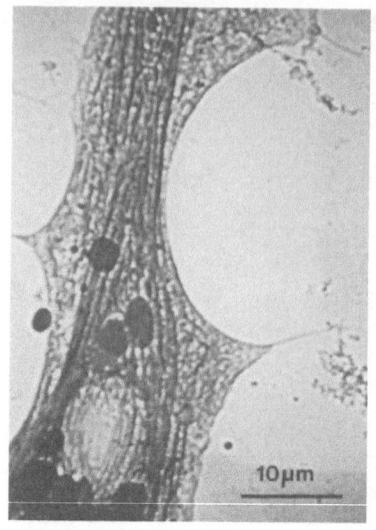

Fig.29.4   3D pair of *Xenopus laevis* muscle cell. Replicated at ±30° with carbon
K radiation. Key: (F) myofibril bundle,
(arrow) yolk granule from another cell
which has attached to surface of the
cell being viewed

10µm

29.4 Commercial Sub-Microsecond Flash X-Ray Sources

Aiming mainly at the x-ray lithography market in the microfabrication industry, several firms are bringing out moderate-cost short-pulse soft x-ray
sources suitable for laboratory use. These sources are not ideal for soft
x-ray microscopy, because of their complex spectra and their tendency to be
somewhat underpowered for one-pulse exposures, but as reported above we have
made successful exposures with one of them.

Two firms (Maxwell Laboratories, 8835 Balboa Avenue, San Diego, CA 92123;
and Physics International, 2700 Merced Street, San Leandro, CA 94577) have
announced sources in which x rays are emitted by a briefly confined hot compressed plasma formed by electrical discharge through a gas. Pulse lengths
are approximately 200 ns, and x-ray power is of the order of 10 joules per
pulse, with the Maxwell source apparently giving slightly more power. Price
of the units is in the $150K range.

Fig.29.5   Osmium fixation, *Xenopus laevis* fibroblast-like cell. Cultured on
Formvar film, fixed in 1% glutaraldehyde, post-fixed in 0.5% OsO_4,
dehydrated in EtOH, and critical-point dried. Vanadium L (24Å)
imaging. Key: (Y) yolk granule, (S) stress fiber, (double arrow)
x-ray dense cell margin, (N) nucleus, (No) nucleolus, (er) endo-
plasmic reticulum, (m) mitochondria

Fig.29.6   Higher magnification view, showing nuclear region

Fig.29.7   Osmium fixation, *Xenopus laevis* muscle cell. Cultured on Formvar
film, fixed in glutaraldehyde (1%), post-fixed in OsO_4 (0.5%),
dehydrated in EtOH, and critical-point dried. Vanadium L (24Å)
imaging. Key: (A) A band of sarcomere, (I)I band, (Z)Z line

Fig.29.8   Higher magnification view of Fig.29.7

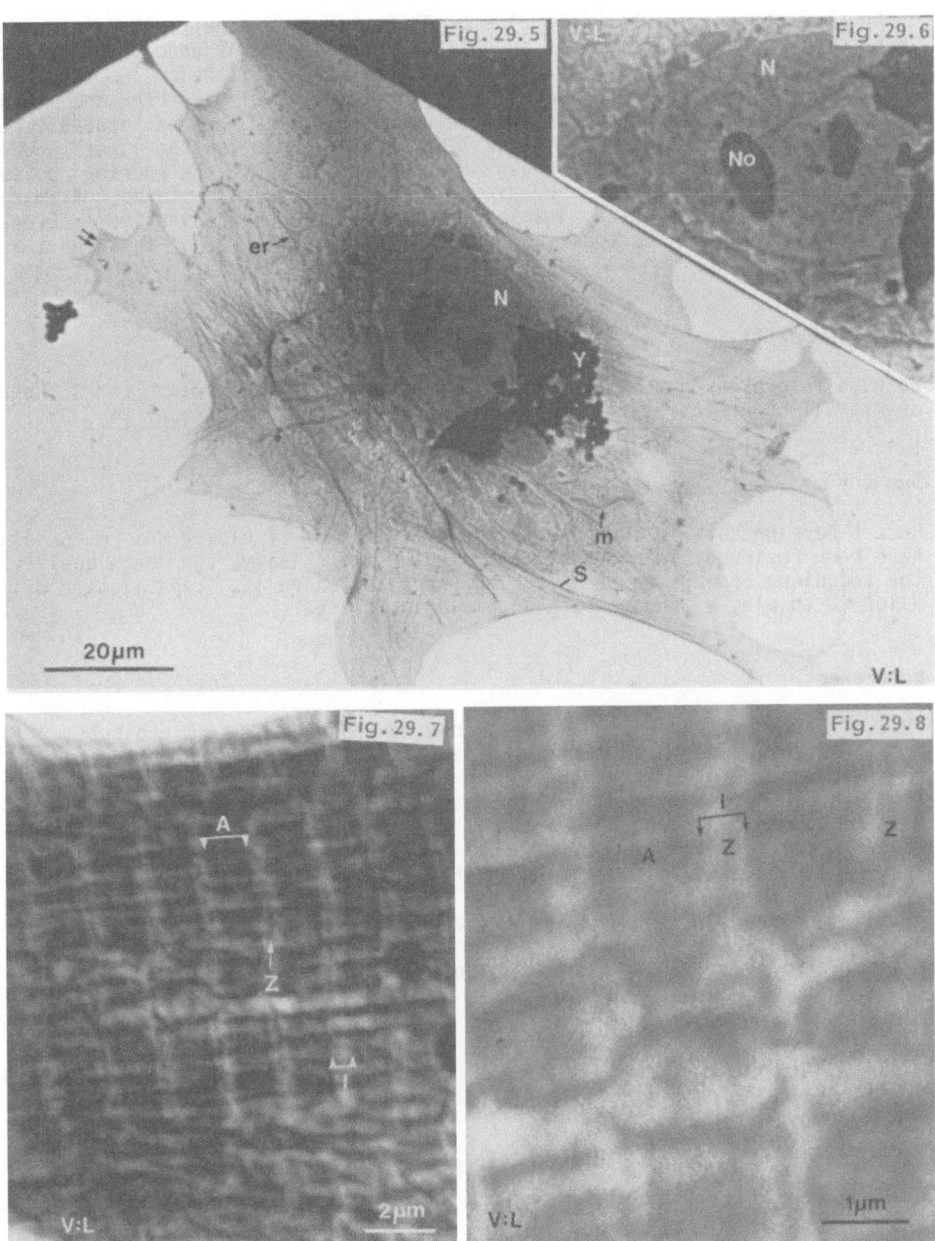

Figs.29.5-8. Captions see opposite page

Another source, in which the plasma is formed by the heating of a solid surface by the pulsed output of an optical power laser, is under development by XMR Inc., 3350 Scott Boulevard, Santa Clara, CA 95051. Pulse length is in the 1-10 ns range, and x-ray power is approximately 2-4 joules for the model currently being developed. It should be noted that specimens can probably be brought closer to the source in the laser-heated case, so the lower power may not be significant. It should also be noted that the shorter pulse length cannot automatically be taken to be preferable in terms of image sharpness, as the faster imaging must be balanced against more severe shock-wave motion of the specimen structures [29.6]. Price of this unit is in the $100K range.

Acknowledgments

We wish to express our thanks to Drs. J. Pearlman, J. Riordan, and S. Robb of Maxwell Laboratories for their help in making the exposure of Fig. 29.1.

Summary

Recent developments in soft x-ray contact imaging of biological materials have been reviewed. In practicality, imaging flexibility, and image quality, the technique appears to be reaching the point where its capabilities will allow it to play a genuine role in biological research.

References

29.1 D. Sayre and R. Feder: "Status Report on Contact X-Ray Microscopy", in SPIE Vol. 316 (High Resolution Soft X-Ray Optics, 1981), E. Spiller, ed., pp. 56-61
29.2 P.C. Cheng, K.H. Tan, J.Wm. McGowan, R. Feder, H.B. Peng, D.M. Shinozaki: this volume, no.30
29.3 R. Feder et al.: in preparation
29.4 H.B. Peng, J.J. Wolosewick, and P.C. Cheng: Devel. Biol. 88, 121-136 (1981)
29.5 P.C. Cheng, H.B. Peng, R. Feder, and J.W. McGowan: Electron Microscopy, 461-462 (1982)
29.6 J.C. Solem and G.C. Baldwin: Science 218, 229-235 (1982)

30. Soft X-Ray Contact Microscopy and Microchemical Analysis of Biological Specimens

P. C. Cheng and H. B. Peng Department of Anatomy, University of Illinois at Chicago, Chicago, IL 60612, USA
K. H. Tan Canadian Synchrotron Radiation Facility (CSRF), Physical Sci. Lab., University of Wisconsin, Stoughton, WI 53589, USA
J. Wm. McGowan Physics Department and Center for Chemical Physics University of Western Ontario, London, Ontario N6A 3K7, Canada
R. Feder IBM Research Center, Yorktown Heights, NY 10598, USA
D. M. Shinozaki Faculty of Engineering Sci., University of Western Ontario, London, Ontario N6A 5B9, Canada

30.1 Introduction

Since the development of high resolution x-ray contact microscopy [30.1], many works on the x-ray imaging of biological samples have been carried out and published [30.2-11]. However, in order to make this technology a useful tool in biological research, detailed interpretation of biological images and correlations of fine structures with conventional TEM (transmission electron microscope) images of the original specimens are needed. Furthermore, the effect of different energies on the resulting images is not clear. In this paper, we will present contact images of cultured cells taken with both monochromatic and polychromatic synchrotron radiation, stationary target source and a submicrosecond flash x-ray source.

The development of x-ray resists (e.g., polymethyl methacrylate, PMMA) and an intense tuneable x-ray source (synchrotron radiation) have generated a great deal of interest in microchemical analysis using x-ray contact microscopy on both side of an absorption edge [30.3]. Therefore, knowledge of the properties of the recording media and specimens are needed before any meaningful attempt can be made on above and below edge imaging. A start has been made in the studies of absorption properties of x-ray resists, supporting films and biological materials. Furthermore, localization of phosphorus by imaging on either side of the phosphorus L absorption edge has been conducted.

The ultimate goal in x-ray microscopy is to image wet specimens or live cells. A paper at this symposium [30.12] reports recent results in wet cell imaging by contact microscopy.

30.2 Material and Methods

Cells (muscle cell and fibroblast-like cell) were isolated from stage 20-22 *Xenopus laevis* embryos according to the methods previously described [30.13]. Cells were cultured on UV-sterilized Formvar-carbon coated gold or nickle index grids. Cells were also cultured directly on the surface of PMMA resists. Mouse 3T3 fibroblast cells were also used; these were cultured in Dulbecco's modified Eagle's medium under 5% CO_2 at 37 C.

Cells were fixed in 1% glutaraldehyde (in water) for 1 hr. After fixation, the grids were dehydrated in a multigrid holder through an ethanol series and finally critical-point dried.

The sources used in this study were stationary target (IBM and UWO), pulsed plasma (Maxwell Laboratories, 8835 Balboa Ave., San Diego, CA 92123) and synchrotron radiation sources (Tantalus storage ring and Brookhaven VUV storage ring). Both monochromatic and white light radiation were used. Monochromatic x rays 140 eV and 128 eV were used, being on either side of the phosphorus edge. To insure uniformity, the images were taken at equal radiation dosages, developed together, and viewed under a scanning electron microscope (SEM) operated under identical conditions.

The x-ray recording media (x-ray resist) used in this study were either a 1.0 μm PMMA or copolymer which was spun uniformly onto a Si_3N_4 window [30.8] or onto a silicon wafer. Cell bearing index grids were placed in intimate contact with the resist-coated window (or wafer) in a spring-loaded holder. Images were also taken by growing the cells directly on the resist surface; after exposure, the cells were removed with a 1% Chlorox solution before development. The second method provided the best intimate contact between the specimens and the resist.

For x-ray absorption studies, self-standing Formvar, polystyrene and protein films were used. The Formvar and polystyrene films were prepared as previously reported [30.14]. A thin protein film was made by dipping a 1.5 cm diameter wire loop into a 10% gelatin solution and pulling out slowly, cooling of the liquid film causing gelation; subsequent drying produced a clear protein film. Various biological compounds could be mixed into the gelatin solution and thin films prepared. PMMA and PBS (poly(butene-1 sulfone)) thin films were prepared by the spinning method. X-ray absorption studies were conducted at the Tantalus storage ring with a Mark IV grasshopper monochromator (CSRF).

30.3 X-Ray Absorption Properties of PMMA and PBS Resists, Formvar and Polystyrene Supporting Film, and Protein

Extensive studies on the absorption properties of various biological compounds, resists and supporting film are in progress. In this paper, examples of the works will be presented.

The x-ray absorption properties of PMMA and PBS resists (Fig.30.1) were studied by HOWE [30.15]. The absorption spectrum of the PMMA resist is characterized by the pronounced carbon edge and the high absorption in long wavelengths. In contrast, PBS exhibits a very steep sulfur L edge in addition to the carbon edge. Detailed studies revealed several components in the sulfur edge (Fig. 30.1, insert). It is clear that PBS will not be useful as a recording medium for the work at wavelengths near the sulfur edge since the resist has great variations in its absorption properties at this region.

Formvar is used routinely as the supporting matrix for biological specimens (e.g., tissue culture cells). Figure 30.2 shows the x-ray absorption spectrum of a Formvar film in the range of 3-10 nm. The spectrum shows a prominent carbon K edge at 4.3 nm. Below the carbon edge is a relatively transparent window. Based on the mass fraction calculation, Formvar shows a significant oxygen absorption edge. Therefore, for the work near the oxygen edge, polystyrene film is more suitable than the Formvar film: since polystyrene consists only of carbon and hydrogen, no oxygen absorption occurs. However, at wavelengths longer than the carbon edge, the polystyrene film shows no significant difference from the Formvar film (Fig. 30.3). In comparison with Formvar, the polystyrene film is quite brittle, which causes some handling difficulties, but the film does provide excellent chemical stability under culture conditions. A thin layer of carbon coating was usually applied in order to facilitate the attachment of cells on both Formvar and

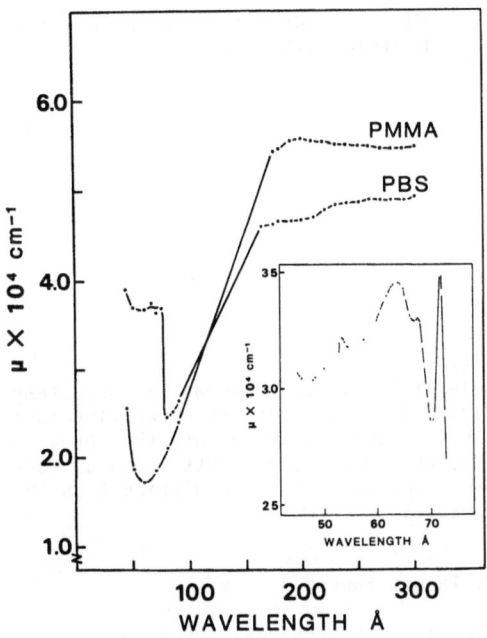

Fig.30.1 Absorption spectra of PMMA and PBS resists. Both resists exhibit carbon edges and high absorption properties at long wavelengths, furthermore, the sulfur L edge is evident in the PBS spectrum. Detailed examination of the sulfur L edge revealed fine structures (insert) (adopted from HOWE [30.15])

polystyrene films. For imaging above and below the absorption edge, one should consider not only the absorption characteristics of the photo resist for equal radiation deposition, but also the absorption characteristics of the supporting film (if there is any) which could differentially attenuate the x-ray transmittance at various wavelengths. From Fig.30.2 and 30.3, it is clear that both Formvar and polystyrene are quite suitable for the work at 4.5-10 nm region. If the two chosen wavelengths are quite close to each other, very little exposure compensation for the difference in absorption of the supporting film has to be made.

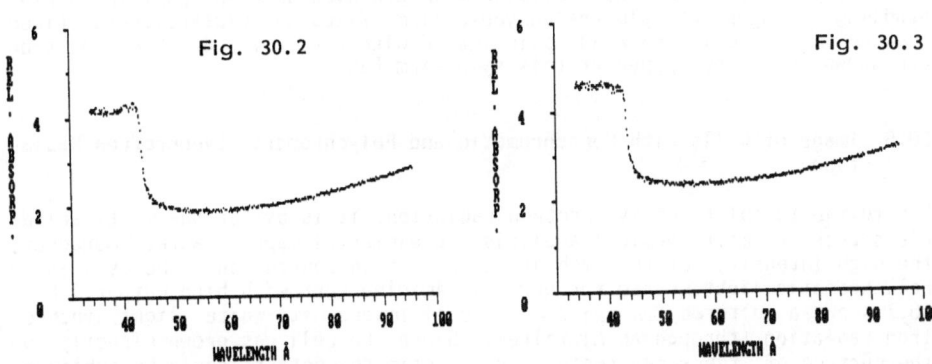

Fig.30.2   Relative x-ray absorption of a Formvar film

Fig.30.3   Relative x-ray absorption of a polystyrene film. Note the similarity to the Formvar film

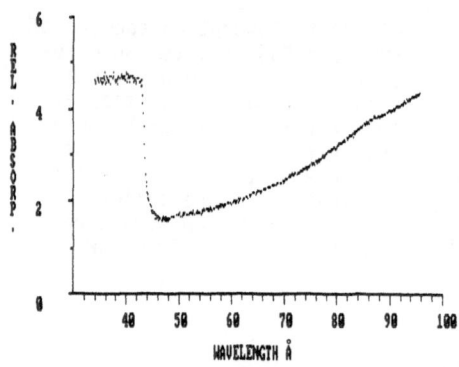

Fig.30.4 Relative absorption of a
protein film (gelatin)

The absorption spectrum of a protein film (gelatin) shows a very steep
carbon K edge located at 4.3 nm (Fig.30.4). The relative absorption just
below the carbon K edge is low; however, at longer wavelengths, the protein
film becomes quite opaque. Therefore, detecting trace elements (such as mag-
nesium) embedded in the film by absorption becomes more difficult as the
background absorption continues to increase.

30.4 Image of Cells with a Stationary Target Source

A stationary target source has been used in many previously published works,
and in general, the resulting contact images were magnified by a SEM
[30.3-7,10]. However, the image obtained from SEM is rather difficult to
interpret and the resolution is generally limited. The use of Si_3N_4 window-
supported resist for TEM viewing was first introduced by FEDER et al. in
1981 [30.8]; the methodology became an important step in the advancement of
x-ray contact microscopy. Some x-ray images using this technique have been
published since then [30.8, 10]. Figure 30.5 shows a portion of a *Xenopus
laevis* muscle cell imaged with C:K radiation and magnified with a TEM. Figure 30.6 shows the corresponding original cell shown in Fig. 30.5. By comparing the two images, one can begin to make structural correlations. Some
cellular structures have been labelled such as mitochrondria (m) and stress
fibers (s). Most of the yolk granules (Y) are from other broken cells in the
culture, many of these extracellular yolk granules shifted position during
handling. Images of glutaraldehyde-osmium fixed and glutaraldehyde fixed
Xenopus laevis cells and a 3D pair imaged with either V:L or C:K radiation
are shown in another paper at this symposium [30.12].

30.5 Image of Cells with Monochromatic and Polychromatic Synchrotron Radia-
tion

Due to the tunablity of synchrotron radiation, it is by far the most suitable source for microchemical analysis and elemental mapping work. Moreover,
the high intensity of the sychrotron radiation source could be used as a
polychromatic light source for general imaging work with high output. Fig.
30.7 shows a cultured *Xenopus laevis* cell imaged with white light synchrotron radiation (through an Al filter). Since the cell was grown directly on
the surface of the x-ray resist, very intimate contact could be achieved.
Many cellular features can be recognized such as stress fiber (black arrow)
and yolk granules (Y). The insert is a phase contrast image of the original
cell before x-ray irradiation. Figure 30.8 shows an image of a *Xenopus
laevis* fibroblast-like cell imaged with 130 eV monochromatic radiation (a

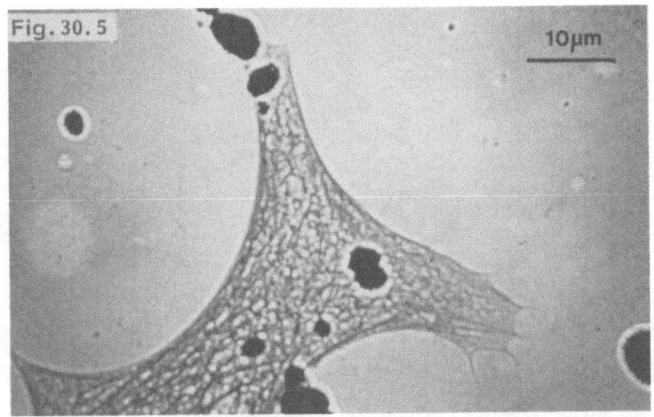

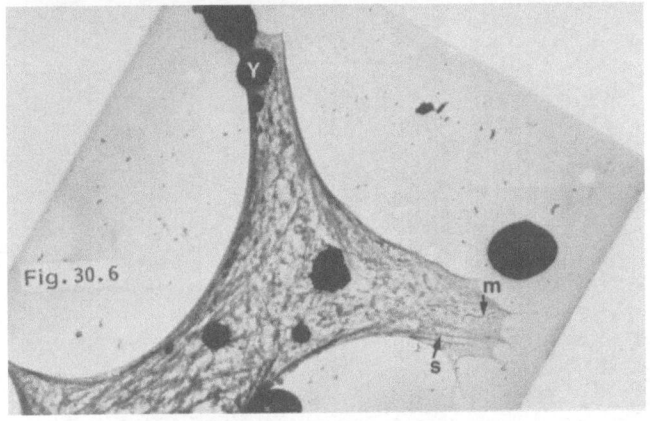

Fig.30.5 Soft x-ray image of a *Xenopus laevis* muscle cell imaged with C:K
          radiation. Yolk granule (Y), mitochondria (m), stress fiber (s).
          Resist: PMMA
Fig.30.6 TEM image of the original cell shown in Fig. 30.11. Some of the
          unmatched yolk granules are of extracellular origin and were dis-
          located during specimen handling

TEM magnified contact image). Due to the long wavelength used, diffraction
becomes very pronounced (double arrow). Figure 30.11 shows a SEM magnified
contact image of a *Xenopus laevis* fibroblast-like cell with 140 eV monochromatic light. The high x-ray density of the nucleolus is evident.

30.6 Image of Cells with a Pulsed Plasma Source (Lexis II)

Live cell imaging requires an x-ray source of high intensity which allows
one to make very short exposures. A pulsed plasma source, put on the market
by Maxwell Labs, meets this requirements. The first wet cell image produced
by such a source is reported in another paper at this symposium [30.12]. The
images shown in this paper are cells which have been fixed in glutaraldehyde
and critical-point dried. The cells were cultured on a Formvar film. During
exposure, a 100 nm Si_3N_4 window was placed in front of the specimen, therefore, the effective radiation was the filtered light through the window and

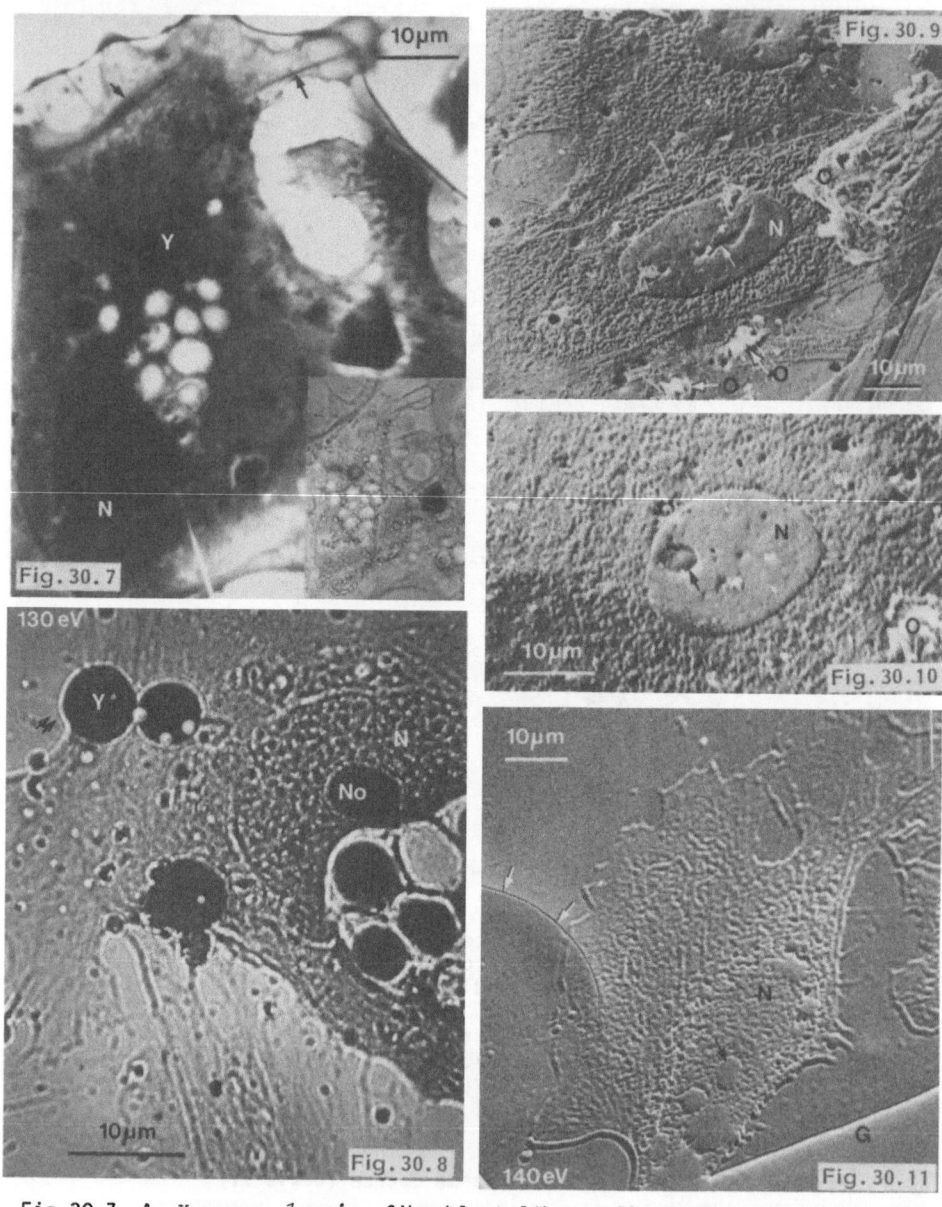

Fig.30.7 A *Xenopus laevis* fibroblast-like cell imaged with white light (>170 eV) synchrotron radiation. The cell was cultured directly on the surface of PMMA. After development, the resist was floated off the silicon wafer with 1% HF [30.11], and the free-standing resist was viewed under a TEM. Yolk granule (Y), nucleus (N), stress fiber (arrow). Insert: Phase contrast image of the original cell before x-ray irradiation

Figs.30.8-11. Caption see opposite page

Formvar film. Figures 30.9 and 30.10 show the images of mouse 3T3 cells imaged by this method. Note the high resist profile of the cell nucleus which
suggests a high x-ray density; however, the nucleolus shows much lower x-ray
density with respect to the nucleus. In comparison to the images obtained
with white light synchrotron radiation (Fig. 30.7) and monochromatic radiation (Fig. 30.8,11) as well as the stationary target source (C:K) (Fig.
30.5), one can observe that the relative x-ray density between nucleus and
nucleolus is reversed compared to the image taken with the pulsed plasma
source. The reason for such reversal is presently under investigation.

30.7 Images of a Cell Above and Below P:L Edge

We have begun a series of investigations on imaging identical cells at two
energies just above and below certain elemental absorption edges. One of the
elements chosen was phosphorus. In considering the possible edge shift from
its elemental state, we also obtained absorption spectra of phosphorus-containing biological compounds which were used to provide a general reference
for the selection of exposure energies. Figure 30.12 and Fig.30.13 represent a set of images (*Xenopus laevis* muscle cell) which were exposed at
140 eV (above L edge) and 128 eV (below L edge) synchrotron radiation. One
can clearly see that the cytoplasm of the cell has a much coarser texture
and lower resist profile when imaged at 128 eV (see also Fig. 30.14). Therefore, the image shows lower x-ray absorption by the cytoplasm. In contrast,
the 140 eV image (see also Fig. 30.15) shows much higher x-ray absorption.
It is believed that the difference in x-ray density is due to the presence
of phosphorus-containing compounds in the cytoplasm. Figures 30.12-15 are
contact images magnified by a SEM. The images are useful for providing a
visual impression of the differences in x-ray absorption at the two energies; however, the required specimen tilt presents difficulties in image
digitizing and subsequent image subtraction. Therefore, it was necessary to
magnify the contact images by a TEM. (Not shown here, but shown in the talk,
is a set of contact images taken at 459 and 590 eV, and magnified with a
TEM). The computer program necessary for subtracting the two images above
and below an absorption edge is under development.

Fig.30.8 *Xenopus laevis* fibroblast cell imaged with 130 eV monochromatic
synchrotron radiation. (resist: PMMA). Nucleus (N), Nucleolus
(No), Yolk granule (Y). Note the diffraction fringes (double arrow)
Fig.30.9 and 30.10 Mouse 3T3 fibroblast. Fixed in 1% glutaraldehyde, dehydrated in EtOH and critical-point dried. High-speed (200 ns) imaging with Lexis II source. In addition to the Formvar film which
supported the cell, a 100 nm thick Si_3N_4 window was used as a filter. Note the high x-ray density of the nucleus (N) and relative
low absorbance of the nucleolus (arrow). Some fragments of the
original cell remain on the surface of the resist (O). Resist:
copolymer
Fig.30.11 Monochromatic soft x-ray image of *Xenopus laevis* fibroblast-like
cell. This contact image was magnified by a SEM. Note the relatively high absorbance of the nucleolus (black arrow) with respect
to the nucleus (N). Grid bar (G), crack on the resist (white arrows). Resist: PMMA

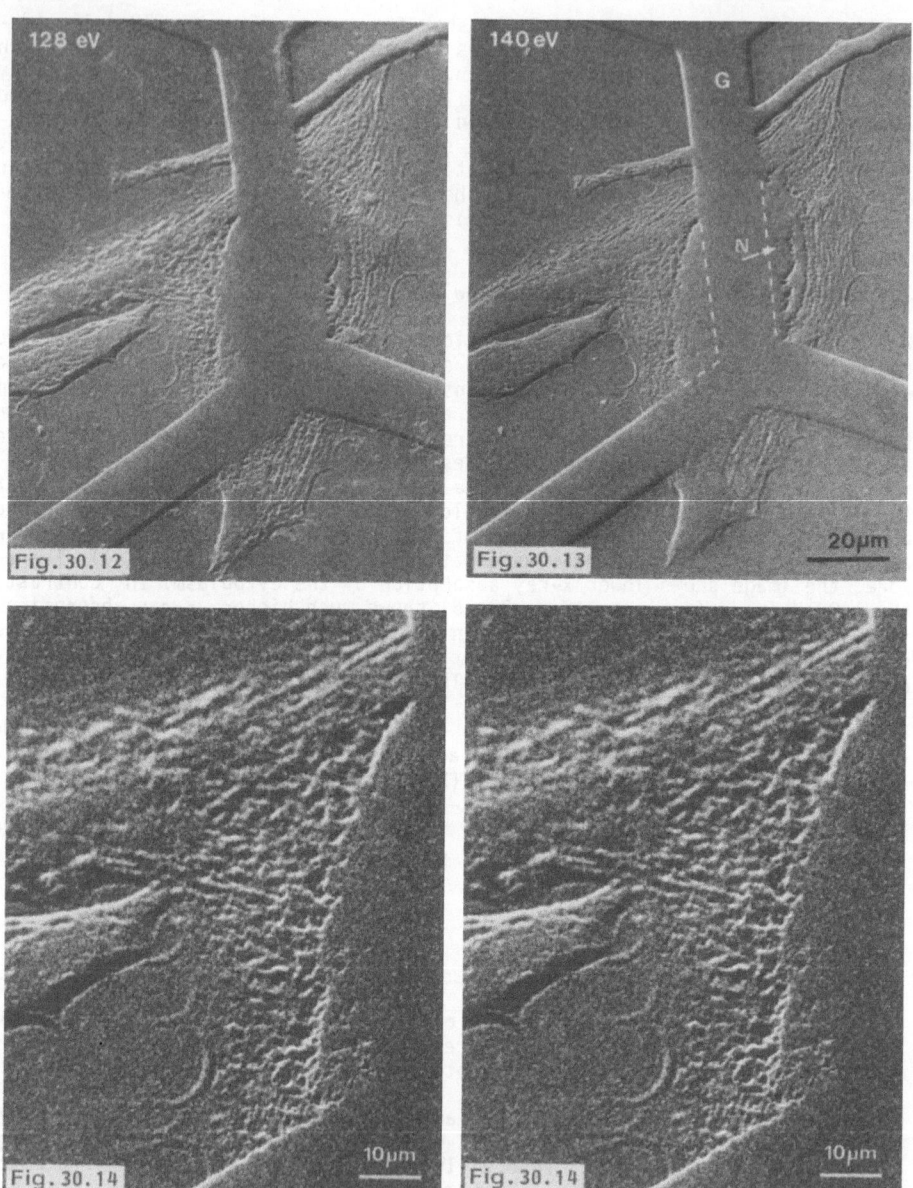

Fig.30.12 Monochromatic soft x-ray images of a *Xenopus laevis* muscle cell.
and   Note the coarser appearance of the cell cytoplasm in   Fig.30.14
Fig.30.13 (128 eV: below P:L edge) in contrast to Fig.30.15 (140 eV: above
       P:L edge). The cytoplasm of the cell has a higher absorbance   at
       140 eV than at 130 eV; this difference is due to the presence   of
       phosphorus-containing compounds. The nucleus (N) is too thick   to
       be imaged in this replication. Grid bar (G)
Fig.30.14 Same as Fig.30.12 and 30.13 but at a higher magnification to show
and   details of the cell cytoplasm
Fig.30.15

Summary

The present status of soft x-ray contact microscopy is reminiscent of electron microscopy during the 40's when most of the related techniques were under development. Considerable refinement is needed before this technique can become a useful tool in modern biological research. We believe that three important areas are worth concentrating on: the continued development of wet cell imaging techniques, the development of proper specimen handling and preparation techniques, and the establishment of basic rules for image interpretation.

Acknowledgments

We are very grateful to the Canadian Medical Research Council, the Natural Science and Engineering Research Council and the Veeco Fund for financial support. We wish to express our thanks to R.J. Walter and P.K. Gaetano of University of Illinois for the 3T3 cell, J. Perlman, J. Riordan and S. Robb of Maxwell Laboratories for their help on LEXIS-II operation, and R.J. Howe for providing Fig.30.1.

References

30.1   R. Feder, E. Spiller, J. Topalian, A.N. Broers, W. Gudat, B.J. Panessa, Z.A. Zadunaisky and J. Sedat: Science 197, 259 (1977)

30.2   R. Feder and D. Sayre: Ann. NY Acad. Sci. 342, 213-234 (1980)

30.3   J.Wm. McGowan, B. Borwein, J.A. Medeiros, T. Beveridge, J.D. Brown, E. Spiller, R. Feder, J. Topalian and W. Gudat: J. Cell Biol. 80, 732-735 (1980)

30.4   B.J. Panessa-Warren and J.B. Warren: Ann. NY Acad. Sci. 342, 350-367 (1980)

30.5   B.J. Panessa, J.B. Warren, P. Hoffman and R. Feder: Ultramicroscopy 5, 267-274 (1980)

30.6   E. Spiller, R. Feder, J. Topalian, D.E. Eastman, W. Gudat and D. Sayre: Science 191, 1172-1174 (1976)

30.7   L. Manuelidis, J. Sedat and R. Feder: Ann. NY Acad. Sci. 342, 304-325 (1980)

30.8   R. Feder, J.L. Costa, P. Chandhari and D. Sayre: Science 212, 1398-1400 (1981)

30.9   K.H. Tan, P.C. Cheng, G.M. Bancroft and J.Wm. McGowan: submitted to Can. J. Spectroscopy (1983)

30.10 P.C. Cheng, and H.B. Peng: Science Monthly (Taipei, Republic of China) 12-9, 11-12 (1981)

30.11 P.C. Cheng, H.B. Peng, R. Feder and J.Wm. McGowan: Electron Microscopy (10th Int. Electron Microscopy Congress Proc.) 1, 461-462 (1982)

30.12 R. Feder, V. Mayne-Banton, D. Sayre, J. Costa, B.K. Kim, M.G. Baldini and P.C. Cheng: this volume no.29

30.13 H.B. Peng, J.J. Wolosewick and P.C. Cheng: Develop. Biol. 88, 121-136 (1981)

30.14 P.C. Cheng and M.I. Lin: Natl. Sci. Council Monthly, Rep. of China (ISSN 0250-1651) 9, 15-22

30.15 R.J. Howe: Absorption Properties of Polymers using Synchrotron Radiation. MSc thesis, Univ. of Western Ontario (1983)

31. X-Ray Microscopy as a Possible Tool for the Investigation of Plant Cells

V. Sarafis

Department of Applied and Environmental Science, School of Food Sciences, Hawkesbury Agricultural College, Richmond, 2753 N.S.W., Australia

31.1 Introductory Remarks

The scope of this article is to highlight the potential use of x-ray microscopy in the investigation of the functional morphology of the plant cell.

A particular focus on soft x rays in the region 1 - 10 nm has been made both because elements of interest in plant cells have appropriate absorption edges and emission lines in this wavelength region [31.1] and several x-ray microscopes are already available which use soft x rays.

In this article the possible use of x-ray microscopy is emphasized only where it would appear to be the method of choice.

In a brief article such as this it is impossible to list all the potential uses, therefore a few examples have been chosen and extensive references made to the review literature to enable the reader to explore further.

From the study of KIRZ and SAYRE [31.2] resolutions of the order of 10 nm can be expected for wet specimens of 10 μm at doses of 10^4J/G. Doses of 10^4J/G are considered [31.3] to be the limit before structural damage occurs although it is likely that cell death will have occurred before then.

If one allows that an upper dose of 10 J/G is lethal [31.3], useful resolution in vivo could occur in many cells and organelles at 100 nm resolution. Even allowing for cell death there would still be considerable advantage in looking at fixed wet cells.

One must however point out the pressing need to conduct suitable viability studies on specimens which have undergone x-ray observation to determine the point of lethality. Useful tests could be the fluorescein diacetate test [31.4] which can be performed on cells and plasmalemma bound cytoplasm fragments and tests the retention of semipermeability or assessment of photosynthesis and respiration.

31.2 Specimen Preparation for Observation

Live specimens for observation may be prepared by using poly-L-lysine as a "bioglue" for immobilization. A 1 % solution of poly-L-lysine is made in water and a drop is placed on the substrate. After 10 minutes it is drained off and the cell or organelle suspension applied and left for 10 minutes. Sticking of a proportion of cells or organelles usually occurs. A gentle rinsing with the suspension medium removes non-adhering components. Alternatively one may lightly flame the substrate where it is possible to do this without damage before carrying out the procedure above.

It is also possible to use lower concentrations of poly-L-lysine and/or to allow a thin layer of poly-L-Lysine solution to evaporate to near or complete dryness before applying the suspension of cells or organelles.

Light pressure by a coverslip may also be applied to enhance adhesion. Some experimentation is generally needed to optimise the sticking. This technique allows cells to survive, is non-toxic and can be used for walled and naked cells as well as for organelles such as chloroplasts.

If one wishes to view cells whose dimensions are not ideal one can prepare protoplasts by any of a number of methods [31.5] using cell wall degrading enzymes. The protoplasts can be compressed to give the desired thickness and if desired some plants such as *Bryopsis*, *Caulerpa*, *Codium* and *Acetabularia* which are siphonaceous algae can be cut and microscopic droplets of cytoplasm are made.

It might also be possible to prepare germlings for examination in situ by attaching spores to substrates and flooding with culture solution to allow growth followed by removal of the medium while the cell systems remain attached.

31.3 Isolated Organelle Studies

The highly visible and functional chloroplast has attracted great attention to its structure and function. The ultrastructure has been observed with the electron microscope in fixed dehydrated embedded cut and stained specimens [31.5].

However is what one sees in the electron microscope (EM) really what one sees in vivo? Instead of the lens-like profiles shown by the EM [31.6] the in vivo appearance of chloroplasts in higher plant cells for example, spinach, is dinner plate-like with several mobile extensions from a mobile outer envelope [31.7,8]. Chloroplasts isolated from spinach retain their shape and mobile outer envelope under appropriate isolation criteria [31.8]. It would be interesting to formulate an in vivo three-dimensional map of the structure of an undamaged chloroplast which was still capable of carrying out the full range of photosynthetic activities. Such isolated chloroplasts when fixed as for electron microscopy maintain their morphological structure when seen under the light microscope. It is only in freeze-etched chloroplasts that one can retain internal structural integrity. However, it would be impossible to make a three-dimensional map of such an organelle because a fracture plane would occur only once through a chloroplast.

The tantalizing location of the complex chloroplast genome and the manner of chloroplast replication would also be rewarding to study [31.9].

It would be particularly interesting to observe stages in starch grain formation within chloroplasts as also in amyloplasts [31.10].

Other modification of chloroplasts are chromoplasts in such things as capsicums, tomatoes, carrots and rosehips [31.11]. These are high in lipids and are unlikely to be well preserved by the traditional methods used in electron microscopy. Phase separation, which may occur during lipid freezing (as in preparation for freeze fracture) means that x-ray microscopy is a likely technique to yield a useful result because freezing is avoided.

31.4 Cell Walls

The higher plant cell wall contains both fibrillar components such as cellulose embedded in a water-rich region which contains amorphous polysaccharides. Investigations of cell wall structure have often involved drastic preparation techniques [31.12]. In TEM, cellulose is a very low contrast material which is extremely difficult to stain by the standard electronmicroscopical stains. However, cellulose which because of its molecular structure binds very little water in its immediate vicinity would be a high contrast material within the aqueous parts of the wall due to the H_2O x-ray window. Thus, the true three-dimensional configuration of cellulose fibrils could be apparent in wet walls. The development of (a) lignification [31.13], (b) cutinisation [31.14] and (c) suberization [31.14] could be followed in suitable material by cutting hydrated material.

Several types of plant cures are involved in secreting specialized cell wall components. These include scales, silica shells and mucilages [31.15-18]. It is the Golgi apparatus which is responsible for their manufacture and the structure and/or the dynamics of secretory product formation would repay investigation.

31.5 Plant Vacuoles

The plant cell vacuole is the potentiator of turgor as well as a store for acids, ions, pigments, enzymes, protein "storage", crystals and inorganic crystals [31.5,19]. Physiological activity might be followed, particularly phases of ion accumulation or druse formation. Developmental studies might well be performed on crystallites of calcium oxalate and carbonate, some of which are known to develop in vacuoles [31.20].

31.6 Whole Cells

By preparing protoplasts [31.5] and/or cytoplasmic droplets [31.21], the following would become accessible to study. Cell wall synthesis as for example the de novo formation of cell wall by protoplasts which have had the medium changes from a preparative one to an osmotically protected culture medium [31.5]. An interesting problem which has been addressed has been the gravity detection of *Chara* rhizoids [31.23]. In these barium sulphate crystals are used as statoliths. The kinetics of statolith function might be approached directly using x-ray microscopy.

Complex interrelationships between plant cell organelles have been noted from optical microscopical investigations such as by HONDA & WILDMAN [31.23]. High-resolution scanning of such cells would be revealing because fixed cells are altered in their structure during the process of fixation-dehydration and embedding.

In some red algae,e.g. *Bonnemaisonia*,there are special cells which accumulate iodine and bromine. An x-ray microscopical study may well be able to follow accumulation in such cells.

During germination of spores, hydration of the cytoplasm takes place, vacuoles form, ribosomes aggregate into polysomes and cell wall biosynthesis commences [31.24]. These changes are difficult to observe under optical microscopy due to the spherical shape of many spores and due to the presence of highly refractile granules which renders observation very difficult. X-ray high resolution study of synchronous stages in spore germination, e.g. in mosses, would yield valuable insights into the events accompanying germination, particularly a comparison between the ultrastructure of dry spores and germinating ones.

Studies of the interaction of gametes during syzygy suffer from the difficulty of preserving morphological integrity of quite labile membrane interaction. Unfixed gametes "doing their thing" might throw a particular light on the initial interaction between gamete envelopes.

Budding in *Saccharomyces* involves complex cell activity at the bud site [31.25]. The mode of budding in yeast cells would be amenable to x-ray microscopical investigation unhindered by the need in phase-constrast microscopy for matching the refractive index and the problems caused by the birefringent walls obtained with Nomarski differential interference microscopy.

Fungal filament growth and cytoplasmic organisation were studied in vivo and by EM [31.26]. The material is suitable for x-ray microscopy and could be used to study the complex nuclear and wall formation events in crozier development described by GIRBARDT [31.26] in *Polystictus*.

An extremely exciting activity in botanical sciences today is gene transfer [31.27]. *Agrobacterium tumefaciens*, the causative organism of crown gall disease, is a vector for gene transfer. A similar vector is *Agrobacterium rhizogenes* which causes multiple rooting. In both these cases it has been impossible to see the actual process of plasmic transfer. It has had to be deduced. However, using the model of *Pylaisiella* which is induced to form buds or rhizoids by these organisms respectively [31.28-30], the possibility of actually seeing plasmid transfer due to the appropriate dimensions would be feasible, as it is known that specific adherence of the agrobacteria to the moss protenema is required [31.31].

Analogously, the formation infection threads during nodulation in legumes [31.31] would also be amenable to kinetic studies using x-ray microscopy for comparison with those done by optical means [31.33].

Stomatal function is a characteristic of land plants where ion fluxes, particularly potassium , play an important light-dependent part in opening [31.34]. The formation of active guard cell protoplasts [31.5,34] would allow the dynamics of the process to be followed at a high resolution by absorption-edge imaging or x-ray fluorescence.

31.7 Some Concluding Remarks

The x-ray water window means that for organic molecules the resolution cannot theoretically exceed about nm, however one has one additional degree of freedom in the scanning confocal methods of SHEPPARD [31.35],where increased resolution is demonstrated. Improvements using special apodized screens are obtainable in the light domain [31.35] which could be translated into the x-ray scanning systems using zone plates. Resolution can be doubled by these methods.

The possibility of modifying in vivo stains to include x-ray absorbing atoms of high atomic number would also expand the range of possibilities for ultrastructural study.

While it will not be useful to try to use the x-ray microscope to do the same as the EM specially using sections prepared in a similar way as for the TEM, it is evident in this brief review that the x-ray microscopy offers new possibilities in investigating plant cell structure and function where the brutal EM techniques yield more mysteries that they appear to solve [31.37].

References

31.1 E. Spiller : "The Scanning X-Ray Microscope - Potential Realisations and Applications" in Scanned Image Microscopy, pp. 365-391 (1980)

31.2 G. Schmahl, D. Rudolph, B. Niemann and O. Christ: Quart. Rev. Biophys. 13, 297 (1980)

31.3 J. Kirz and D. Sayre: "Soft X-Ray Microscopy of Biological Materials" in Synchrotron Radiation Research. pp. 277-322 (1980)

31.4 J. Heslop-Harrison and Y. Heslop-Harrison: "Evaluation of Pollen Viability by Enzymatically Induced Fluorescence Intracellular Hydrolysis of Fluroescein Diacetate", Stain Tech. 45, 115 (1970)

31.5 E. Galun: Ann. Rev. Plant Physiol. 32, 237 (1981)

31.6 J.A. Schiff: "Development, Inheritance & Evolution of Plastids and Mitochondria", in the Biochemistry of Plants, A Comprehensive Treatise, 1, pp. 209-265 (1980)

31.7 S.G. Wildman, C.A. Jope and B.A. Atchison: Bot. Gaz. 141, 24 (1980)

31.8 D. Spencer and H. Unt: Aust. J. Biol. Sci. 18, 197 (1965)

31.9 J.V. Possingham: Ann. Rev. Plant Physiol. 31, 113 (1980)

31.10 W. Banks and D.D. Muir: "Structure and Chemistry of the Starch Granule" in The Biochemistry of Plants 3, 321-366 (1980)

31.11 J.T.O. Kirk and R.A.E. Tilney-Bassett: The Plastids: Their Chemistry, Structure, Growth and Inheritance (Elsevier/ Nth Holland 1978)

31.12 J.R. Colvin: "Ultrastructure of the Plant Cell Wall: Biophysical Viewpoint", in Plant Carbohydrates 11, Encyclopedia of Plant Physiology, 138, (1981)

31.13 T. Higuchi: "Biosynthesis of Lignin" in Plant Carbohydrates 11, Encyclopedia of Plant Physiology, 13 B, 194-221 (1981)

31.14 P.E. Kolattukudy, K.E. Espelie and C.L. Soliday: "Hydrophobic Layers Attached to Cell Walls, Cutin, Suberin and Associated Waxes" in Plant Carbohydrates 11, Encyclopedia of Plant Physiology, 13 B, 225-248 (1981)

31.15 D.G. Robinson: "Algal Walls - Cytology of Formation" in Plant Carbohydrates 11, Encyclopedia of Plant Physiology, 13 B, 317-329 (1981)

31.16 E.L. Duke and B.E.F. Reimann: "The Ultrastructure of the Diatom Cell", in The Biology of Diatoms (Blackwell Sc. Pub. Oxford 1977) pp. 65-109

31.17 H.H. Mollenahauer and D.J. Morre: "The Golgi Apparatus", in the Biochemistry of Plants, a Comprehensive Treatise, Vol. 1 The Plant Cell, 438-483 (1980)

31.18 W.P.D. Stewart: Algal Physiology and Biochemistry (Blackwell Scientific Publications, Oxford 1974)

31.19 R.A. Leigh, F. Marty and D. Branton: "Plant Vacuoles", in the Biochemistry of Plants, a Comprehensive Treatise, 1, 625-655 (1980)

31.20 D.A. Larson: J. Ultrastruct. Res. 16, 55 (1966)

31.21 M. Tatanaki and K. Nagata: J. Physiol. 6, 401 (1970)

31.22 K. Schroter, A. Lauchli and A. Sievers: Planta 122, 213-225 (1975)

31.23 R. Honda and S. Wildman: Organelles in Living Plant Cells, (28 min. 16mm) (International Telefilm Enterprises, Toronto, Canada) (1962)

31.24 P. Olesen and G.S. Mogensen: Bryologist 81, 493 (1978)

31.25 E. Cabib: "Chitin: Structure, Metabolism, and Regulation of Biosynthesis" in Plant Carbohydrates 11, Encyclopedia of Plant Physiology, 138, 394-411 (1981)

31.26 M. Girbardt: Chap. XV, die Pilzzelle, in Grundlagen der Cytologie, (Gustav. Fischer Verlag, Stuttgart 1973) pp. 441-460

31.27 S.H. Howell: Rev. Plant Phys. $\underline{33}$, 609 (1982)

31.28 L.C. Spiess, B.B. Lippincott and J.A. Lippincott: Am. J. Bot. $\underline{59}$, 726 (1971)

31.29 L.D. Spiess: Bot. Gaz. $\underline{138}$, 35 (1977)

31.30 L.D. Spiess: Physiol. Plantarum $\underline{51}$, 99 (1981)

31.31 L.D. Spiess, J. Turner, P. Mahlberg, B.B. Lippincott and J.A. Lippincott: Amer. J. Bot. $\underline{64}$, 1200 (1977)

31.32 W.D. Bauer: Ann. Rev. Plant Physiol. $\underline{32}$, 407 (1981)

31.33 P.S. Nutman, C.C. Doncaster and P.J. Dart: Infection of clover by root nodule bacteria. (Made under the auspices of Rothamstead Experimental Station England. Produced by British Universal Film, London) 1973

31.34 T.C. Hsiao: "Stomatal Ion Transport" in Transport in Plants 11, Encyclopedia of Plant Physiology, $\underline{2\ B}$, 195-217 (1976)

31.35 C.J.R. Sheppard: "Imaging Modes of Scanning Optical Microscopy", in Scanned Image Microscopy, 201-225 (1980)

31.36 D. Yansen: Diversified Research, Lexington, Mass. U.S.A. Personal Communication

31.37 T.P. O'Brien and M.E. McCully: The Study of Plant Structure - Principles and Selected Methods (Temarcarphi Pty., Ltd., Melbourne, Vic. Aust. 1981)

32. Possible Applications of X-Ray Microscopy in Pathology

F. Pfannkuch and D. Hoder

Universitätsklinikum Charlottenburg, Institut für Pathologie, Freie
Universität Berlin, Spandauer Damm 130, D-1000 Berlin 19

H. Baumgärtel

Institut für Physikalische Chemie der Freien Universität Berlin
Takusstraße 3, D-1000 Berlin 33

The greatest achievement of R. VIRCHOW was his application of the microscopic technique in the diagnosis of diseases. This development revolutionized diagnosis in medicine and led to a new "bioptical" dimension in the field. In our institute - which is considered to be medium sized - the number of histological and cytological investigations has reached 25 000 cases or 80 000 preparations per year.

In the area of basic research (and to a lesser extent routine diagnostics) electron microscopy, with its advanced technological development, is of continual use to us.

32.1 Light Microscopy

For routine diagnostics in pathology, light microscopy plays the most important role. The useful range of magnification for our practical medicinal diagnostics is achieved with objectives between 4x and 63x, whereby the usual ocular of 12.5x yields magnifications of 60x - 600x. This corresponds to a resolution of 1 μm. Further magnification has basically no advantage because of the increasing probability of artefacts. This is the greatest problem of histopathology today: this area is in essence fully standardized, but in practice, all the more so by its artefacts. These are a result of

1. Fixation (formaldehyde, alcohol, etc.)
2. Dehydration by increasing alcohol concentrations
3. Paraffining
4. Staining (haematoxylin, etc.).

Methods which avoid these shortcomings, such as dark-field, interference or phase-contrast microscopy, are quite restricted in their reliability and are of no importance in pathological-histological diagnostics.

32.2 Electron Microscopy

With transmission electron microscopy (TEM) a new dimension in the structural analysis of the cell has been revealed. The theoretical point resolution of our electron microscope (EM 10, Zeiss, Oberkochen, West-Germany) is 0.24 nm, corresponding to a magnification of up to 500 000x. A resolution of better than 1 nm is, however, possible only by use of extremely sensitive and time-consuming techniques of preparation and optical processing (UNWIN and HENDERSON [32.1]). Here, too, there is for us a resulting practicable range of magnification of 4 000x and 25 000x, and in exceptional cases 100 000x. Some reasons for these restrictions are as follows:

1. Denaturation of protein due to fixation (glutaraldehyde,etc.)
2. Embedding in resins (Araldite; Epon,etc.)
3. Contrasting with heavy metals (osmium tetroxide, uranyl acetate).
   Resolution is limited by the particle size of the substance used
   (1-2 nm).
4. Decreased slice thickness (60 - 100 nm).

Finally, an essential limitation of methods to be mentioned is that the analyses can be carried out only under high vacuum. Many of the attempts to enable such investigations at atmospheric pressure proved to be useless. Whereas use of light microscopy can enable observation of living conditions and the judgment of dynamic cell functions and metabolism, electron microscopy practically cannot.

Since the electron permeability of the samples to be investigated requires extremely thin slices, complicated sectioning and analysis are necessary in order to reconstitute the complete picture - for example of one cell. Scanning electron microscopy (SEM) in medicinal-diagnostic research enables resolution of down to 10 nm and yields a description of surface features with high depth of field.

32.3 X-Ray Microscopy

The development of zone plate x-ray microcopy (SCHMAHL et al. [32.2-4]) is of great importance, as far as the practical histo-pathological aspects and the demands of pathological-anatomical basic research are concerned. For example:

- Resolution is in the order of magnitude of previously described light and simple electron microscopies (70-10 nm).

- Depth of field, i.e., judgment of objects within several μm makes possible a new dimension in research of spatial organisation of the cell.

- The range of 2.3 - 4.5 nm is the absorption optimum for proteins, i.e., for the cell components of greatest interest to us. Contrasting can in many cases be omitted here.

- Since investigations can be carried out at atmospheric pressure, observations of vital functions are also possible.

- Adverse affects due to heat of the electron beam are much less and can be decreased even more by scanning x-ray microscopy (NIEMANN et al. [32.5]).

(At present the damage of structures under observation cannot be accurately appraised. The long exposure times necessary must be reduced in order to avoid inadequate focussing or picture entirety of living objects).

The following are some possible utilizations of x-ray microscopy for pathology:

1. Observation of living cells and of cell culture monolayers with special regard to their function, such as cell movement, phagocytosis, pinocytosis, cytopempsis, excretion under abnormal conditions (eg.,pathological phagocytosis inhibition, lacking excretion of lysosomal enzymes form pathological granulocytes,etc.).

2. Investigations of metabolism in the cell under conditions of normal synthesis of structure and function (incorporation of particles, formation of metabolites such as pigments and lipids; activation of metabolism under especially demanding conditions or pathological conditions).

3. Observation of growth and degeneration occurrences of cells, especially in regard to their cytoarchitecture (filament formation, filament distribution, origin of cell contacts such as adhesion; desmosomes,etc., autolysis).

4. Correlation of results of x-ray microscopy with hitherto unknown conceptions of function and design of cells and cell organelles. Ordinary slices can in any case be used for x-ray microscopy as well, whereby embedding and contrasting are not necessary.

5. Elemental analysis through spectroscopy (eg., recognition of substances pathologically stored intracellularly; incorporation of vital dyes,etc.).

Results of x-ray contact lithography techniques (SPILLER et al. [32.6]; McGOWAN et al. [32.7]; PANESSA et al. [32.8]; PANESSA et al. [32.9]; FEDER [32.10]; CHENG et al. [32.11]) have already shown that new structures are to be found during the investigation of biological samples. Using this technique, a resolution of better than 6 nm was stated. Flush x-ray microscopy allowed the exposure time to be decreased to about 60 ns.

In contrast to zone plate x-ray microscopy, this is,however,an indirect method, one which is technically more complicated and at the same time less flexible.

In the past we have not had the chance of gaining experience in x-ray microscopy. The above-mentioned utilisations therefore correspond solely to the conceptions we have had concerning light and electron microscopic endeavours in our research and routine projects.

In regard to practical considerations which must be met upon use of x-ray microscopy, the following items are of interest:

1. The microscope must be installed in the vicinity of the user (an intensive source of radiation with several different wavelengths must be available, but taking up little space).

2. The objective chamber must be quick to load and unload,whereby the horizontal position is to be preferred.

3. Observation of the picture (on the screen) must be possible in sitting position, so that the object in the chamber can simultaneously be shifted in the x/y plane. (This problem is easily solved with the scanner and picture screen).

4. The objective field must be at least 1 mm^2.

References

32.1 P.N.T. Unwin, R. Henderson: Molecular structure determination by electron microscopy of unstained crystalline specimens. J. Mol. Biol. $\underline{94}$, 425-440 (1975)
32.2 G. Schmahl, D. Rudolph, B. Niemann, O. Christ: X-ray microscopy of biological specimens with a zone plate microscope. Ann. NY. Acad. Sci. $\underline{342}$, 368-386 (1980)

32.3 G. Schmahl, D. Rudolph, B. Niemann, O. Christ: Zone-plate x-ray micro-scopy. Quart. Rev. Biophys. $\underline{13}$, 297-315 (1980)

32.4 G. Schmahl: X-ray microscopy. Nuclear Instruments and Methods $\underline{208}$, 361-365 (1983)

32.5 B. Niemann, D. Rudolph, G. Schmahl: The Göttingen x-ray microscopes. Nuclear Instruments and Methods $\underline{208}$, 367-371 (1983)

32.6 E. Spiller, R. Feder, J. Topalian, D. Eastman, W. Gudat, D. Sayre: X-ray microscopy of biological objects with carbon K_α and with syn-chrotron radiation. Science $\underline{191}$, 1172-1174 (1976)

32.7 J.Wm. McGowan, B. Borwein, J.A. Medeiros, T. Beveridge, J.D. Brown, E. Spiller, R. Feder, J. Topalian, W. Gudat: High resolution microche-mical analysis using soft x-ray lithographic techniques. J. Cell. Biol. $\underline{80}$, 732-735 (1979)

32.8 B.J. Panessa, J.B. Warren, R. Feder, D. Sayre, P. Hoffman: Ultrastruc-tural and elemental imaging of biological specimens by soft x-ray con-tact microscopy. Scanning Electron Microscopy $\underline{2}$, 107-116 (1980)

32.9 B.J. Panessa, G. Colemann, R.A. McCorkle, J.B. Warren: A wetcell for x-ray imaging of hydrated biological specimens. Scanning $\underline{4}$, 21-26 (1981)

32.10 R. Feder: Improved details in biological soft x-ray microscopy: study of blood platelets. Science $\underline{212}$, 1398-1400 (1981)

32.11 P. Cheng, H.B. Peng, R. Feder, J.Wm. McGowan: The use of transmission electron microscope as a viewing tool for high resolution soft x-ray contact microscopy. Electr. Microscopy 1982 (10th Int. Congr. Electron Microscopy), 1, 461-462 (1982)

33. Time Resolved X-Ray Spectroscopy in Biology

R. Rigler

Department of Medical Biophysics, Karolinska Institutet, Box 60400
S-104 01 Stockholm, Sweden

33.1 Introduction

The time structure as well as the high intensity of synchrotron radiation
offer new opportunities to study dynamic properties of complex biological
molecules. While the use of synchrotron radiation in the x-ray region for
stationary investigations of biopolymers has become widespread (for reviews)
see [33.1-3]) its applications to follow dynamic events is still limited to
a few cases. These, however, clearly indicate the potentialities available
for elucidating the structural dynamics and functional properties of biomolecules.

33.2 Dynamics of Molecular Structures

In order to follow kinetic structural events synchrotron radiation can be
used in two ways: (a) making use of its own time structure in a stroboscopic
detection of fast events; (b) using the high beam intensity for following
the time course of structural changes with stationary illumination.

Since electrons or positrons are circulating within the storage ring in
bunches spaced at regular intervals, synchrotron radiation appears to the
observer to consist of repetitive light pulses. Dependent on the beam optics
the full width half-maximum (FWHM) varies between 150 to 50 ps and pulse
widths down to 10 ps appear to be feasible. Dependent on the number of electron bunches circulating in the ring as well as on the ring size the interval between successive pulses varies between 1 µs to 10 ns.

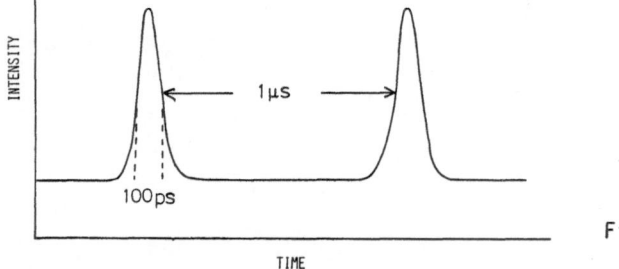

Fig.33.1

At present the time structure has been used for experiments in the visible and VUV region in order to probe the dynamics of excited states
[33.4]. From measurements of the time dependence of fluorescence polarization information on segmental mobility and internal motion in proteins
[33.4] and nucleic acids [33.5,6] is available.

Similar experiments could in principle be performed when observing x rays scattered or diffracted by biological structures provided adequate time and spatial resolution of x-ray detectors. Currently electron density maps refer to the time and space average of electron positions as described by their temperature factors [33.7,8]. The positions of the scattering electrons however could be frozen in a ps time scale with the pulse width available and would allow discrimination between dynamic and static disorder of crystalline structure, which at present is available only for special situations [33.7,9,10].

Time-resolved studies of molecular motions, however, on a much slower scale (ms and longer) can be performed by perturbing structural relaxations and measuring stationary diffraction, scattering or absorption of x-ray radiation.

Pioneering experiments on time-resolved x-ray diffraction of myosin during muscle contraction have been performed by HUXLEY and collaborators [33.11] showing a close correlation between the development of muscular tension and contraction due to sliding motion between overlapping actin and myosin filaments. These studies have been possible due to the development of position-sensitive detectors with high count rates [33.12,13] in addition to high intensity and well-collimated x-ray beams as provided by synchrotron radiation. Analogous experiments can be performed with crystalline structures, e.g. in studying temperature relaxations of diffraction patterns [33.14].

Due to the high intensity of synchrotron radiation also time-resolved x-ray scattering studies of randomly oriented molecules have been possible, a notable example being the kinetics of microtubuli assembly [33.15] which has provided information on the formation of microtubli from the tubulin subunits.

Optimization of the x-ray beam geometry and the development of circular position-sensitive detectors [33.13] as well as of mixing equipment allows one to follow structural transitions in small proteins such as ACTase [33.16] as well as in nucleic acids such as tRNA [33.17].

Structural information on the immediate vicinity of x-ray absorbing atoms can be obtained from their time structure above the absorption edge [33.18] and time-resolved studies are being carried out [33.19].

33.3 Real-Time Imaging

An important development in real-time imaging has been the digital subtraction angiography initiated by HUGHES [33.20] and KULIPANOV [33.21]. Here the differential absorption before and after the K edge of iodine is used to follow the flow in heart and blood vessels. These approaches could in principle be transferred into the microscale. The dynamics and motions of cellular constituents, such as cell organelles, as well as of macromolecules of various degrees of organization could be followed in real space and would be complementary to time-resolved scattering and diffraction experiments.

References

33.1   D. Winick and S. Doniach: Synchrotron Radiation Research. Plenum Press, New York (1980)
33.2   A. Castellani and I.F. Quercia: Synchrotron Radiation Applied to Biophysical and Biochemical Research. Nato Advanced Study Institutes Series. Series A: Life Sciences. Plenum Press, New York (1979)

33.3 H.B. Stuhrmann: Uses of Synchrotron Radiation in Biology. Academic Press, London (1982)

33.4 I. Pecht: Application of synchrotron radiation to biochemical fluorescence spectroscopy. In Uses of Synchrotron Radiation in Biology, ed. H.B. Stuhrmann, pp. 71-82. Academic Press, London (1982)

33.5 D.P. Millar, R.J. Robbins, and A.H. Zewail: Direct observation of the torsional dynamics of DNA and RNA using picosecond spectroscopy. Proc. Natl. Acad. Sci. USA 77, 5593-5597 (1980)

33.6 M. Ehrenberg, R. Rigler and W. Wintermeyer: On the structure and conformational dynamics of yeast phenylalanine-accepting transfer ribonucleic acid in solution. Biochemistry 18, 4588-4599 (1979)

33.7 H. Frauenfelder, G.A. Petsko and D. Tsernoglou: Temperature-dependent X-ray diffraction as a probe of protein structural dynamics. Nature 280, 558-63 (1979)

33.8 P.J. Artymiuk, C.C.F. Blake, D.E.P. Grace, L.J. Oatley, D.C. Phillips and M.J.E. Sternberg: Crystallographic studies of the dynamic properties of lysozyme. Nature 280, 563-68 (1979)

33.9 H. Hartmann, F. Parak, W. Steigemann, G.A. Petsko, D. Ringe Ponzi and H. Frauenfelder: Conformational substates in a protein: Structure and dynamics of metmyoglobin at 80 K. Proc. Natl. Acad. Sci. USA 79, 4967-4971 (1982)

33.10 F. Parak, E.N. Frolow, R.L. Mössbauer and V.I. Goldanskii: Dynamics of Metmyoglobin Crystals Investigated by Nuclear Gamma Resonance Absorption. J. Mol. Biol. 145, 825-833 (1981)

33.11 H.E. Huxley, A.R. Faruqi, J. Bordas, M.H.J. Koch and J.R. Milch: The use of synchrotron radiation in time-resolved X-ray diffraction studies of myosin layer-line reflections during muscle contraction. Nature 284, 140-143 (1980)

33.12 A. Gabriel: Position sensitive X-ray detector. Rev. Sci Instrum. 48, 1303-1305 (1977)

33.13 J. Hendrix: Position-sensitive X-ray detectors. In Uses of Synchrotron Biology, ed. H.B. Stuhrmann, pp. 285-320. Academic Press, London (1982)

33.14 W.A. Gilbert, J. Kuriyan, G.A. Petsko and D. Ringe Ponzi: Mapping the spatial distribution of protein fluctuations by X-ray diffraction. In Structure and Dynamics: Nucleic Acids and Proteins. Eds. E. Clementi and R.H. Sarma, pp. 405-420. Adenine Press, New York (1983)

33.15 E.-M. Mandelkow, A. Harmsen, E. Mandelkow and J. Bordas: X-ray kinetic studies of microtubule assembly using synchrotron radiation. Nature 287, 597-599 (1980)

33.16 M.F. Moody, P. Vachette, A.M. Foote, A. Tardieu, M.H.J. Koch and J. Bordas: Stopped-flow X-ray scattering: The dissociation of aspartate transcarbamylase. Proc. Natl. Acad. Sci. USA 77, 4040-4043 (1980)

33.17 L. Nilsson, R. Rigler and P. Laggner: Structural variability of tRNA: Small-angle X-ray scattering of the yeast tRNA^{Phe}-Escherichia coli tRNA$^{Glu}_2$ complex. Proc. Natl. Acad Sci. USA 79, 5891-5895 (1982)

33.18 S. Doniach, P. Eisenberg and K.O. Hodgson: X-ray absorption spectroscopy of biological molecules. In Synchrotron Radiation Research. Eds. H. Winick and S. Doniach, pp. 425-458. Plenum Press, New York (1980)

33.19 B. Chance: Personal communications. (1983)

33.20 E.B. Hughes, H.D. Zeman, L.E. Campbell, R. Hofstadter, U. Meyer-Berkhout, J.N. Otis, J. Rolfe, J.P. Stone, S. Wilson, E. Rubinstein, D.C. Harrison, R.S. Kernoff, A.C. Thompson and G.S. Brown: K-edge digital subtraction angiography with synchrotron X-rays. In Medical Applications of Synchrotron Radiation, Suppl. Acta Radiologica Scandinavica, eds. R. Rigler, R. Prins and R. Walstam, in press (1983)

33.21 G.N. Kulipanov, N.A. Mezentsev, V.F. Pindyurin, A.N. Skrinskiy, M.A. Sheromov, A.P. Ogirenko and V.M. Ogimov: Synchrotron radiation for examination of the circulatory system. In Medical Applications of Synchrotron Radiation, Suppl. Acta Radiologica Scandinavica, eds. R. Rigler, R. Prins and R. Walstam, in press (1983)

34. Quantitative Microanalysis with High Resolution Using Soft X-Rays – Possible Applications

J. Kirschner

Institut für Grenzflächenforschung und Vakuumphysik
Kernforschungsanlage Jülich, Postfach 1913
D-5170 Jülich, Fed. Rep. of Germany

The feasibility of a Scanning Photoelectron Microscope (SPM) is studied. Comparisons to scanning Auger microscopy are made, which indicate substantial advantages of SPM for the analysis of sensitive materials. Estimates of the attainable lateral resolution and signal intensities are made on the basis of a synchrotron light source and zone plate x-ray optics.

34.1 Introduction

The present contribution tries to explore the possibilities of quantitative microanalysis using soft x rays and to make comparisons to existing techniques. The emphasis is on photoelectron spectroscopy, as electron spectroscopies hold the largest potential for microanalysis due to the short inelastic mean free path of electrons. With a probe diameter of $\sim$ 30 nm the volume analyzed is of the order of 10^{-6} μm^3 or 10^{-18} cm^3. As the depth of information is of the order of 1 nm, the information is essentially two-dimensional, confined to the near-surface region of a sample. For a discussion of bulk microanalysis, e.g. by means of x-ray fluorescence, [34.1] is suggested. Among the large number of surface analysis techniques developed during the last decade, only Auger spectroscopy has become a mature microanalysis tool. Any new technique will therefore have to be compared to Scanning Auger Microscopy (SAM). The next section contains a brief description of the present state of SAM, of its merits and drawbacks. Then the prospects for photoelectron microanalysis are discussed in comparison to SAM and some possible applications are indicated.

34.2 Scanning Auger Microscopy (SAM)

The primary excitation source in SAM is a finely focused electron beam of some ten keV energy. The electrons leaving the surface are analyzed by an electrostatic analyzer, most efficiently by one of the CMA type. For optimum lateral resolution the beam forming optics are of the magnetic type, and modern SAM's therefore resemble ultrahigh vacuum scanning electron microscopes. High primary voltages are chosen because of the compromise between beam size and beam current, though they are not optimum for the excitation of Auger electrons. For acceptable signal-to-noise ratios the beam currents have to be of the order of several nA, corresponding to about 10^{10} to 10^{11} electrons/s. Using a field emission cathode, an edge resolution around 30 nm has been reported [34.2]. Even if brighter cathodes were available, not much improvement would be possible, as there is a contribution to the Auger current from electrons scattered back from the bulk. These cause a halo around the impact point, thus deteriorating the lateral resolution and sometimes falsifying Auger micrographs [34.3]. An 'Auger resolution' of 50 to 100 nm therefore may be considered a practical limit to SAM, depending on the accuracy required in the quantitative evaluation [34.4]. Quantitative analysis is generally being performed by comparison to standards of known composition. An accuracy of the order of 10% can be expected for homogeneous samples, which can be improved to a few percent under favourable conditions. The development of efficient procedures, also dealing with inhomogeneous samples, is the subject of current research.

Scanning auger microscopy is a satisfactory surface-sensitive technique for many applications, such as semiconductor characterization, analysis of VLSI circuits, materials research or corrosion. Its most important limitation is electron-beam damage to 'fragile' specimens, such as organics, glasses, adsorbates and some compounds. The most obvious cause of damage is the thermal load. With a beam of say 1 μm^2 cross section at 10^{-8} A current and 10 keV energy the power density amounts to 0.1 mW/μm^2 or 100 W/mm^2. Though this load is in general not harmful to metallic specimens, with insulators or particles in bad thermal contact with the substrate it may lead to thermal decomposition or even evaporation. While this problem may eventually be solved by appropriate sample preparation, the damage due to electronic excitations is unavoidable. It may only be reduced by lowering the dose. An example for the decomposition of a relatively stable material by electron irradiation is shown in Fig.34.1 for oxidized aluminium [34.5].

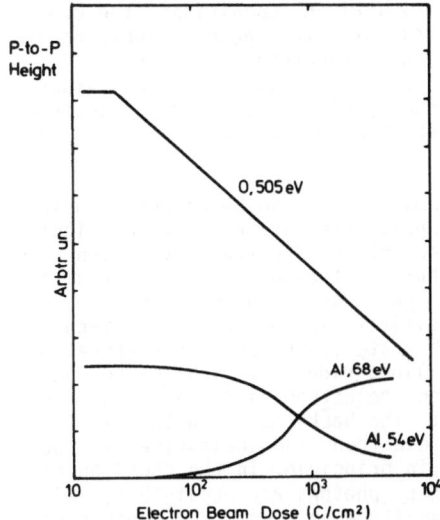

Fig.34.1 Electron-beam induced decomposition of oxidized Al at 5 keV. The Auger signals of oxygen (505 eV), oxidized Al (54 eV) and metallic Al (68 eV) are plotted versus the electron-beam dose

Figure 34.1 shows relative intensities of oxygen, oxidized Al and metallic Al as a function of electron-beam dose. After a dose of $\sim$ 50 Cb/cm^2 the oxygen and oxide signals begin to fall while the metallic Al signal rises. This dose corresponds to applying a beam of 10 nA with $\sim$ 1 μm diameter for 50 sec to the sample, a time that is hardly sufficient for taking a complete Auger spectrum. (A multichannel detection system that is compatible with the count rate requirements has not yet been implemented.) For adsorbates, the situation may be even more serious. For recent reviews of electron-beam damage in AES see [34.6,7]. This limitation of SAM is the main motivation to search for a less damaging technique offering similar lateral resolution.

34.3 X-Ray Photoelectron Spectroscopy (ESCA)

In many circumstances ESCA in the conventional manner using line radiation from Al or Mg anodes has been found to be less damaging than standard Auger spectroscopy without spatial resolution. Primarily, this is due to the low photon fluxes in conventional ESCA spectrometers, being of the order of 10^{12} photons sec^{-1} over a useful sample area of the order of 10 mm^2 [34.7]. On the basis of electronic damage per incident particle, however, the differences are not very large [34.8]. For protein the degradation was found to be roughly equal for equal fluxes of electrons and photons. In organic polymers photons of 1.5 keV produced even 3 times more damage than electrons of 10 keV [34.9,10]. The desorption of chemisorbed species on metals goes to a large extent via the generation of core holes, where the nature of the particle producing the hole is unimportant [34.11,12]. Thus one should expect photons and electrons of the commonly used energies to be about equally damaging, perhaps with a slight advantage for photons as they produce not as much kinetic secondary electrons as primary electrons of 10 keV energy. The differences may lie within a factor of three or so.

The photoionization cross sections of the most useful lines are in the range of 10^{-19} cm^2 to 10^{-18} cm^2 using Mg K$_\alpha$ radiation [34.13]. As the energy dependence of the ionization cross section is different for photons and for electrons, no direct comparison can be made. The maximum cross section for electrons occurs at about 4 times the binding energy, decreasing slowly towards higher energies [34.14]. Thus, for typical SAM conditions the cross sections are of the order of 10^{-18} cm^2. A slight advantage of electrons is compensated, however, by the fact that the Auger electrons originating from a given core hole are distributed over a number of fairly broad lines. Therefore, the signal intensities per incident particle at a given electron analyzer energy are roughly equal in Auger spectroscopy and ESCA, again within a factor of three or so.

From the above considerations, there seems not much to be gained by using a fine photon beam instead of an electron beam. There is, however, a significant difference between Auger and photoelectron spectroscopy, which is outlined in Fig.34.2. The upper panel shows a typical Auger spectrum of slightly contaminated Cu, while the lower panel shows an ESCA spectrum of clean Pt. Both elements are of average sensitivity in the respective techniques, and both spectra have been taken with state-of-the-art instrumentation [34.15,16]. It is apparent that the Auger lines ride on a high background, much larger than the signal, while the photoelectron lines are on a low background, much smaller than the signal. The background in the electron excited Auger spectrum is due to secondary electrons and inelastically scattered primary electrons. It is unavoidable in principle. In the ESCA spectrum, secondary electrons directly generated by photons are negligible, while those from photoelectrons are of similar magnitude as those from Auger electrons. The signal-to-background ratio S/B defined by the peak intensity minus the background intensity above the line divided by the background intensity is thus much more favourable in ESCA than in AES. In Auger spectroscopy S/B is generally of the order of 0.1, approaching 1 in favourable cases. In photoelectron spectroscopy S/B is of the order of 10, approaching 100 in favourable cases. This difference by roughly two orders of magnitude is of great significance. In SAM, a background measurement is mandatory, as its variation at different locations of a sample may be of the same magnitude as the signal variation itself. In ESCA, it may be neglected in many cases. Considering the counting statistics the presence of a large background means that for a given statistical uncertainty of the signal a much larger number of counts must be aquired than if the background were negligible. Assuming

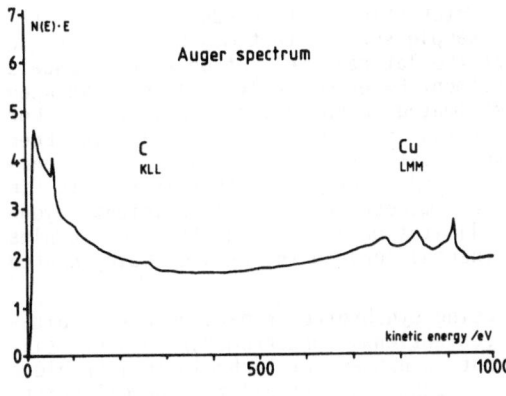

Auger spectrum

ESCA spectrum

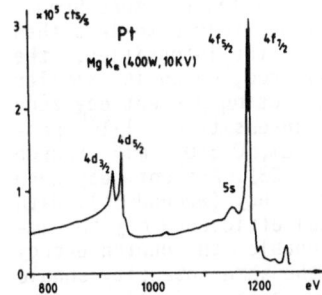

Fig.34.2 Typical spectra of Auger analysis (upper panel) and ESCA (lower panel) taken by pulse counting. Note the large differences in the signal-to-background ratios of the two techniques

the efficiency of Auger electron production by electrons to be roughly the same as that of photoelectron generation by x rays, it turns out that about one to two orders of magnitude more primary electrons than photons are required for a given statistical uncertainty. For a given lateral resolution this means that the irradiation (and with it the damage) may be reduced by the same amount when using photons instead of electrons. This is what I consider the most significant advantage of a future 'scanning photoelectron microscope'.

There is a further, more technical advantage in using photons. With primary electrons, insulating samples frequently charge up negatively, depending on the secondary electron emission coefficient being smaller or larger than 1. A negative charge is difficult to remove or to stabilize, and there is no standard procedure to deal with this problem. With photons, the charge-up is always positive and by providing a cloud of thermal electrons close to the sample it should be much easier to remove or at least to stabilize the charge-up.

34.4 Scanning Photoelectron Microscopy (SPM)

With conventional x-ray generators it appears rather difficult to produce a sufficiently intense photon beam of, say, 1 μm diameter [34.17]. An alternative approach has been first suggested by CAZAUX [34.18] and demonstrated by HOVLAND [34.19]. A sample of thickness less than the absorption length of soft x rays is placed on a thin metal foil, which is bombarded from the rear by a rastered electron beam. The characteristic x rays produced in the metal foil generate photoelectrons in the sample surface that may be analyzed in the usual manner. It was found that the lateral resolution roughly equals the total thickness of foil plus specimen. Experimentally, ~ 10 μm have been obtained, the theoretical limit is estimated to be around 1 μm [34.18]. For a thin Au layer on an Al foil count rates in excess of 10^4 sec^{-1} have been obtained on the Au $N_{6,7}$ doublet with an estimated x-ray flux of $3 \cdot 10^9$ photons/sec [34.20]. The main advantage of this approach is that it results in a laboratory-type apparatus, with very modest additional equipment beyond the electron spectrometer. The main limitation lies in the type of samples to be used (thin films), and in the thermal load to the sample, being heated by the converter foil.

Both problems could be solved by using synchrotron radiation with suitable focussing monochromators. With conventional monochromators a spot size of ~ 1 μm seems feasible [34.17], but a number of technological problems have yet to be explored and solved (mechanical stability, thermal drift, scanning). A very promising technique originated through the development of zone plates for the soft x-ray regime [34.21,22], that allow for monochromatization as well as for demagnifying the synchrotron source to less than 1 μm spot size. For an order-of-magnitude estimate of the intensities, the dedicated storage ring BESSY, Berlin, shall be considered, which is similar in performance to the Brookhaven UV storage ring. Using present day zone plate technology, with a two-plate set-up a photon intensity of $1 \cdot 10^9$ photons sec^{-1} into a 1 μm^2 spot is estimated (100 mA beam, 0.5x0.5 mm^2 source size, wavelength 4.5 nm, resolution $\lambda/\Delta\lambda = 250$) [34.23]. Present day zone plates of the absorber type have rather low efficiency (around 5%). With improved plates of the phase type the overall optical efficiency may be improved by an order of magnitude. Thus, for double the photon energy (~ 550 eV) and half the energy width (~ 1 eV) about the same intensity should be expected. As the photon energy can be optimized to core levels of low binding energy (around ~ 100 eV), count rates of the order of 10^4 cps may be anticipated. The detection efficiency of electron spectrometers can be improved by two orders of magnitude using existing multichannel detectors, which in turn could be used to reduce the spot size correspondingly. The scanning photoelectron microscope operating at the 50 nm level seems therefore feasible.

There are yet some technological problems to be solved to bring the SPM into reality: UHV technology has to be implemented into x-ray microscopy, problems of mechanical stability and reproducibility have to be considered, and the scanning technique needs to be adapted. In the SPM as a routine tool, scanning the sample would be a nuisance, so that the zone plates should be scanned.

All these problems do not seem to be insurmountable, and the SPM appears as a promising alternative to the scanning Auger microprobe. Relative to SAM the beam-induced damage could be reduced by one to two orders of magnitude. The thermal load to the sample could be reduced by three orders of magnitude (two from the reduction of the flux density, one from the use of lower energy photons). The SPM would therefore be most useful in areas that are hardly accessible to SAM due to the electron-beam effects. Glasses, ceramics, polymers, ionic crystals, and adsorbate systems could be analyzed on a microscopic scale.

Acknowledgements

Stimulating discussions with G. Schmahl and D. Rudolph are gratefully acknowledged. Thanks are also due to G. Pirug for advice and the provision of the ESCA spectrum of Fig.34.2.

References

34.1 H. Winick, S. Doniach (eds.): Synchrotron Radiation Research (Plenum Press, New York, London, 1980)

34.2 A.P. Janssen, J.A. Venables: Surface Sci. 77, 351 (1978)

34.3 J. Kirschner: Scanning Electron Microsc. 1976 I, 215

34.4 J. Cazaux: Surface Sci. 125, 335 (1983)

34.5 A. van Oostrom: Surface Sci. 89, 615 (1979)

34.6 C.G. Pantano, T.E. Madey: Appl. Surf. Science 7, 115 (1981)

34.7 R.G. Copperthwaite: Surface Interface Anal. 2, 17 (1980)

34.8 M.S. Jsaacson, in: EXAFS Spectroscopy, B.K. Teo, D.C. Joy (eds.) (Plenum Press, New York, London, 1981) 269

34.9 J. Kirz, D. Sayre, in: Synchrotron Radiation Research, H. Winick, S. Doniach (eds.) (Plenum Press, New York, London, 1980) chapter 8

34.10 M. Utlaut in: EXAFS Spectroscopy, B.K. Teo, D.C. Joy (eds.) (Plenum Press, New York, London, 1981) 255

34.11 J. Kirschner, D. Menzel, P. Staib: Surface Sci. 87, L267 (1979)

34.12 R. Franchy, D. Menzel: Phys. Rev. Lett. 43, 865 (1979)

34.13 J.H. Scofield: J. Electron Spectr. Rel. Phen. 8, 129 (1976)

34.14 J. Kirschner, in: Topics in Current Physics, Vol. 4 , H. Ibach (ed.) (Springer Verlag, Berlin, Heidelberg, New York, 1977) 63

34.15 Perking Elmer Model 560 ESCA/SAM

34.16 Leybold Heraeus, Model EA 10

34.17 T.A. Carlson: Surface Interface Anal. 4, 125 (1982)

34.18 J. Cazaux: Rév. Phys. Appl. 10, 263 (1975)

34.19 C.T. Hovland, in: Proceedings 7th Int. Vac. Congr. & 3rd Int. Congr. Solid Surf., Vienna, 1977, R. Dobrozensky et al. (eds. and publishers) Vol. 3, p. 2363

34.20 J. Cazaux, D. Mouze, J. Perrin, in: Proc. 8th Intern. Vacuum Congr. Cannes, 1980 (Le Vide, Les Couches Minces 201, Suppl. 1) 291

34.21 G. Schmahl: Nucl. Instr. Meth. 208, 361 (1983)

34.22 B. Niemann, D. Rudolph, G. Schmahl: Nucl. Instr. Meth. 208, 367 (1983)

34.23 G. Schmahl, D. Rudolph: private communication

35. On the Possibility of Imaging Microstructures by Soft X-Ray Diffraction Pattern Analysis

D. Sayre   IBM Th.J. Watson Research Center, P.O. Box 218
Yorktown Heights, NY 10598, USA
R. P. Haelbich   IBM Th. J. Watson Research Center, P.O. Box 218
Yorktown Heights, NY 10598, USA
Permanent address: Zweites Institut für Experimentalphysik
Universität Hamburg, D-2000 Hamburg, Fed. Rep. of Germany
J. Kirz and W. B. Yun   Physics Department, State University of New York
Stony Brook, NY 11794, USA

As two-dimensional x-ray microscopy develops, it is of interest to consider also possible three-dimensional imaging methods, in view of the thick-specimen capability of soft x rays. The highest-quality 3-dimensional images of atomic assemblies today are produced by Fourier inversion of 1A x-ray or neutron diffraction patterns of crystalline specimens. Here we consider whether a similar form of imaging may be obtained with noncrystalline specimens using soft x rays.

As x-ray wavelengths increase from 1A, elastic scattering persists, Compton scattering decreases, and photoelectric absorption increases. The first and third of these give rise to coherent scattering and are effective in diffraction, while the second contributes an incoherent background and does not contribute to diffraction. Thus we should expect that diffraction will exist in the soft x-ray region and with lower intrinsic background than in the shorter wavelength region.

We now consider the expected intensity of the diffraction. For sufficiently small objects that beam attenuation in incident and diffracted beams may be ignored, the diffracting cross section of a specimen in direction $\bar{u}$, with incident beam direction $\bar{u}_0$ and wavelength λ, is given by

$$\left(\frac{d\sigma}{d\Omega}\right)_{\bar{u},\bar{u}_0,\lambda} = r_0^2 \; |F\left(\frac{\bar{u}-\bar{u}_0}{\lambda}\right)|^2 , \qquad (35.1)$$

where the function F is determined by the diffracting structure and is given by

$$F(\bar{h}) = \Sigma \; f_j \; e^{i2\pi\bar{h}\cdot\bar{r}_j} . \qquad (35.2)$$

Here r_0 is the classical radius of the electron, and the $\bar{r}_j$ and f_j are the coordinates and atomic scattering factors of the atoms composing the specimen. Values of the scattering factors for soft x rays for all atomic species have recently been tabulated [35.1]. In the case where the specimen is described by its scattering density $\rho(\bar{r})$,

$$F(\bar{h}) = \int \rho(\bar{r}) \; e^{i2\pi\bar{h}\cdot\bar{r}} \; d^3\bar{r}. \qquad (35.3)$$

In the large-angle region ($|\bar{h}|>1/a$, where a is the edge length of the specimen), (35.2) yields $\langle|F(h)|^2\rangle = \Sigma_j f_j^2$ (see any standard text in crystallography). We therefore have

$$\langle\left(\frac{d\sigma}{d\Omega}\right)_{\bar{u},\bar{u}_0,\lambda}\rangle = r_0^2 \; \Sigma \; f_j^2 \qquad (35.4)$$

for the average differential cross section of diffraction. The total number of photons diffracted into the large-angle pattern per unit time is thus

$$n_{\text{large angle}} = 4\pi\Phi_0\, r_0^2\, \Sigma\, f_j^2, \tag{35.5}$$

where Φ_0 is the incident photon intensity.

We note that the elementary Shannon volume in diffraction space is of order $1/a^3$, while the volume of diffraction space accessible with wavelength λ is of order $1/\lambda^3$. Thus in a complete measurement of the diffraction pattern the number of elementary volumes among which the photons given by (35.5) must be divided is of order a^3/λ^3. The average number of photons diffracted into an elementary volume in unit time is thus

$$n_{\text{elem.vol., av.}} \approx \Phi_0\, r_0^2\, (\lambda/a)^3\, \Sigma\, f_j^2, \tag{35.6}$$

the $\approx$ indicating that a numerical constant has been omitted. The time needed to measure the diffraction pattern to a given level of accuracy will be inversely proportional to this quantity.

For low-Z atoms, f_j can be replaced with sufficient accuracy in the soft x-ray region by the atomic number Z_j, and Σf_j^2 by NZ^2 in the case where the specimen consists of N atoms of atomic number Z. In particular (35.6) then becomes

$$n_{\text{elem.vol., av.}} \approx \Phi_0\, r_0^2\, \lambda^3\, mZ^2, \tag{35.6'}$$

where m is the average number of atoms per unit volume.

Assuming $\Phi_0 = 10^{18}$ photons/cm^2 sec, $\lambda = 30$A, m = $0.05/$A^3, Z = 8, we find $n_{\text{elem. vol., av.}} \approx 10^{-2}$ photons/sec, meaning that a diffraction pattern could be collected with an average of 1000 counts per elementary volume, or 3% accuracy, in approximately a day. 10^{18} photons/cm^2 sec is the predicted value for Φ_0 for soft x-ray undulators now being planned.

The present output of the U15 beam line at Brookhaven is 10^{13} photons/cm^2 sec. If the specimen is kept stationary in the incident beam, the number of elementary volumes among which the large-angle photons are divided decreases to approximately a^2/λ^2. In this case $n_{\text{elem.vol.}}$ increases by a factor of a/λ, i.e.,

$$n_{\text{elem.vol., instantaneous}} \approx \Phi_0\, r_0^2\, \lambda^2\, amZ^2. \tag{35.6''}$$

This increase can be several orders of magnitude, meaning that there is a possibility that regions of the diffraction pattern which are above average in intensity may be seen even on this beam line, if the specimen is kept stationary.

To test this latter possibility we have constructed a simple diffraction apparatus for use on U15. The incident beam is defined by a 10 μm pinhole. About 2 mm further downstream is a larger (80 μm) pinhole, to screen the large-angle pattern from parasitic radiation. The specimen is mounted on an EM grid, which in turn is mounted on the 80 μm pinhole, and about 1/2'' downstream from the specimen is a 1''x1'' photographic detector. The camera is evacuated to approximately 10^{-5} torr. To date one picture possibly showing diffraction from the specimen (in this case a thin section of nerve) has been taken. More work is needed, however, before it can be concluded that diffraction has actually been observed.

315

Should it prove possible to measure a diffraction pattern, it would still be necessary to phase it (obtain the function F from the measured function $|F|$), probably by an adaptation of the isomorphous replacement or anomalous dispersion phasing techniques of protein crystallography. These would require the collection of one or more additional diffraction patterns. The Fourier inverse of (35.3) would then be used to compute the image $\rho(\bar{r})$.

Some characteristics of this image, if it is produced, may be mentioned. First, provided that the full diffraction pattern (to diffraction angles 2θ approaching $180°$) can be measured and phased, the resolution of the image will be $\lambda/2$, or 15A for 30A x rays. The three-dimensionality of the image has already been noted. Finally, the image will in general be complex-valued, owing to the presence of both elastic scattering and photoelectric absorption in the specimen. At a given wavelength, the amounts of real and imaginary image will vary from point to point according to the local atomic composition, and it should be possible to use this effect to estimate the composition in each resolution element of the specimen. The amounts of real and imaginary image will also vary with x-ray wavelength, and this should be of further help in establishing local composition in the specimen.

A few final points should be noted. The diffraction pattern will in general be continuous, not discrete as with crystalline specimens, and this will have a harmful effect on signal/noise, making noise reduction in the experiment especially important. Second, there may be a serious problem of radiation damage to the specimen, arising from (a) the inefficiency of photon absorption as a source of diffraction and (b) the time necessary to explore the full set of directions $\bar{u}, \bar{u}_0$ represented in a three-dimensional diffraction function. It may be necessary in some cases to protect the specimen through the use of chemical fixatives or low temperatures, or to curtail the set of directions studied and accept a reduction in the quality of the three-dimensional imaging. Third, the holographic approach to imaging is closely related to the one described here. Both depend on the existence of the diffraction pattern, but holography phases it differently, and normally does use a curtailed set of $\bar{u}, \bar{u}_0$ directions with resulting loss in the three-dimensional image quality.

We are grateful to M. Howells, J. Kenney, and H. Rarback for their help and suggestions concerning the design and test of the apparatus. The work of WBY has been supported by an IBM predoctoral fellowship.

Summary

We suggest that single microscopic objects (such as a biological cell or cell fragment) are large enough to act as diffraction specimens for soft x rays, and are hence potentially soft x-ray imagable in three dimensions and at resolutions of the order of 15A.

Reference

35.1 B.L. Henke et al.: "The Atomic Scattering Factor, $f_1 + if_2$, for 94 Elements and for the 100 to 2000 eV Photon Energy Region", in American Institute of Physics Conference Proceedings No. 75 (Low Energy X-Ray Diagnostics - 1981), eds. D.T. Attwood and B.L. Henke, pp. 340-388

Part VI

X-Ray Holography

36. Possibilities for X-Ray Holography Using Synchrotron Radiation

M. R. Howells

National Synchrotron Light Source, Brookhaven National Laboratory
Upton, NY 11973, USA

36.1 Introduction

Since the theoretical [36.1] and experimental [36.2,3] demonstrations of the effectiveness of soft x rays in imaging biological material there has been considerably study [36.4,5] given to the prospects for further development of the presently existing techniques. This has been motivated to a large extent by advances in source technology, particularly the use of undulators on electron storage rings and recent improvements in short wavelength lasers. The present authors have carried out theoretical evaluations of the possibilities of holographic imaging and have also recorded a number of holograms using the U15 soft x-ray beam line at the National Synchrotron Light Source (NSLS) 750 MeV storage ring at Brookhaven. These have been successfully reconstructed using He:Cd laser light.

In this paper we first review the physical processes which generate information containing wave fronts when soft x rays interact with matter. We then briefly describe the holographic method which has been highly developed using visible light lasers and identify holographic geometries which are promising for x-ray applications. We discuss some of the practical and theoretical limitations involved in making holographic images and then give the results of our own experiments.

36.2 Soft X-Ray Interactions with Matter

The situation of interest is represented in Fig.36.1. Soft x rays are incident from the left as a plane wave of unit amplitude. They interact with the sample which is represented by a refractive index distribution $n(\underline{r}) = 1 - \delta(\underline{r}) - i\beta(\underline{r})$ and are received by the detector where we wish to make a holographic recording. We know that in free space the amplitude $\psi(\underline{r})$ of the wave must satisfy the Helmholtz equation

$$\nabla^2\psi + k_o{}^2\psi = 0 \tag{36.1}$$

where k_0 is the free space propagation constant. In the sample k_0 becomes nk_0 and (36.1) becomes

$$\nabla^2\psi + k_o{}^2\psi = (1-n^2)k_o{}^2\psi = \upsilon\psi. \tag{36.2}$$

This is a familiar equation and it is a standard result [36.6] that the

---

1) This work was supported by the U.S. Department of Energy under contract number DE-AC02-76-CH00016

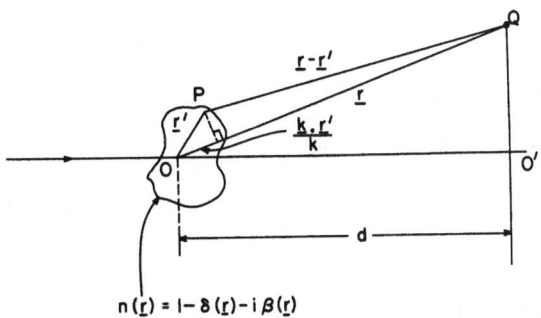

$$n(\underline{r}) = 1 - \delta(\underline{r}) - i\beta(\underline{r})$$

Fig.36.1 Notation for discussion of soft x-ray interactions with a refractive index distribution $n(\underline{r})$

solution is

$$\psi(\underline{r}) = e^{ik_0 z} - \frac{1}{4\pi} \int_{sample} \frac{\exp ik_0|\underline{r}-\underline{r}'|}{|\underline{r}-\underline{r}'|} U(\underline{r}) \, \psi(\underline{r}') \, d^3\underline{r}' . \qquad (36.3)$$

For particle scattering by a potential $V(\underline{r})$, $U(\underline{r})$ would be $(8\pi^2 m/h) V(\underline{r})$. For scattering of x rays by a charge distribution of $\rho(\underline{r})$ electrons per unit volume, $U(\underline{r})$ is $4\pi r_0 \rho(\underline{r})$ where r_0 is the classical electron radius. For the present case $U(\underline{r}) = (1-n^2)k_0^2$. In all these cases (36.3) tells us that at each volume element $d^3 r'$ the wave is depleted by the scattering of a secondary wavelet of strength, $-U(\underline{r}')/4\pi$ times the value of the incident amplitude at that point. In our case $U(\underline{r})$ in general is complex so that both absorption and scattering are expected to occur. For conventional x-ray diffraction (36.3) is interpreted by making two idealizations: (i) the Fraunhofer approximation, $w^2/\lambda d \ll 1$, where w is the width of the sample; and (ii) the Born approximation, the use of the unmodified incoming wave as an approximation to $\psi(\underline{r}')$ in (36.3). It is helpful at this point to define unit vectors $\underline{s}_0$ and $\underline{s}$ in the incoming and outgoing (OQ) directions, respectively. With these assumptions $k_0|\underline{r}-\underline{r}'|$ in the exponent becomes $k_0 r - k_0 \underline{s} \cdot \underline{r}'$ and using the definition $\underline{K} = k_0(\underline{s}_0 - \underline{s})$ (36.3) can be written (dropping the prime)

$$\psi(\underline{r}) = e^{ik_0 z} - \frac{e^{ik_0 r}}{4\pi r} \int_{sample} U(\underline{r}) e^{ik_0 \underline{K} \cdot \underline{r}} \, d^3\underline{r} . \qquad (36.4)$$

(36.4) is of the form of an asymptotic solution $e^{ik_0 z} + f(\theta_x, \theta_y) e^{ik_0 r}/r$ so we can identify $f(\theta_x, \theta_y)$ and see that

$$\frac{d\sigma(K_x, K_y)}{d\Omega} = |f(\theta_x, \theta_y)|^2 = \frac{1}{16\pi^2} |\tilde{U}(K_x, K_y)|^2 \qquad (36.5)$$

where

$$K_x = k_0 \sin\theta_x, \quad K_y = k_0 \sin\theta_y , \quad K_z = -k_0(1 - \sqrt{1 - \sin^2\theta_x - \sin^2\theta_y}) \qquad (36.6)$$

and Q (Fig.36.1) is the point ($r \sin\theta_x$, $r \sin\theta_y$). This defines the intensity at Q (x,y) in terms of the original function $n(\underline{r})$ within the limitations of the Fraunhofer and Born approximations. To see the relation of our refractive index picture to the atomic picture implied in $\rho(\underline{r})$, we note that according to HENKE [36.7] δ and β are related to f_1 and f_2, the real and imaginary parts of the atomic scattering factor, by

$$\delta = \frac{r_o\lambda^2}{2\pi} n_o f_1 \quad , \quad \beta = \frac{r_o\lambda^2}{2\pi} n_o f_2 \tag{36.7}$$

where n_0 is the number of atoms per unit volume. So

$$U(\underline{r}) = k_o^2(1-n^2) \simeq 2k_o^2(\delta+i\beta) \; ; \tag{36.8}$$

inserting (36.7) into (36.8) gives

$$U(\underline{r}) \simeq 4\pi r_o n_o(f_1+if_2) \equiv 4\pi r_o \rho(\underline{r})$$

as it should, where now $\rho(\underline{r})$ is understood as the effective number of electrons per unit volume.

Of course it is not normally possible to obtain $U(\underline{r})$ from $|\tilde{U}(K_x, K_y)|^2$ without prior knowledge of $U(\underline{r})$ and such prior knowledge is not expected to be available in microscopy. In order to obtain $U(\underline{r})$ from $|\tilde{U}(K_x, K_y)|^2$ we need a method of determining the phase of $\psi(\underline{r})$ at each point Q. The holographic approach to this is to beat the waves $\psi(\underline{r})$ against a standard reference wave. The resulting interference fringes give a full record of the phases of $\psi(\underline{r})$ at all points on the detector. In addition, if the recording is suitably made, it can be used as a diffracting structure. In this case illumination by the original reference wave gives the diffraction pattern of the diffraction pattern and reconstructs the original optical density distribution of the sample. We will enlarge on this later. The process is similar to the Bragg microscope [36.8] and to optical transform methods [36.9] used in crystallography. If the holographic recording is inverted optically then it is not subject to the limitations of the Fraunhofer and Born approximations and would in principle allow a perfect reconstruction if all spatial frequencies K_x, K_y present in the sample were recorded. However, as we shall see there are some fundamental difficulties in providing reference waves over the entire Ewald sphere. Generally speaking the holographic method should be complementary to the alternative of measuring the entire diffraction pattern without the use of reference waves. For holography, high-frequency information is lost due to incomplete collection of the diffraction pattern. For the diffraction approach the methods for determining the phases will be the limitation.

36.3 Resolution Limitations Due to Beam Geometry

From (36.6) we see directly that a feature of width δx corresponding roughly to a frequency $2\pi/\delta x$ leads to a diffracted beam at θ_x where

$$K_x = \frac{2\pi}{\delta x} = \frac{2\pi}{\lambda} \sin\theta_x \; . \tag{36.9}$$

We recognize the grating equation $\lambda = \delta x \sin\theta_x$ and we see clearly the relation between transverse resolution and the diffracted beam angle, i.e., for δx to be small, as desired, θ_x must be large.

We could make a similar argument from (36.6) for K_z and get the limit of the longitudinal resolution. However, we prefer to make an optical argument. It is a standard result of geometrical optics that the wave front aberration (δW) for the edge rays of a pencil of half angle θ whose source point is subject to a longitudinal focal shift δz, is given by

$$\delta W \simeq \tfrac{1}{2} \theta^2 \delta z \quad (\theta \text{ small}).\tag{36.10}$$

If we accept that the Rayleigh quarter wave criterion gives the range of wave front deformations which are not detectable at the image then we can set $\delta W = \lambda/4$ and get a δz which is a measure of the longitudinal resolution

$$\delta z \simeq \frac{\lambda}{2\theta^2} .\tag{36.11}$$

To get an idea of scale suppose $\theta = 0.1$ radians and $\lambda = 30$ A. We find $\delta x, \delta y \simeq 300$ A but $\delta z \simeq 1500$ A. Using (36.6) gives the similar result $\delta z \simeq 3000$ A. So we see without even considering how to do holography that its potential for gathering three-dimensional information is strictly limited by the collection angle and that through (36.9) and (36.11) or (36.6) the transverse and longitudinal resolution are related.

36.4 The Holographic Method [36.10-13]

A generalized view of holographic layout is shown in Fig.36.2. The subject is illuminated by waves from a source. At the same time a reference object is illuminated by the same source in a manner which maintains coherence over the whole wave front reaching both objects. The waves are scattered by the objects so that a wave front from the subject arrives at the hologram recording plane, leading to an amplitude distribution $a_S(x,y)$ in that plane. Similarly another wave front from the reference object gives an amplitude $a_R(x,y)$. The amplitudes add and give rise to an intensity distribution $I = (a_S + a_R)(a_S{}^* + a_R{}^*)$. This is recorded, for example, photographically and developed to give an amplitude transmittance (t) proportional to the exposure, i.e.,

$$t \propto I = a_S a_S{}^* + a_R a_R{}^* + a_S a_R{}^* + a_S{}^* a_R .\tag{36.12}$$

Now suppose we illuminate the hologram with the original reference wave. The amplitude just pass the hologram w is given by

$$
\begin{aligned}
w = a_R t \propto\ & a_R |a_S|^2 && \text{intermodulation term} && \left.\begin{array}{l}\text{Zero-}\\\text{order}\\\text{terms}\end{array}\right\} \\
& + a_R |a_R|^2 \\
& + a_S |a_R|^2 && \text{VIRTUAL IMAGE TERM} \\
& + a_R a_R a_S{}^* && \text{Confusing wave .}
\end{aligned}\tag{36.13}
$$

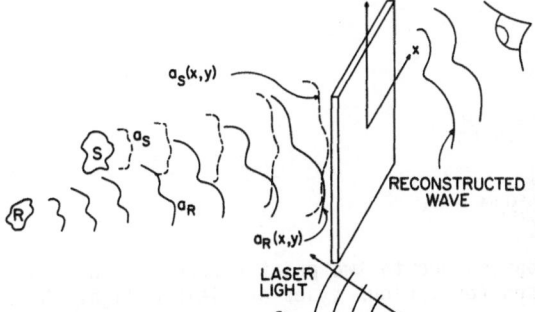

Fig.36.2 Combination of a subject and reference wave front to give an interference pattern. The amplitude functions $a_S(x,y)$ and $a_R(x,y)$ refer to amplitude distributions at the hologram plane

321

The third term is an aberration free reconstruction of the wave front which previously was emitted directly by the subject. The fourth term, can, under suitable conditions, lead to formation of a real image of the subject but from the point of view of observing the virtual image it is a nuisance.

We may note the following general points: (i) (36.12) is symmetrical in a_S and a_R. So illumination of the hologram by either one reconstructs the other. (ii) Although, according to the above, any wave can be the reference wave, it is much more convenient to use plane or spherical waves. These can be recreated anytime without needing to re-use the reference object. (iii) Reconstruction need not be with the original reference wave. If the reference source was a point (including the point at infinity) then any point source will reconstruct an image of the subject. However, if the wavelength or the source to hologram distance or the off-axis angle is changed then, in general, the magnification will no longer be unity and aberrations are introduced. Similarly, if the hologram is scaled up or down in size the same effects occur. This is important for us because we wish to make a recording with x rays and do the reconstruction in the visible in order to get magnification. There are some special cases where aberrations are small or zero and we return to these later.

36.5 Hologram Recording Geometries

If we make the simplification of considering a point subject and point reference source then we can draw a useful representation of the interference fringes that would result if both sources emit coherently in all directions. This is shown in Fig.36.3. We see that there is a rich variety of geometries with which visible light holograms can be recorded. Only three of these are of interest for x-ray use: the Gabor in-line hologram [36.14], the Leith-Upatnieks off-axis hologram [36.15] and the Fourier transform hologram [36.16]. We consider them in turn.

The Gabor in-line hologram uses the light that happens to be transmitted through the sample as the reference beam. For the Gabor hologram, each point

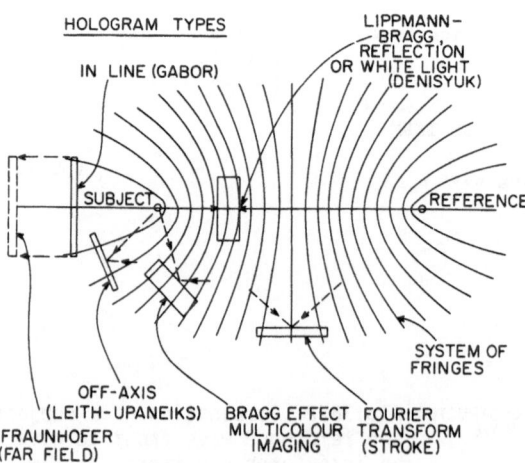

Fig.36.3 Idealized fringe distribution due to two point sources with indication of the regions used for various types of visible light holograms

of the subject contributes an amplitude distribution in the hologram which is a complete circular Fresnel zone plate pattern centered near the axis. The advantage is that no optics are needed except a pinhole. The most obvious disadvantages are that the sample must be chosen to transmit suitably and there is no good solution to the problem of separating the wanted wave front from the confusing wave. This problem can, however, be much alleviated by placing the recording surface in the far field of the sample [36.17]. The far-field or Fraunhofer in-line geometry was the one used by AOKI and KIKUTA [36.30] in recording what are probably the best x-ray holograms made with a microfocus tube.

The Leith-Upatneiks hologram is similar to the Gabor but an off-axis reference beam is used. In this case each point of the subject contributes an off-axis zone plate. The center may be off the recording area and so higher numbered, more closely spaced zone plate rings are utilized. The advantage is that separation of the virtual image wave and the confusing wave can be achieved. The disadvantage is that a higher resolution detector and a source with greater coherence length are needed. Some optics are also required.

The Fourier transform hologram is a radically different approach. Here the reference is typically a point source and must be set at a distance from the hologram equal to that of the subject. The subject-reference distance is chosen about equal to the subject width which in microscopy would be small. The interfering beams thus have only a small angle between them so rather coarsely spaced Youngs fringes are recorded on the detector. Consequently the resolution is not limited by the detector resolution but rather by the actual size of the nominal point source used as a reference or by the considerations of Section 36.3. Detector resolution plays a role in limiting the field of view.

In Fig.36.4 we indicate ways in which the three recording geometries just discussed might be used to make x-ray holograms.

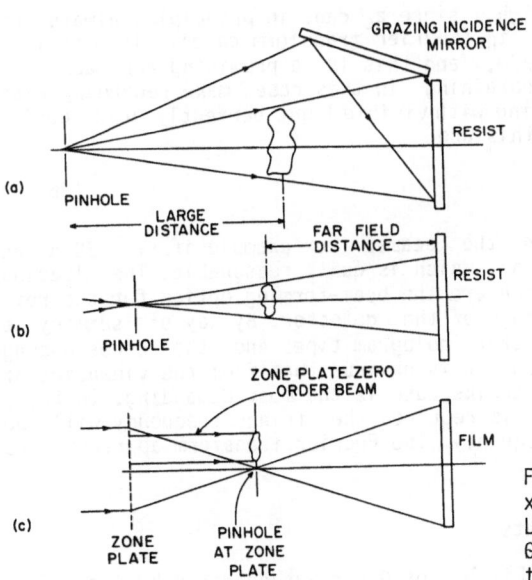

Fig.36.4 Possible geometry for x-ray holography utilizing (a) Leith-Upatneiks geometry, (b) Gabor geometry, (c) Fourier transform geometry

36.6   The Paraxial Optics of Point Source Holograms

Consider a subject point (x_1,y_1,z_1) and a reference point (x_R,y_R,z_R) in a coordinate frame defined so that the recording surface is the plane $z = 0$. Suppose a hologram is formed with light of wavelength λ_1 and reconstructed with light of wavelength λ_2 $(\lambda_2/\lambda_1 = \mu)$. Let the reconstruction source be at (x_c,y_c,z_c) and suppose the real and virtual images are at (x_{3R},y_{3R},z_{3R}) and (x_{3V},y_{3V},z_{3V}) respectively. Furthermore, let the hologram be scaled by a factor or before reconstruction. It is now a standard result [36.10.11] that the positions of the images are given by

$$\frac{1}{z_{3V,R}} = \frac{1}{z_c} \pm \frac{\mu}{m^2 z_1} \mp \frac{\mu}{m^2 z_R} . \tag{36.14}$$

The upper signs refer to z_{3V}, the lower to z_{3R}. Also the lateral magnification M_{LAT} is given by:

$$M_{LAT,V,R} = \frac{m}{1 \pm \frac{m^2 z_1}{\mu z_c} \mp \frac{z_1}{z_R}} = \frac{\mu}{m}\frac{z_{3V,R}}{z_1} \tag{36.15}$$

where a similar sign convention applies.

Since we are interested in microscopy (36.15) is an important formula. It was by special choices of the parameters in (36.15) that Gabor originally intended to achieve high magnification by his 'projection method'. We are also seeking to use a high value for μ and achieve magnification. However, it is clear by inspection of (36.15) that high μ alone does not guarantee a high value of M_{LAT}. There are two ways to achieve this: with $z_1 \neq z_R$ (Gabor and Leith-Upatneiks case) and supposing m = 1 then $M_{LAT,R} \to \infty$ if

$$\frac{z_1}{\mu z_c} + 1 + \frac{z_1}{z_R} . \tag{36.16}$$

This does not really depend on high μ since z_c can, in principle, always be chosen to satisfy (36.16). $z_1 = z_R$ (Fourier transform case). Then setting $z_c = z_1 = z_R$ leads to $M_{LAT,V} = \mu/m$, and this is a promising approach for x-ray work. The coarse fringes obtaining in this case make recording with photographic film and reconstructing with visible light perfectly practicable. We note that the conditions for this are

$$\text{Fringe spacing} = \frac{\lambda_1}{\phi} > \lambda_2 \tag{36.17}$$

where ϕ is the angle between the beams. For example if $\lambda_1 = 30$ A and $\lambda_2 = 6328$ A then we find $\phi > 5$ mrad, which is quite reasonable. The leading problems remaining in this approach are the beam-forming optics for the reference source and the dynamic range of the detector. By way of summary we show in Fig.36.5 the three important hologram types and the corresponding fringe frequencies in the hologram. It is obvious that from the viewpoint of detector resolution the Leith-Upatneiks case is the most demanding. In fact, if separation of the two images is required the fringe frequency will be four times the highest sample frequency. The Fourier transform approach has a great advantage in this respect.

36.7   Source Coherence Requirements

The original 'protection method' [36.14] of Gabor was defeated by lack of a suitable source. Even the visible light work which was successfully carried

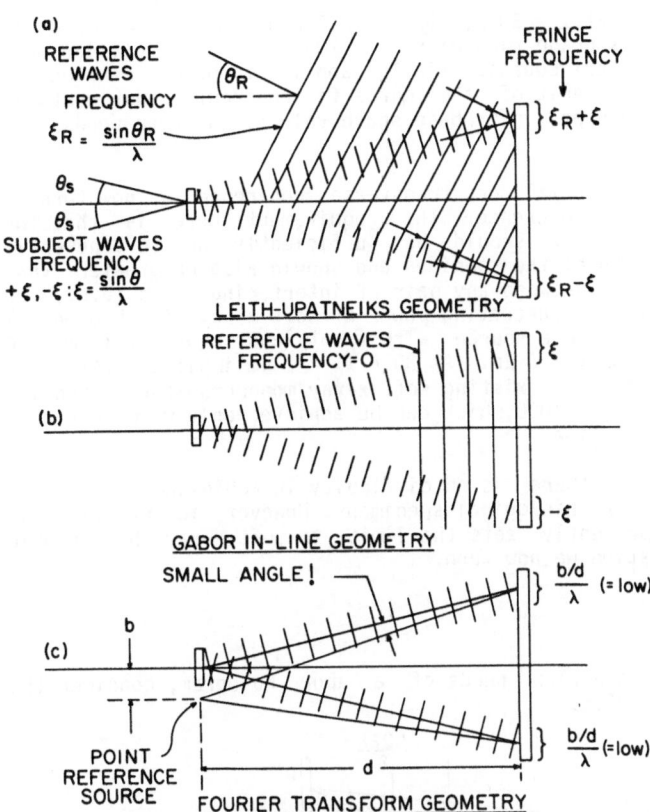

REFERENCE
WAVES
FREQUENCY
$\epsilon_R = \dfrac{\sin\theta_R}{\lambda}$

θ_R

FRINGE
FREQUENCY

$\}\epsilon_R + \epsilon$

θ_S

θ_S

SUBJECT WAVES
FREQUENCY
$+\epsilon, -\epsilon : \epsilon = \dfrac{\sin\theta}{\lambda}$

$\}\epsilon_R - \epsilon$

LEITH-UPATNEIKS GEOMETRY

REFERENCE WAVES
FREQUENCY=0

$\}\epsilon$

(b)

$\}-\epsilon$

GABOR IN-LINE GEOMETRY

SMALL ANGLE !

$\}\dfrac{b/d}{\lambda}$ (= low)

b

(c)

$\}\dfrac{b/d}{\lambda}$ (=low)

POINT
REFERENCE
SOURCE

d

FOURIER TRANSFORM GEOMETRY

Fig.36.5 Fringe frequencies produced by the three main holographic geometries: (a) Leith-Upatneiks, (b) Gabor, (c) Fourier transform

out in the forties and fifties did not make a great impact until the inven-
tion of the laser. Similarly the development of the x-ray holography has
been limited up to the present time by lack of a suitable source. However, we
believe this is changing. So far we do not have any sources which generate
soft x rays in a way which naturally makes all of the radiation coherent in
the manner of a laser. Even on this matter, however, there are possibilities
for progress [36.18]. For the time being we must still use methods which
isolate the coherent part of the output of an incoherent source but much
better sources are becoming available.

According to the Van Cittert-Zernike theorem [36.19] the output of any
source is spatially coherent within a region defined by the first minimum of
the Fraunhofer diffraction pattern of the spatial distribution of the
source. For example, if the source is a circular pinhole of radius x then
spatial coherence exists within the Airy cone (of half angle x'). This sets
an upper limit on the emittance ϵ of a coherent beam given by [36.20]

$$\epsilon_{max}^c = xx' = 0.61\lambda .$$

(36.18)

This is a very small emittance and (36.18) sets a strong limitation on the
fraction of the output of most sources which is useable for coherent

experiments. For example at λ = 30 A, $\epsilon_{max} \simeq 18.0$ A·Radians $\equiv 1.8$ mrad·µm. We see that for a good storage ring with x $\simeq$ 100 µm we can utilize only $2x' \simeq 2\epsilon_{max}/100 = 0.036$ mrad! Equation (36.18) applies to both dimensions of the beam so the figure of merit of the source for coherent experiments is the flux per unit bandwidth per unit phase space volume. In other words the brilliance.[2]

This takes care of spatial or transverse coherence. We now turn to temporal or longitudinal coherence. The requirement here is that the coherence length $\ell_c = \lambda^2/\Delta\lambda$ should be sufficiently long to coherently illuminate the entire depth of the specimen and should also be greater than the largest path difference between any pair of interfering beams (e.g., PQ in Fig.36.6). In view of the penetrating power of soft x rays in biological material we would like to have a source with ℓ_c at least equal to 1 µm and preferably in the region of 3-5 µm. At 30 A ℓ_c = 1 µm implies $\lambda/\Delta\lambda \simeq 300$ which is readily achieved using existing soft x-ray monochromator technology. For ℓ_c > 3 µm then $\lambda/\Delta\lambda$ > 1000, which can be achieved only with considerable effort at the present time.

We see that in principle there is no difficulty in achieving the needed coherence for microscopy of biological specimens. However, in practice, it is lack of flux which presently sets the limit to soft x-ray holographic imaging and to this question we now turn.

36.8 Flux Requirements

In order to understand the flux needs of a Gabor hologram, consider the arrangement shown in Fig.36.6.

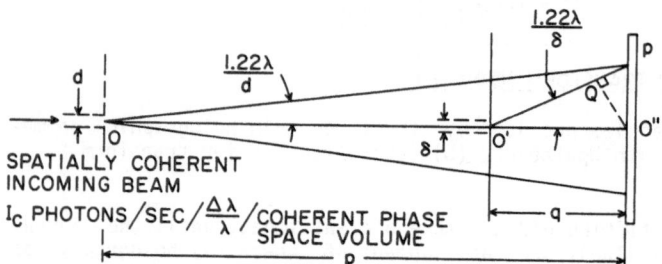

Fig.36.6 Layout of imaginary experiment to compute the flux needs of a Gabor hologram

A spatially coherent beam of flux I_c [photons/sec/$\Delta\lambda$/λ] is focussed on to a pinhole of diameter d at O distance p from the detector. By definition the beam will just fill the Airy cone of the pinhole. This beam is used to illuminate a subject consisting of only one element: a circular hole of diameter δ. The latter is surrounded by an absorbing screen which transmits just enough to provide a reference beam. The Airy cone of δ illuminates an area of the detector of radius r_n = 1.22·λq/δ, where q is the working distance. The result is the creation of a Fresnel zone plate pattern of radius r_n. According to (36.9) we must utilize the full Airy cone of δ if we are to reconstruct it without loss of resolution. We therefore choose a large value of p which satisfies

---

2) known to optics specialists for many years as brightness

$$\frac{p}{d} = \frac{q}{\delta}$$ (36.19)

so that the coherent beam exactly fills the circle of radius r_n. The focal length f of the zone plate is actually given by

$$\frac{1}{f} = \frac{1}{q} + \frac{1}{p}$$ (36.20)

but we use the approximation $f \simeq q$.

Now in order for the zone plate to work properly in reconstruction of the subject δ we need each distinguishable element to be registered with adequate signal to noise ratio. Suppose signal/noise = 5 is considered adequate. Then each element of detector of area Δr^2 (Δr is the outer ring spacing of the zone plate) should end up registering at least 25 photons from the subject. This will be achieved if

$$I_c \left(\frac{\delta^2}{4r_n^2}\right)\left(\frac{\Delta r^2}{\pi r_n^2}\right).2Qt \geq 25$$ (36.21)

where Q is the detective quantum efficiency of the detector, t is the time, and the two arises because only the bright fringes need to receive light. Now we know from zone plate theory that $\delta \simeq 2.44 \cdot \Delta r$[3] and for both film and resist Q = 0.2 is a reasonable estimate, so we end up with an estimate of the fluence $I_c t$ needed to record the elementary hologram. (Of course if the sample had additional elements these would also contribute zone plate amplitude patterns without needing any more flux.)

$$I_c t \geq K_1 \frac{\lambda^4 q^4}{\delta^8}$$ (36.22)

where $K_1 = 1.0 \times 10^4$. This may also be written

$$I_c t \geq K_2 n^4$$

where n is the number of zones and $K_2 = 2.0 \times 10^3$.

So far we have regarded q as an arbitrary parameter but we see from (36.22) that it has a strong effect on the fluence. The choice of q in a real hologram recording or a real attempt to manufacture a zone plate would depend on a number of issues which we prefer not to open in the present discussion. Instead we simply plot some sample values of the fluence for interesting values of δ and q (see Fig.36.7). We note that these figures apply to a high contrast subject feature.

Now suppose that we are dealing with Fourier transform geometry. For this case the sample plane in Fig.36.6 will contain two pinholes of diameter δ, separated by a distance b (say). The illuminated circle must now be of diameter b and (36.21) becomes

$$I_c \left(\frac{\delta^2}{b^2}\right)\left(\frac{\lambda^2 q^2}{b^2} \middle/ \frac{\pi \times 1.22^2 \lambda^2 q^2}{\delta^2}\right) 2Qt \geq 25 .$$ (36.23)

If we now write b as $N\delta$ then we get

---

3) We are dealing with the full width of the Airy disk

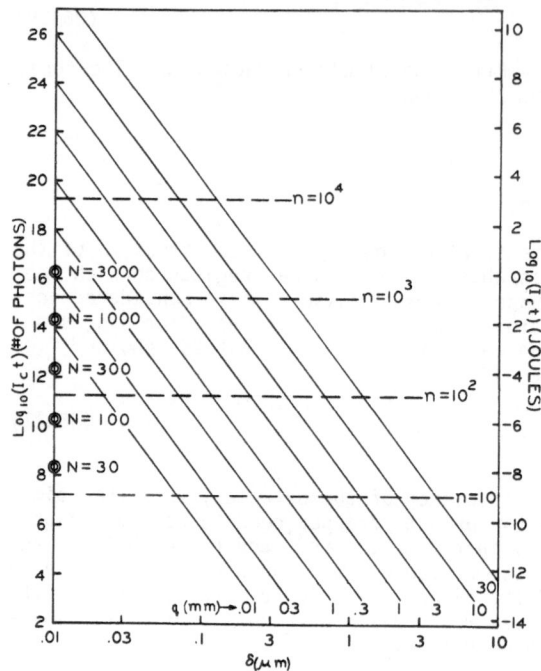

Fig. 36.7 Calculated fluence needs of Gabor holograms with various parameters. For all cases $\lambda = 30$ A. q is the sample-hologram distance, $I_c t$ is the fluence, δ is the resolution element size in the sample, n is the number of rings in the recorded zone plate pattern. The fluences needed for Fourier transform holograms of δ are shown as double circles labelled with N values, where N^2 is the total number of resolution elements in the sample

$$I_c t \geq K_3 N^4 \qquad (36.24)$$

where $K_3 = 2.9 \times 10^2$. We notice that the required fluence is independent of δ, q and even λ and depends only on N which is a measure of the field of view. Some fluence values from (36.24) are entered on Fig.36.7. We notice that the Fourier transform values are quite encouraging and are without the obligation to miniaturize which is implied in the zone plate cases. The lower information content of Fourier transform holograms arises because formation and reconstruction are with spherical waves and therefore it is not necessary to store 'lens' information in the hologram.

In order to give a feeling for the extent to which real sources can approach the flux needs expressed in Fig.36.7, we present in Fig.36.8 some estimates of the coherent power of various synchrotron radiation sources and some other sources which have been used for x-ray holography. One can easily see that the older sources were far short of achieving the beams needed for high-resolution holography but that the newest ones offer great promise.

36.9 Discussion

We have considered many of the limitations involved in holographic imaging and we can now make some judgements on the best strategies for approaching our goal, which is three-dimensional imaging of biological material with a resolution of 100-200 A. First we note that there are a number of reviews [36.21-24] of this general question in the literature and there is no consensus as to the preferred approach. There are different presumptions about available sources and different interests in regard to wavelength. Some authors regard good x-ray lenses as unavailable by definition. Others regard them as something to work toward.

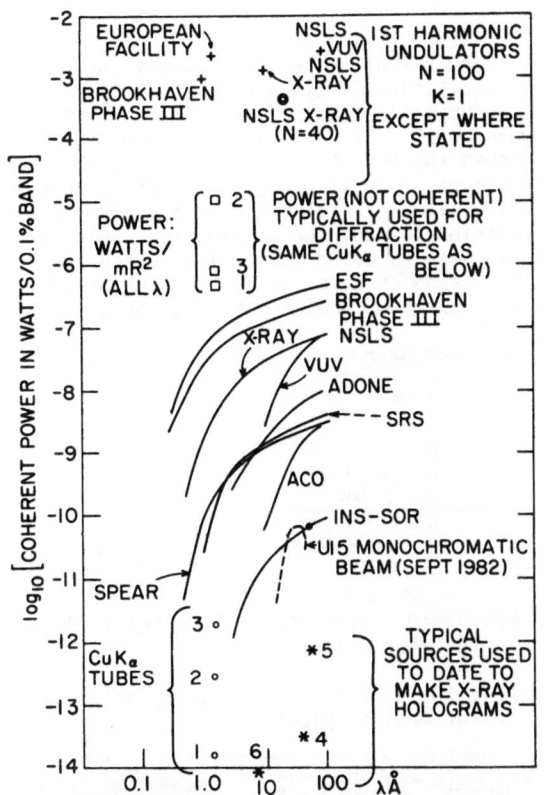

Fig. 35.8 Coherent power output of various sources. The numerical labels are to be interpreted as follows: (1) sealed CuK tube, 2kW, 50 kV, (2) high power rotating anode CuK tube, 50 kW, 50 kV, (3) microfocus rotating anode CuK tube, 3.5 kW, 50 kV, (4,5 and 6) x-ray holography experiments due to REUTER and MAHR [36.29], AOKI and KIKUTA [36.30] and AOKI et al. [36.31), respectively. The double circle represents an undulator presently under construction at Brookhaven for use on the NSLS x-ray ring

We take the availability of both synchrotron radiation and undulator radiation as given. We take the 30 A region as the one of prime interest for biology. We also assume that some zone plates are available with resolution around 0.1 μm and that better ones will be available soon.

In light of these conditions we choose to make the following first steps: (i) to proceed immediately to make what holograms we can with techniques that are on hand, (ii) to move toward the high-resolution regime via the Fourier transform approach. (i) involves making Gabor holograms because these need the least technology. (ii) gives us the following advantages: (a) it is possible to use photographic recording and reconstruct in the visible (b) one can move ahead readily using existing zone plates as beam—forming optics while still looking for an improved reference source through development of special pinholes, etc. (c) future detector developments involving computer reconstruction can be carried out using devices that read directly to the computer in spite of the relatively low resolution of such devices (d) aberrations [36.25,26] are much less in Fourier transform holograms than in other types.

We consider our plans for the future in detail in another publication [36.25]. For the present we will report our experimental work to date and give some ideas on how we are beginning to understand it.

36.10 Experimental Results

We have recorded a number of Gabor holograms using the arrangement shown in Figs.36.9 and 36.10. Light from the National Synchrotron Light Source 750 MeV storage ring was roughly monochromatized using the U15 beam-line system [36.27]. The light (of wavelength 31 A) fully illuminated at 200 x 200 μm^2 Si_3N_4 window which formed the end of the U15 vacuum system. In the holography chamber a 2 μm pinhole was illuminated and used as a source of spatially coherent radiation. This illuminated the sample and also provided the reference beam by transmission through the sample. The holograms were recorded on Kodak type 131-02 high-speed holographic film, which has a quoted resolution of 2500 ℓ/mm. Our samples were small objects of width 10 μm or less, so we see that in all cases recording was made in the far

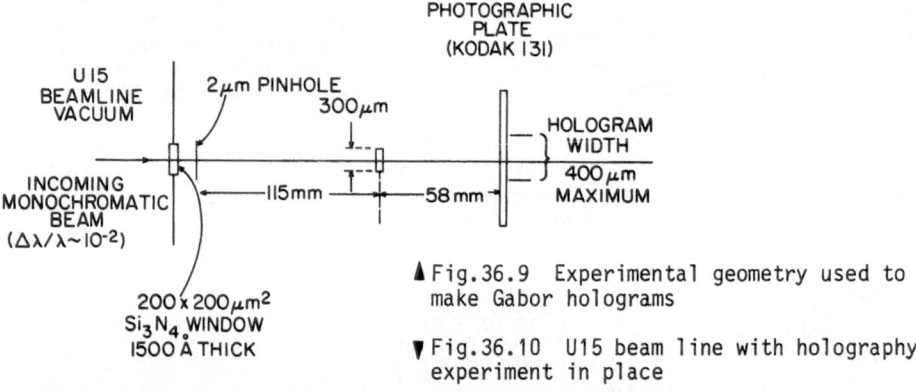

▲Fig.36.9 Experimental geometry used to make Gabor holograms

▼Fig.36.10 U15 beam line with holography experiment in place

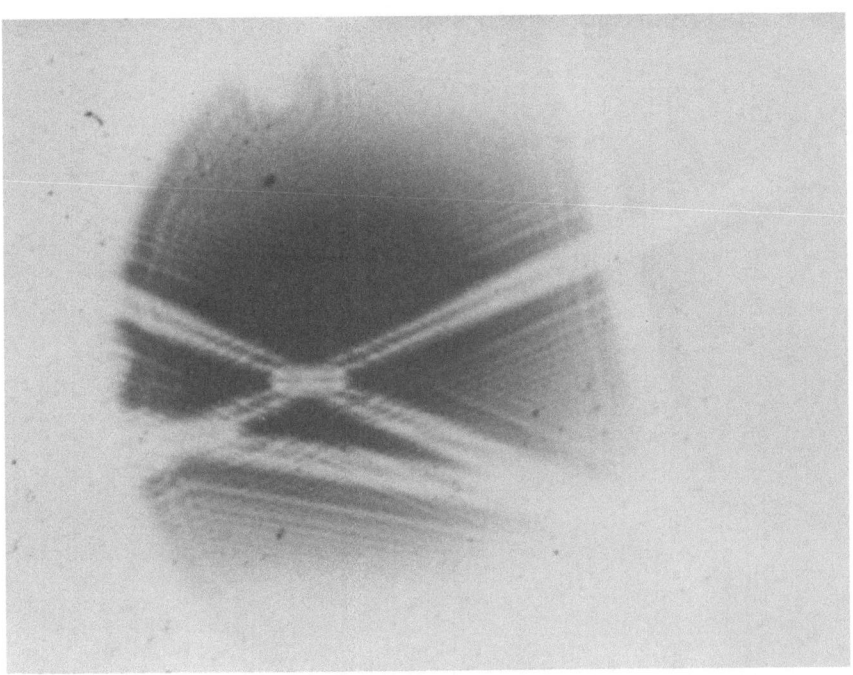

Fig.36.11 Gabor hologram of cross wires 10 μm diameter recorded on Kodak 131-02 film (quoted resolution ∿2500 lines per mm) using an exposure of 3 minutes at 100 mA beam current (about 50 μJoules/cm²). Magnification ≃ 250 X

field of the sample. BAEZ [36.32] has given a prescription for calculating the resolution of the reconstruction of a Gabor hologram using a finite size source and a recording medium of finite resolution. Using this method for our arrangement predicts a resolution of about 0.8 μm.

Figure 36.11 shows the hologram of several 10 μm diameter wires. We see a rich fringe structure superimposed on the central peak of the Airy disk of the 2 μm pinhole. Figure 36.12 shows a similar recording of a group of 3-5 μm diameter spherical, glass beads mounted on a Si_3N_4 window. Again, a complex fringe structure is seen, this time resembling more closely the 'blotchy' appearance of highly magnified visible light holograms. Figure 36.13 shows the hologram of a single wire. It is very reminiscent of the analogous pictures taken in visible light by THOMPSON and TYLER [36.17] and of the famous first-ever hologram taken in 1932 by Kellstrom.

We have attempted to calculate the intensity distribution $I(x)$ expected in a hologram such as Fig.36.13. According to Thompson and Tyler this should be given by

$$I(x) = 1 - \frac{4a}{\sqrt{\lambda z}} \cos\left(\frac{\pi x^2}{\lambda z} - \frac{\pi}{4}\right) \, \text{sinc}\left(\frac{2\pi a x}{\lambda z}\right) + \frac{4a^2}{\lambda z} \, \text{sinc}^2\left(\frac{2\pi a x}{\lambda z}\right) \qquad (36.25)$$

where a is the radius of the wire, z the sample to hologram distance and x the coordinate in the detector perpendicular to the wire. This formula

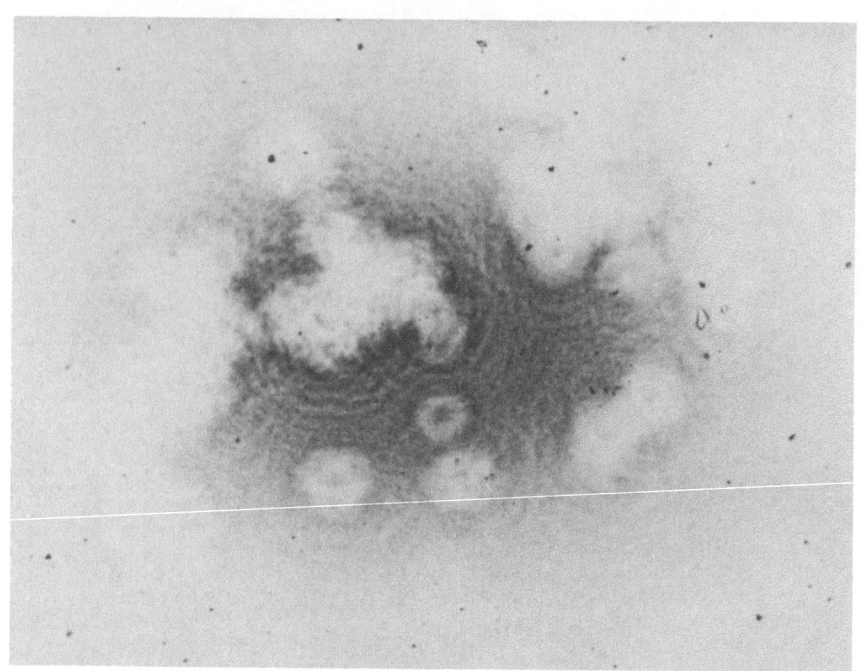

Fig.36.12 Hologram of glass spheres 3-5 μm diameter taken under similar conditions to Fig.36.11

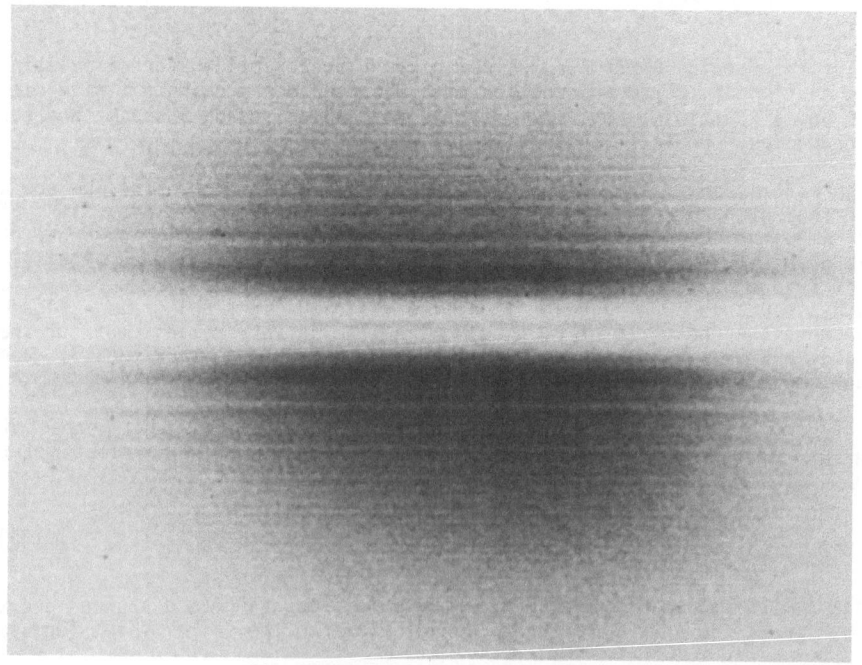

Fig.36.13 Similar to Fig.36.11 using just a single wire

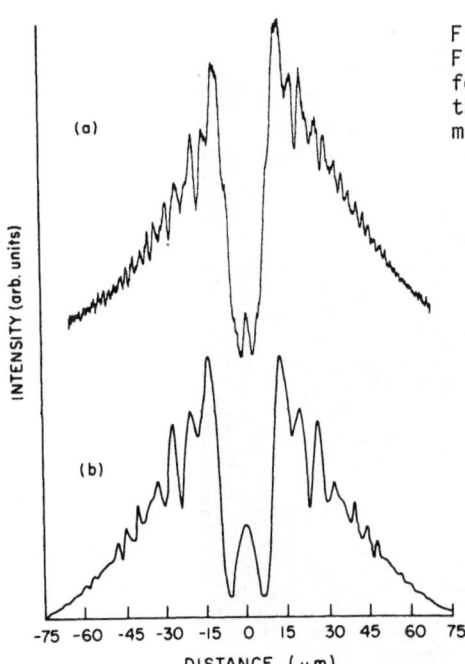

assumes uniform plane wave illumination which is not really appropriate for our case. To make some allowance for this we have folded a Gaussian of roughly the right width into (36.25). The resulting calculated form is shown in Fig.36.14 together with a microdensitometer trace of Fig.36.13. We see that without any further attempts at a more realistic model we obtain reasonable qualitative agreement. A more careful analysis of this question is presently in progress.

In addition to attempting model calculations we are also trying to reconstruct our x-ray holograms using He:Cd laser light (λ = 4166 A). These efforts are just beginning but we have had some successes and one is illustrated in Fig.36.15. We see a fairly good reconstruction of the crossed wires used to make the hologram in Fig.36.11.

36.11 Conclusion

We have reviewed the physics of soft x-ray interactions with matter that could lead to imaging applications. We have given an outline of the operation of the holographic method especially emphasizing the geometries with promise for x-ray applications. We have considered the degree of spatial and temporal coherence and the flux needed for successful imaging and have pointed out the promise of high brightness undulator sources in this connection. In the light of the general characteristics of the hologram forming methods we have concluded that for immediate use the Gabor geometry is best and for the medium term the Fourier transform offers the best chance of approaching the goal of high-resolution biological imaging in three dimensions.

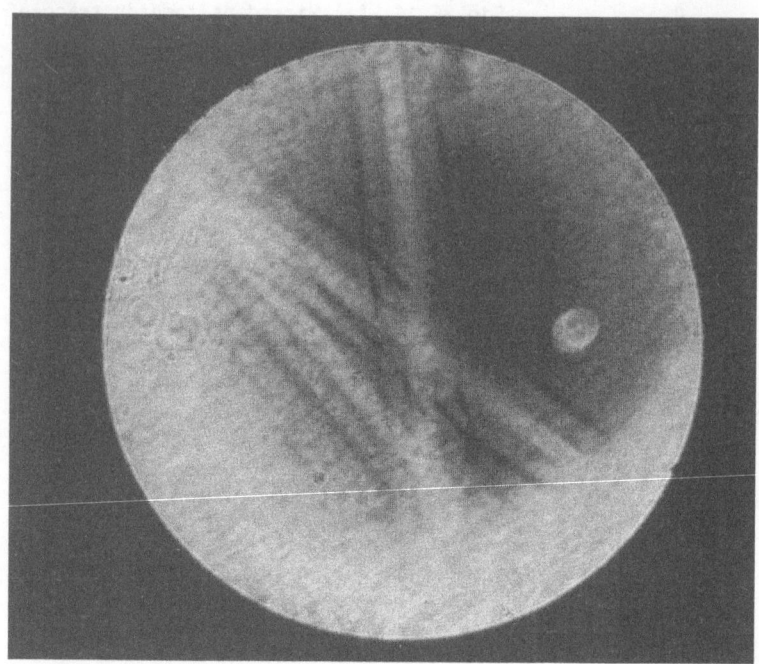

Fig.36.15 Reconstruction of the hologram shown in Fig.36.11 using He:Cd laser light (λ = 4166 A)

We have reported some results of our first attempts to make holograms with soft x rays. These are of objects which, although uninteresting in themselves, allow rehearsal of the technique. Being simple they also allow comparison with theory and we have been qualitatively successful in achieving agreement with the calculated hologram intensity and with the expected geometry of the reconstructed image.

Acknowledgments

The author wishes to express his indebtedness to M. Iarocci of Brookhaven National Laboratory for creating the holographic experimental system, to H. Rarback and J. Kenney of the State University of New York at Stony Brook for providing the U15 beam-line and to J. Kirz also of Stony Brook for continuing discussions, advice and intellectual stimulation.

References

36.1 J. Kirz and D. Sayre: Soft X-ray microscopy of biological specimens in <u>Synchrotron Radiation Research</u>, H. Winick and S. Doniach, Eds. Plenum, New York (1980)
36.2 G. Schmahl et al.: X-ray microscopy in <u>Uses of Synchrotron Radiation in Biology</u>, H.B. Stuhrmann, Academic, London, 1982
36.3 J. Kenney et al.: Proceedings of the Brookhaven Conf. on Synchrotron Radiation Instrumentation, Sept. 1983
36.4 D. Sayre: Prospects for long wavelength x-ray microscopy and diffraction in <u>Imaging Processes and Coherence in Physics</u>, Ed.

M. Schlenker et al., Proceedings Les Houches 1979, Springer Verlag, New York (1979)

36.5 J.C. Solem and G.C. Baldwin: Science, 218, 4569, 229 (1982)

36.6 See for example N.F. Mott and H. Massey: The Theory of Atomic Collisions, 2nd Edition, Oxford, 1950, p. 116

36.7 B.L. Henke: Monterey Conf. on Low Energy X-Ray Diagnostics, Proc. AIP 75, 146 (1981)

36.8 B.L. Bragg: Nature, 143, 678 (1939)

36.9 See for example: Optical Transforms, H. Lipson, Academic, London (1972)

36.10 R.J. Collier, C.B. Burckhardt and L.H. Lin: Optical Holography, Academic, 1971

36.11 H.M. Smith: Principles of Holography, Wiley, New York, 1975

36.12 F.T.S. Yu: Optical Information Processing, Wiley, New York, (1983)

36.13 J.W. Goodman: Introduction to Fourier Optics, McGraw Hill, San Francisco (1968)

36.14 D. Gabor: Proc. Roy. Soc. A197, 454 (1949)

36.15 E.N. Leith, J. Upatnieks: J.O.S.A., 52, 1123 (1962)

36.16 G.W. Stroke: Appl. Phys. Lett. 6, 201 (1965)

36.17 G.A. Tyler, B.J. Thompson: Optica Acta 23, 685 (1976)

36.18 C. Pellegrini: Brookhaven National Laboratory Report #31684 (1982); also Optical Society of America, Topical Meeting on Free Electron Generation of Extreme Ultraviolet Coherent Radiation, Brookhaven Sept. 19, 1983; and Department of Energy Field Task Proposal (1983) by C. Pellegrini, et al.

36.19 M. Born and E. Wolf: Principles of Optics, 3rd Ed., Pergamon, Oxford, p. 508 (1965)

36.20 M.R. Howells, J. Kirz, S. Krinsky: A beamline for experiments with coherent soft x-rays, Brookhaven Laboratory Report #32519 (1982)

36.21 V.V. Aristov and G.A. Ivanova. J. Appl. Cryst. 12, 19 (1979)

36.22 G.L. Rogers and J. Palmer: J. Microsc. (Oxford), 89, 125 (1969)

36.23 J. Solem: High intensity x-ray holography: an approach to high resolution snapshot imaging of biological specimens. Los Alamos National Laboratory Report LA-9508-MS (1982)

36.24 A.M. Kondratenko, A.N. Skrinsky: Opt. Spectrosc. (USSR) 42, 189 (1977)

36.25 SPIE Meeting on Advances in Soft X-Ray Science and Technology, Brookhaven, Oct. 17, 1983. to be published as Proc. SPIE, Vol. 447

36.26 R.W. Meier, J.O.S.A. 56, 219 (1966)

36.27 J. Kirz et al, this volume, no.27

36.28 G. Kellstrom: Nov. Acta Reg. Soc. Sci., Uppsaliensis 8, 5 (1932)

36.29 B. Reuter and H. Mahr: J. Phys. E 9, 746 (1976)

36.30 S. Aoki, S. Kikuta: Jap. J. Appl. Phys. 11, 1857 (1972)

36.31 S. Aoki, et al.: Jap. J. Appl. Phys. 13, 9, 1385 (1974)

36.32 A. Baez: J.O.S.A. 42, 756 (1952)

37. The Study of the Helical Undulator Parameters Installed in the Storage Ring VEPP-2M as a Source of X-Ray Microscopy and Holography

E. S. Gluskin, G. N. Kulipanow, G. Ya. Kezerashvili, V. F. Pindyurin, A. N. Skrinsky, and A. S. Sokolov

Institute of Nuclear Physics, Siberian Division of the USSR Academy of Sciences, SU-630090 Novosibirsk, USSR

P. P. Ilyinsky

Novosibirsk State University, SU-630090 Novosibirsk, USSR

In the work presented here the possibilities of using undulator radiation (UR) in holographic microscopy are considered. Absolute measurements of the spectral-angular density of UR and observations of the UR spatial coherence have been performed aiming at the study of the real undulator parameters.

1. The development of various techniques of x-ray microscopy was stimulated with the advent of powerful sources of synchrotron radiation (SR). The most impressive results are obtained in the field of contact microscopy [37.1]. Microscopy using x-ray optical elements as Fresnel zone plates [37.2,3] and multilayer interference mirrors [37.4] is in good progress. In recent years many discussions were devoted to holographic x-ray microscopy [37.5,6] the possibility of which has already been demonstrated in principle [37.7-10].

An important advantage of holographic microscopy is that the holographic picture contrast (with selection of the optimum reference light) does not depend directly on the sample contrast; it can have high contrast even for very weak absorption and for weak phase shifting samples, as the central part of the passed beam carries only a small fraction of holographic information (of the form of the object under study) and this fraction can often be cancelled in detection. In addition, the hologram enables one to detect also the longitudinal position of the object components at the same time.

An analysis of the requirements of the sources of x-radiation for various schemes of x-ray microscopy (including the holographic one) shows [37.6] that the total number of the "effective" quanta from the source is determined only by the spectral brightness of the source B_λ

$$\dot{N}_\lambda = B_\lambda \cdot \lambda^2 \cdot \frac{\Delta\lambda}{\lambda} . \qquad (37.1.1)$$

But the necessity of having monochromatic radiation imposes limits on the possibilities to use the SR beams from the bending magnets, since "as a rule" the monochromators "spoil" the phase density of the beam of radiation quanta, as the monochromator band pass $\Delta\lambda$ determines an angular spread $\Delta\theta \sim \Delta\lambda/\lambda$ introduced into the quantum flux. Just this circumstance determines the fact that it is of principal importance for x-ray holography to use the super bright beams of x radiation from the undulators which do not need the preliminary monochromatization.

As is well known [37.11-15] the undulator radiation (UR) has some new properties compared to the SR properties.

The radiation spectral brightness of the N-period undulator is by N to N^2 times higher than that from the bending magnet. In addition, the UR is a quasimonochromatic radiation with $\Delta\lambda/\lambda \sim 1/N$ and a sharp angular distribution $\Delta\Theta_\lambda \sim (\gamma\sqrt{N})^{-1}$. These properties enable one to get the high intensity without substantial deterioration of the UR beam monochromaticity only by selection of the optimum diaphragm size without use of the monochromator. The effective dimensions of the UR source are determined by the electron beam parameters in the storage ring.

One should also note that mainly due to the development of x-ray lithography, quite a large number of x-ray resists appeared recently - the films of various materials (mostly of organic nature) whose etching rate depends on the x-ray radiation dose absorbed in the unit of the x-ray resist volume (the typical value ranges from 10 mJ/cm^3 to 500 J/cm^3). The spatial profile obtained on the x-ray resist after etching can be read by an electron microscope. In the soft x-ray wavelength range ($\lambda \simeq 50$ to 100 A) the x-ray resists have an excellent spatial resolution (50 to 200 A) and a sensitivity near to that of the ideal detector sensitivity when the minimum detected dose corresponds to the absorption of 1 to 10 quanta per unit cell of the resist volume whose linear dimensions are equal to the resist spatial resolution.

The use of undulators as the sources of x radiation and x-ray resists as detectors open up new possibilities for x-ray microholography studies. In the work presented here the experimental study is carried out in the parameters of the UR from the helical undulator, installed in the storage ring VEPP-2M, which are studying its use in the holographic schemes.

2. Without going into details of the theoretical analysis of the UR characteristics performed in a number of papers (see [37.11,15]), let us mention some characteristic UR spectral and angular properties. As it is known, the UR n-th harmonic from the individual electron (observed at an angle Θ with respect to the helical undulator axis) has a wavelength λ_n and a spectral width $(\Delta\lambda/\lambda)_n$ determined by the following relations:

$$\lambda_n = \frac{\lambda_0}{2n\gamma^2}\cdot(1+k^2+\gamma^2\Theta^2) \; ; \quad (\frac{\Delta\lambda}{\lambda})_n = \frac{1}{Nn} \tag{37.1}$$

where $\gamma= E/(mc^2)$ is the electron relativistic factor, $k=0.0934\lambda_0 H_0$ [cm·kG] is the deflection parameter, λ_0 is the helical undulator magnetic period length, H_0 is the magnetic field amplitude, N is the total number of magnetic periods.

For further considerations of the possibilities of undulator radiation usage in holography let us take a certain example scheme of the Gabor in line axial holography (an example of the small - angular x-ray holography using the SR beam is considered in [37.17]). The scheme for getting the axial x-ray holographic picture is shown in Fig.37.1. The main relations for the resolution capability of this holographic scheme are the following:

a) transverse resolution $\delta_\perp$, determined by the radiation source transverse dimensions σ_x, σ_y

$$\delta_\perp \simeq \frac{q\sigma}{p} \; ; \quad \sigma = \max[\sigma_x,\sigma_y] \tag{37.2}$$

b) transverse resolution determined by the screen-detector capability resolution δ_R

$$\delta_\perp \simeq \delta_R \; , \quad p \gg q \tag{37.3}$$

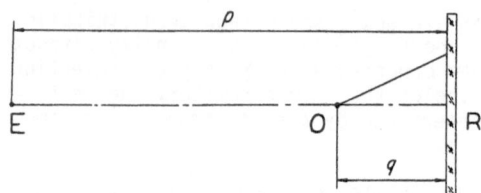

Fig.37.1 Axial hologram registration scheme
p is a distance between
radiation source and the
screen (R), q is a dis-
tance between the
object (0) and the
screen (R)

c) transverse resolution determined by the monochromaticity of radiation $\frac{\Delta\lambda}{\lambda}$:

$$\delta_\perp \simeq \frac{1}{2}\sqrt{q\lambda\cdot\frac{\Delta\lambda}{\lambda}} \quad , \quad p \gg q \tag{37.4}$$

d) longitudinal resolution determined by the radiation monochromaticity:

$$\delta_\parallel \simeq \frac{q}{4}\cdot\frac{\Delta\lambda}{\lambda} \simeq \frac{\delta_\perp^2}{\lambda} \quad , \quad p \gg q \ . \tag{37.5}$$

The organic x-ray resist is used in the scheme as a screen detector with high resolution. In this case, the holographic picture exposure time t with the source spectral brightness B_λ and monochromaticity $\Delta\lambda/\lambda$ given is deter-mined by the following relation:

$$t \simeq \frac{D}{\mu}\cdot\frac{S}{B_\lambda\cdot\lambda^2\cdot\frac{\Delta\lambda}{\lambda}} \simeq \frac{D}{\mu}\cdot\frac{q^2}{B_\lambda\cdot\delta_\perp^2\cdot\frac{\Delta\lambda}{\lambda}} \tag{37.6}$$

where D is a volume dose (J/cm^3), μ is the resist linear absorption coeffi-cient (cm^{-1}), S is the hologram area.

Let us get the numerical value for using the undulator as a radiation source installed in the straight section of the storage ring VEPP-2M [37.16]. This is a helical undulator with a magnetic period length λ_0 = 2.4 cm and the total number of periods N = 10. The magnetic field maxi-mum amplitude on the undulator axis is H_0 = 1.3 kG. Taking into account the electron-beam parameters where the undulator is located in the stor-age ring VEPP-2M at an electron energy E_e = 0.67 GeV, $\sigma_{x'}$ = 0.82 mrad, $\sigma_{y'}$ = 0.22 mrad, σ_x = 0.035 cm, σ_y = 0.00067 cm, the undulator spectral brightness B_λ with the field H_0 = 1.3 kG on the UR 1-st harmonic λ_1 = 79 A with a current of 50 mA has the following value B_λ =7.7·10^8 $W/(cm^2,s,rad,\Delta\lambda/\lambda)$ with the total UR spectral width $(\Delta\lambda/\lambda)_1$ = 0.20. On the other hand, estimates of the source brightness required for an x-ray hologram with a resolution of $\sim$ 500 A and an exposure time of one second give $B_\lambda \gtrsim 10^7$ $W/(cm^2,s,rad,\Delta\lambda/\lambda)$. This evaluation performed does not take into account the additional radiation loss in the sample, substrate, etc. The comparison of this value with those obtained when using SR beams from bending magnets of the storage ring in x-ray holography [37.17] (taking into account the necessary monochromaticity) emphasizes even more the advantages of the UR usage for the solution of such problems.

It is also worth noting the principal possibility to get x-ray holograph-ic pictures with a spatial resolution of $\sim$ 50 A using the UR wavelength of $\sim$ 50 A. Diffractional diffusion and the photoelectron free path in the available x-ray resists put the limits on getting higher spatial resolution values ($\leq$ 50 A) [37.18].

338

3. The absolute measurements of the UR parameters from the helical undulator installed in the storage ring VEPP-2M were carried out at an electron energy E_e = 0.39 GeV and 0.51 GeV within the wavelength range 100 to 300 A. The magnetic field amplitude on the undulator axis was 1.0 to 1.1 kG. This corresponds to values K = 0.224 to 0.247. The main goal of these studies was to check the correspondence of the UR characteristics to the requirements of x-ray holography.

The absolute measurements of the UR spectral angular characteristics have been performed with the grazing incidence monochromator (see Fig.37.2) with a gold-coated spherical grating G with a radius of 2m, p=600 g/mm.

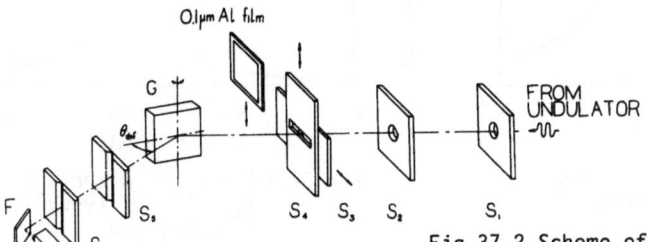

Fig.37.2 Scheme of absolute measurements of the UR spectral-angular characteristics

The diffraction angle $\Theta_{dif} \simeq 5^0$ has been selected so that at a smaller diffraction angle the scattered background increases and at larger angle the diffraction grating reflection coefficient is substantially decreased in the first order.

The monochromator calibration over the wavelengths has been performed on the Al $L_{2,3}$ absorption edge by introducing an aluminium foil $\sim$ 1000 A thick. The radiation was detected with the channeltron (C) VEU-6 with the CsI(F) photocathode.

The monochromator can be operated without an entrance slit because of the small dimensions of the radiation source $\sigma_x \cdot \sigma_y$ = 0.35 x 0.067 mm². The collimators S_1S_2 and S_5S_6 are necessary for suppressing the background caused by radiation from the ends of bending magnets. As spectral measurements have shown, the radiation intensity from the bending magnets within the wavelength range 100 to 300 A is not in excess of 5% of the UR intensity at Θ=0.

The observation angular aperture 0.036 mrad was determined by the entrance diaphragm (S_3S_4) 0.2 x 0.2 mm² and a distance to the source of 555 cm. An effective angular dimension of the observation area for the first harmonic was estimated by $\Delta\Theta \sim 1/(\gamma\sqrt{N})$, thus, at E_e = 0.51 GeV, $\Delta\Theta \sim 0.32$ mrad and at E = 0.39 GeV, $\Delta\Theta \sim 0.41$ mrad. As it is seen, an angular aperture after the diaphragm is ten times smaller than the angular size of the first harmonic. The small size of the diaphragm enabled studying directly the UR spectral angular density.

The electron beam angular spread for VEPP-2M is $\sigma_{x'}$ = 0.628 mrad, $\sigma_{y'}$ = 0.167 mrad at an energy E = 0.51 GeV. The calculated curve for the UR spectral angular density (taking into account the electron angular spread and the beam transverse dimensions) is drawn in Fig.37.3. The UR 1-st harmonic wavelength is λ_1 = 132.9 A, the spectral width is $(\Delta\lambda/\lambda)_1$ = 0.144, absolute spectral angular density is $9.3 \cdot 10^{11}$ phot/(s,mA,mrad²,1% bandwidth)

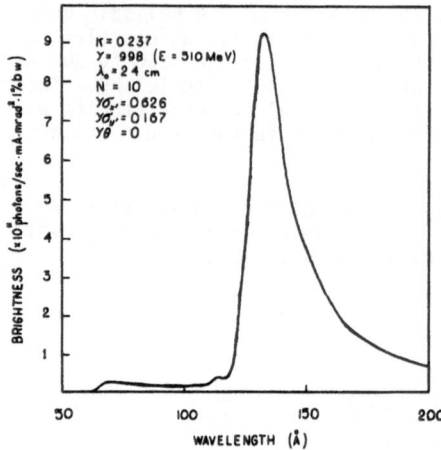

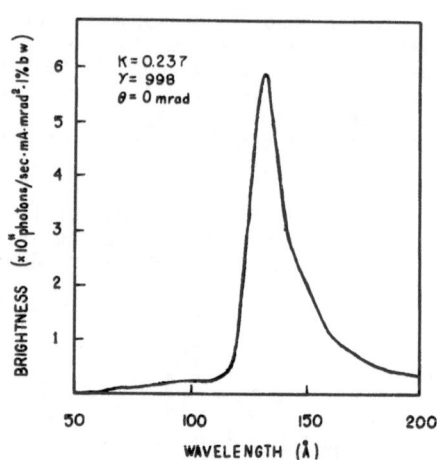

Fig.37.3 Calculated UR spectrum
E_e=0.51 GeV (γ=998),
k=0.237, θ=0

Fig.37.4 Observed UR spectrum
E_e=0.51 GeV (γ=998),
k=0.237, θ=0

which is 1.43 times higher compared to the observed value (Fig.37.4) of
$6.5 \cdot 10^{11}$. The SR from the bending magnet of the storage ring VEPP-2M at
E=0.51 GeV emits $3.1 \cdot 10^{10}$ photons at the same wavelength. This is 21 times
lower than the observed UR. As seen from the measured spectrum, the UR
second harmonic is not observed. This fact is connected with the small
reflection coefficient of the diffraction grating in the corresponding
wavelength range. In addition, in this case, the spectral angular density of
the UR 2-nd harmonic is 50 times lower than that for the 1-st harmonic. For
simultaneous observation of the 1-st and 2-nd UR harmonics the measurements
have been carried out with an electron energy E = 0.39 GeV. In this case,
the wavelength of the two UR harmonics are in the spectral window of the
used monochromator: 100 to 300 A.

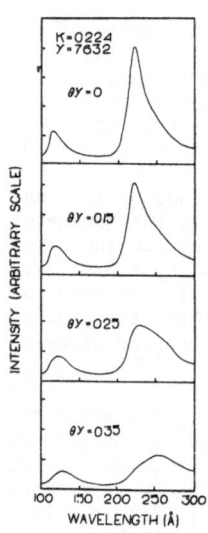

The UR spectra were obtained at E = 0.39 GeV
(γ = 763) k = 0.224, for $\gamma\theta$ = 0, 0.15, 0.25, 0.35
(Fig.37.5). The spectra are not normalized over the
grating reflection coefficient, therefore, the
harmonic ratio on the spectrum is not correct. Figure 37.6 shows the wavelength for both UR harmonics as
a function of the observation angle θ. The electron
angular spread and the UR spectral width
$(\Delta\lambda/\lambda)_n = 1/N \cdot n$ lead to the displacement of the
wavelength for the first and second UR harmonics
compared to that given by relation (37.1) and also
lead to the violation of the dependence on θ^2 in the
region of the observation of small angles
($\theta \ll \sigma_{x'}, \sigma_{y'}$). The difference between the observed
and calculated angular dependences is caused by an
inaccuracy in angular calibration of θ values. The
same reason causes the difference in the UR spectral
angular density dependence on the observation angle
values. The curves obtained are normalized over the
calculated value at θ = 0.

Fig.37.5 Observed UR spectra E_e=0.39 GeV (γ=763), k=0.224, $\gamma\theta$=0, 0.15, 0.25,
0.35 or θ=0, 0,20, 0.33, 0.46 mrad

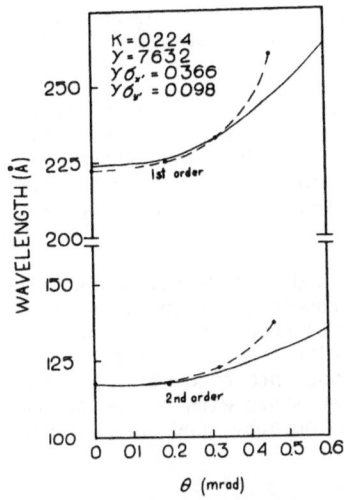

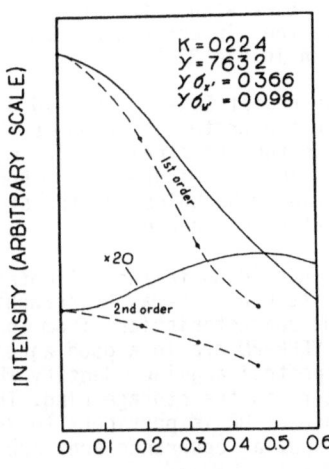

Fig.37.6 Angular dependence of the UR wavelengths on 1-st and 2-nd harmonics (n=1) and (n=2) respectively :
---- observed curve (see Fig.37.5);
—— calculated curve ($\gamma\sigma_{x'} = 0.366, \gamma\sigma_{y'} = 0.098$)

Fig.37.7 Angular dependence of the UR spectral densities of the 1-st(n=1) and 2-nd(n=2) harmonics :
---- observed curve (see Fig.37.5);
—— calculated curve ($\gamma\sigma_{x'} = 0.366, \gamma\sigma_{y'} = 0.098$)

4. The UR spectral angular characteristic measurements do not supply the direct information on the effective transverse size of radiation source which is important in setting up the holographic experiments. According to the electron-beam size in the storage ring VEPP-2M, the helical undulator radiation should have a sufficiently high level of spatial coherence. For an electron energy E = 0.51 GeV at the transverse dimensions of the electron beam $d_x = 2.36\cdot\sigma_x = 0.059$ cm, $d_y = 2.36\cdot\sigma_y = 0.019$ cm the coherence region size at a distance from the source P = 520 cm are the following $C_x = \lambda_1\cdot P/d_x = 110$ µm, $C_y = \lambda_1\cdot P/d_y = 360$ µm at the wavelength of UR $\lambda_1 = 130$ A. The experiments on the real spatial coherence observation of the VEPP-2M undulator radiation have been performed by studying the interference pictures from two small ($\simeq 6$ µm) holes in 2 µm thick Si membrane coated with

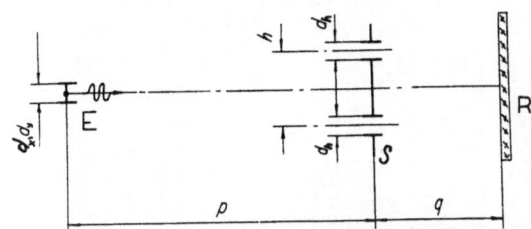

Fig.37.8 The spatial coherence registration scheme :
d_x, d_y - transvers dimensions of the radiation source (E), (d_x=0.059 cm, d_y=0.019 cm), p - distance between the source (E) and absorption screen(S), (p=520cm); h,d_h - distance between holes and their diameters in the screen(S), ($d_h \sim$ 6µm); q - distance between the screen (S) and detector screen(R), (q=31cm)

341

0.3 μm Au and increasing the difference between the holes until disappearance of the interference bands. The optical scheme of the experiment is given in Fig.37.8.

The negative x-ray resist ELN-200 ($D \simeq 10$ J/cm^3, $\mu \simeq 10^5$ cm^{-1}, 0.2 μm thick film on a silicon substrate) is used as the detector screen. The exposure time, according to the spectral-angular measurements described above, is estimated as $I \cdot t \simeq 10$ A·s. Figure 37.9 shows the interference pictures for some distances between the holes in the study of both the vertical (a) and horizontal (b) spatial coherence.

5. Conclusion. According to the estimates obtained, the UR satisfies the requirements for its use in x-ray holographic microscopy. The absolute measurements of the UR characteristics from the helical undulator installed in the storage ring VEPP-2M are in a good agreement with the calculation estimates of the UR spectral angular density which took into account the electron-beam parameters in the storage ring. The experiments with the UR demonstrate that the use of UR is pertinent to x-ray holography, and show that the estimates of the spatial coherence area are correct.

(a) Vertical Disposition of Holes

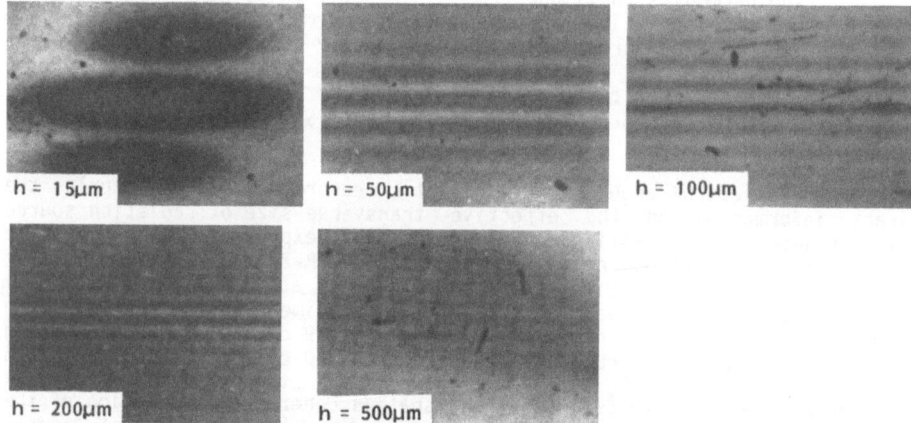

(b) Horizontal Disposition of Holes

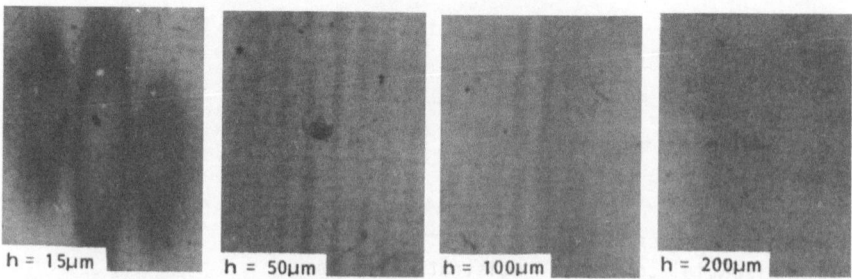

Fig.37.9 Interference pictures obtained for different distances between the holes h during measurements of vertical (a) and horizontal (b) UR spatial coherence. Typical exposure $I \cdot t \simeq 20$ As, I [A] is the storage ring current

References

17.1 R. Feder et al.: Science <u>197</u>, 259 (1977)
37.2 B. Niemann et al.: Appl. Opt. <u>15</u>, 1882 (1976)
37.3 B. Niemann et al.: NIM <u>208</u>, 367 (1983)
37.4 R.P. Haelbich et al.: A Scanning Ultrasoft X-Ray Microscope with Large Aperture Reflection Optics for Use with Synchrotron Radiation: Preprint DESY SR-79/190, Hamburg, 1979
37.5 A.M. Kondratenko, A.N. Skrinsky: Avtometria <u>2</u>, 3 (1977)
37.6 G.N. Kulipanov, A.N. Skrinsky: UPhN, 122, Iss. 3, 369 (1977)
37.7 S. Kikuta et al.: Opt. Commun. <u>5</u>, 86 (1972)
37.8 O.J. Sasoccio: J. Opt. Soc. Am. <u>57</u>, 966 (1967)
37.9 S. Hoki et al.: Jpn. J. Appl. Phys. <u>13</u>, 1385 (1974)
37.10 V.V. Aristov et al.: Opt. Commun. <u>34</u>, N3, 332 (1980)
37.11 D.F. Alferov et al.: Part. Acceler. <u>9</u>, 223 (1970)
37.12 A.N. Didenko et al.: Sov. Phys. <u>49</u>, 973 (1979)
37.13 A.S. Artamonov et al.: NIM <u>177</u>, 239 (1980)
37.14 H. Maezawa et al.: NIM <u>208</u>, 151 (1983)
37.15 D.F. Alferov, Yu.A. Bashmakov: "Spectral-angular characteristics of radiation of the beam of relativistic charged particle in undulator", Preprint, FIAN, SSSR, N77, 1983
37.16 G.Ya. Kezerashvili et al.: Proc. All Union Conf. on Utilis. of SR (SR-82), Novosibirsk, 1982
37.17 A.M. Kondratenko, A.N. Skrinsky: "The use of radiation from the storage rings in x-ray holography of microobjects", Preprint INP 75-102, Novosibirsk, 1975
37.18 E.S. Gluskin et al.: NIM <u>208</u>, 393 (1983)

List of Contributors

X-Ray Optics

Applications to Solids

Editor: **H.-J. Queisser**
1977. 133 figures, 14 tables. XI, 227 pages
(Topics in Applied Physics, Volume 22)
ISBN 3-540-08462-2

Contents: *H.-J. Queisser:* Introduction: Structure and
Structuring of Solids. - *S. Kozaki, M. Yoshimatsu:* High
Brilliance X-Ray Sources. - *R. Feder, E. Spiller:* X-Ray
Lithography. - *U. Bonse, W. Graeff:* X-Ray and Neutron
Interferometry. - *A. Authier:* Section Topography. -
W. Hartmann: Live Topography.

B. K. Agarwal

X-Ray Spectroscopy

An Introduction

1979. 188 figures, 31 tables. XIII, 418 pages
(Springer Series in Optical Sciences, Volume 15)
ISBN 3-540-09268-4

Contents: Continuous X-Rays. - Characteristic X-Rays.
- Interaction of X-Rays with Matter. - Secondary
Spectra and Satellites. - Scattering of X-Rays. -
Chemical Shifts and Fine Structure. - Soft X-Ray Spec-
troscopy. - Experimental Methods. - Appendices. -
Wavelength Tables. - References. - Auhor Index. -
Subject Index.

Springer-Verlag
Berlin
Heidelberg
NewYork
Tokyo

Synchrotron Radiation

Techniques and Applications

Editor: **C. Kunz**
1979. 162 figures, 28 tables. XVI, 442 pages
(Topics in Current Physics, Volume 10)
ISBN 3-540-09149-1

Contents: *C. Kunz:* Introduction - Properties of
Synchrotron Radiation. - *E. M. Rowe:* The Synchrotron
Radiation Source. - *W. Gudat, C. Kunz:* Instrumentation
for Spectroscopy and other Applications. - *A. Kotani,
Y. Toyozawa:* Theoretical Aspects of Inner-Level Spec-
troscopy. - *K. Codling:* Atomic Spectroscopy. -
E. E. Koch, B. F. Sonntag: Molecular Spectroscopy. -
D. W. Lynch: Solid-State Spectroscopy.

Very Large Scale Integration (VLSI)

Fundamentals and Applications

Editor: **D. F. Barbe**
2nd corrected and updated edition. 1982. 147 figures.
XI, 302 pages (Springer Series in Electrophysics,
Volume 5)
ISBN 3-540-11368-1

Contents: *D. F. Barbe:* Introduction. – *J. L. Prince:* VLSI
Device Fundamentals. – *R. K. Watts:* Advanced Litho-
graphy. – *P. Losleben:* Computer Aided Design for VLSI.
– *R. C. Eden, B. M. Welch:* GaAs Digital Integrated
Circuits for Ultra High Speed LSI/VLSI. – *E. E. Swartz-
lander:* VLSI Architecture. – *B. H. Whalen:* VLSI Appli-
cations and Testing. – *D. F. Barbe, E. C. Urban:* VHSIC
Technology and Systems. – *R. I. Scace:* VLSI in Other
Countries. – Addenda. – Subject Index.

Surface Studies with Lasers

Proceedings of the International Conference
Mauterndorf, Austria, March 9–11, 1983
Editors: **F. R. Aussenegg, A. Leitner, M. E. Lippitsch**
1983. 146 figures. IX, 241 pages
(Springer Series in Chemical Physics, Volume 33)
ISBN 3-540-12598-1

Contents: General Surface Spectroscopy. – Surface
Enhanced Optical Processes. – Laser Surface Spectro-
scopy. – Laser Induced Surface Processes. – Index of
Contributors.

Excimer Lasers

Editor: **C. K. Rhodes**
2nd revised and enlarged edition. 1984. 100 figures.
XII, 271 pages (Topics in Applied Physics, Volume 30)
ISBN 3-540-13013-6

Contents: *P. W. Hoff, C. K. Rhodes:* Introduction. –
M. Krauss, F. H. Mies: Electronic Structure and Radiative
Transitions of Excimer Systems. – *M. V. McCusker:* The
Rare Gas Excimers. – *C. A. Brau:* Rare Gas Halogen
Excimers. – *A. Gallagher:* Metal Vapor Excimers. –
D. L. Huestis, G. Marowsky, F. K. Tittel: Triatomic Rare-
Gas-Halide Excimers. – *H. Pummer, H. Egger,
C. K. Rhodes:* High-Spectral-Brightness Excimer Systems.
– *K. Hohla, H. Pummer, C. K. Rhodes:* Application of
Excimer Systems. – Subject Index. – List of Figures. –
List of Tables.

Springer-Verlag
Berlin
Heidelberg
New York
Tokyo

Springer Series in Optical Sciences

Editorial Board: J.M. Enoch D.L. MacAdam A.L. Schawlow K. Shimoda T. Tamir